HIGH-SPEED
CABLE MODEMS

THE McGRAW-HILL SERIES ON COMPUTER COMMUNICATIONS (SELECTED TITLES)

High-Speed Cable Modems

Including IEEE 802.14 Standards

Albert A. Azzam

McGraw-Hill
New York • San Francisco • Washington, D.C. • Auckland • Bogotá
Caracas • Lisbon • London • Madrid • Mexico City • Milan
Montreal • New Delhi • San Juan • Singapore
Sydney • Tokyo • Toronto

Library of Congress Cataloging-in-Publication Data

Azzam, Albert.
   High speed cable modems : including IEEE 802.14 standards / Albert
Azzam.
     p.   cm.
  Includes index.
  ISBN 0-07-006417-2
  1. Modems. I. Title.
TK7887.8.M63A896 1997
384.3'2—dc21   97-14635
                                              CIP

McGraw-Hill

A Division of The **McGraw·Hill** *Companies*

1 2 3 4 5 6 7 8 9 0 DOC/DOC 9 0 2 1 0 9 8 7

ISBN 0-07-006417-2

*The sponsoring editor for this book was Steve Chapman and the production supervisor was
Pamela Pelton. It was set in Vendome by North Market Street Graphics.*

Printed and bound by R. R. Donnelley & Sons Company.

McGraw-Hill books are available at special quantity discounts to use as premiums
and sales promotions, or for use in corporate training programs. For more information, please write to Director of Special Sales, McGraw-Hill, 11 West 19th Street,
New York, NY 10011. Or contact your local bookstore.

 This book is printed on recycled, acid-free paper containing a minimum of
50% recycled de-inked fiber.

I dedicate this book to my family: my wife, Elizabeth, and my children JoAnna, Kathleen, Greg, and Michael.

CONTENTS

Contents

x

Contents

Contents

Contents

PREFACE

The information age is upon us, and multimedia is becoming part of our lives, not only in the workplace but also at home. Integrating voice and data, as ISDN attempted to do last decade but failed, is no longer a matter of preference but a market-driven service. Video is also becoming an important factor with the phenomenal market growth of the World Wide Web (WWW), computer entertainment, and multimedia. These factors, no doubt, played a major role in the recent Telecommunication Act that was passed by the United States Congress in 1996. This reform act, in short, was designed to promote competition among the major network operators for providing these services.

Superhighway is the term used to define high-speed integrated access, and it has become a national goal spearheaded by vice president Al Gore under the National Information Infrastructure Act.

One can classify today's worldwide network as service-specific, more or less. Telecommunication networks were designed and deployed to handle voice traffic. Both platform and fabric were optimized in the design to switch voice traffic efficiently. Cable TV networks were optimized for one-way video broadcasting, and the Internet network was initially developed and optimized for data transport. In light of this, network operators are dealing with more complex issues when planning network evolution. Network *revolution* is more akin to what the industry will be dealing with in order to provide the integrated services, meet the stiff competition, and stay viable in the marketplace.

ATM (asynchronous transfer mode) was conceived in the 1980s with a technique designed to provide a switching fabric that was independent of the underlining services, be they voice, video, or data. For example, a single wire might contain various QOS applications using the same switching fabric. ATM, unlike the low-speed circuit-switch ISDN technology, uniquely addressed the market demand without being technology-driven. Most if not all network operators believe that ATM will be the switching technology of the future communication revolution. ATM, described in Chap. 3, accommodates the video market and also resolves the annoying low-speed access that is becoming more and more of a nuisance when the video applications on the WWW are loaded.

The rivalry between the telco and cable operators can be traced back to 1982, when the MFJ under U.S. Federal Judge Harold H. Greene became a reality. It was the birthday of the seven regional holding companies (RHCs) and their regulated Bell Operating Companies (RBOCs) from Ma Bell (AT&T). Since then, both BOCs and cable operators lobbied the U.S. Congress effectively and maintained their turf. That, of course, changed with the passing of the Telecommunication Act of 1996. With few exceptions, the BOCs can now provide video programming, and the cable operators can also market voice telephony over cable.

The purpose of the book, in light of the above, is to cover all technical aspects of the high-speed cable modem (sometimes called the cable modem). The high-speed modem is the only tool available to cable operators to:

1. Maintain the revenue from the existing customer base by providing them superhighway access

2. Compete and/or fend off the RBOCs, who are very eager to provide video programming, Internet access, video telephony, and such

3. Expand their share of the market

Organization of the Book

The book is composed of fourteen chapters. The first three chapters will describe the HFC environment that the high-speed cable modem must accommodate: the architecture, broadband services, and transport medium.

Chapter 1 deals with HFC architecture and the reference model as defined by the industry today. It also maps the reference model to the CATV network of today and identifies the barriers and rationale for the modernization plan of the CATV network and economic model.

Chapter 2 describes the broadband services as defined in ITU and other organizations. Key services and applications are analyzed and mapped into the cable modem requirements. The limitation and degradation of services is also described in terms of deployment in a shared-medium environment.

Chapter 3 describes the role of ATM in the HFC network. The advantages and disadvantages of taking ATM to the home are examined in terms of maintaining the quality of service for voice, video, and data. Other transport mechanisms (IP/MPEG packets) are also described.

Chapter 4 gives a general description and history of the high-speed modem and its evolution in the digital environment. A long-term architecture of the telecommunication network is described to understand better future network convergence. A timeline for the high-speed cable modem is provided.

Chapter 5 describes the high-speed cable modem's physical layer for both downstream and upstream channels. QAM and QPSK, accepted in the industry, are described. Other modulation techniques, proposed by cable modem vendors in IEEE 802.14, are also fully described. Future work on multiple physical layers and/or advanced physical layers are explored.

Chapter 6 describes the medium access control (MAC) layer and its function in a high-speed modem. Reservations versus collision algorithm schemes are evaluated. All 14 proposals made to the IEEE 802.14 standard committee by the industry are fully described. Performance evaluations are analyzed in detail in terms of complexity, fairness, ATM compatibility, cell delay variation, and so on. Performance analysis is described in great detail, and finally the MAC selected by IEEE 802.14 is described.

Chapter 7 analyzes all the unwanted noises that inevitably find their way in the CATV network. Detailed noise diagrams and channel models are examined, giving the reader an appreciation of the complexities needed for developing a high-speed cable modem. Both upstream and downstream noise characteristics are described.

Chapter 8 deals with HFC management aspects of the cable modem, such as initialization, registration, and parameter settings. ABR traffic operation in the shared medium environment is also analyzed.

Chapter 9 describes the architecture and other factors needed to provide POTS (plain old telephone service) in an HFC architecture. Limitations and opportunities such as PCS facing the cable operators in providing a voice-over-cable modem.

Chapter 10 describes the network services in terms of HFC security. The security needs of each of the services (voice, home shopping, Internet, VOD, and so on) are examined. Modes of cryptography to combat eavesdropping and service theft, pronounced occurrences in the shared-medium network, are also described.

Chapter 11 deals with standard bodies involved with the development and standardization of the cable modem. The views of each standard body are described. A guide is provided on how to obtain specifications and recommendations from the various organizations.

Chapter 12 describes ADSL technology. The deployment strategies of telco and cable operators are explored. Both cable modem and ADSL technologies are compared based on today's service offering. Alternative technology is also explored.

Chapter 13 takes a look at the European cable network operators, architecture, planned services, and field trials both proposed and existing, using the high-speed modem.

Chapter 14 contains a survey of the ongoing high-speed modem field trials. The chapter also contains a cable modem vendor list, a list of MSO players, and a vendor list of cable modems developed by vendors.

Albert Azzam

ACKNOWLEDGMENTS

I especially wish to thank the Alcatel management for providing me with the opportunity to write this book. The book content does not necessarily reflect Alcatel position. Ideas and opinions in this book are my own and do not reflect Alcatel policy in any way.

I also wish to thank my colleagues who contributed directly to this edition of the book: Nada Golmie of NIST for her contribution to the part on MAC performance analysis and the standard IEEE 802.14 MAC description in Chap. 6; and JoAnna Azzam for her contribution to the introduction and preparation of other research materials.

In addition, I would also like to mention my friends who helped in shaping ideas for this book or indirectly contributed to it, namely Emmanuel Desmet, Paul Spruyt, Chris Sierens, John Angelopoulous, and Dieter Beller.

Finally, this book would not have been possible without the support and cooperation of the organizations who submitted proposals and solutions to IEEE 802.14. Each proposal contained the usual disclaimer statements. As their secretary, I would like to note their excellent professional mannerism during difficult times. Their effort should be applauded by the industry.

INTRODUCTION

Global Connectivity

The rapid transformation of the telecommunications industry is a reflection of today's growing market for global connectivity. The information age has created new barriers to overcome and has produced endless technological alternatives. In 1995, worldwide telecommunications was a $600 billion industry; in the year 2000, it is projected to expand to more than a trillion dollars. Though not the largest industry, it is certainly the fastest growing. This introduction will provide a brief history of the telecommunications industry and the effects of government regulation and deregulation on the market. It will examine the telephone and cable companies and the rapidly developing competition in the marketplace. In addition, the profound effects of the phenomenal growth of the Internet will also be discussed.

Only few years ago, telecommunications was virtually synonymous with plain old telephone service (POTS). Technology mainly consisted of copper wires and electromagnetic switches. Other telecommunications services during this era included telex, telegraphy, and data and facsimile transfer. Television and radio services were considered separate. Now, through digitalization and technological convergence, the connotation of *telecommunications* involves the transfer of information—digital, data, voice, image, video, and sound. Although this industry has been regulated by the government, market forces initiated the development and application of new technologies. Now the boundaries that define what is public and what is private are rapidly being redefined.

The accelerated growth rate of the telecommunications industry illustrates a founding capitalist principle that consumers are better served in an open system of free markets rather than through a regulated economy. Rapid technological improvements helped spark the breakup of the American Telephone and Telegraph Company (AT&T). Once a government-protected monopoly, AT&T was regulated by a state public utility commission that imposed financial restrictions. However, on January 1, 1984, that all changed when AT&T settled an antitrust suit with the Justice Department by initiating the breaking up of the Bell system. This breakup included the creation of seven regional holding companies

(RHCs) and their original twenty-two subsidiaries (now there are twenty) called the Bell Operating Companies (BOCs). These seven RHCs or "Baby Bells" include: Ameritech, Bell Atlantic, BellSouth, Nynex, Pacific Telesis, SBC Communications (which was known as Southwestern Bell prior to 1995), and U.S. West. See the following map for the service regions. The acronym *RBOC* (regional Bell Operating Company) is used to define both the regional holding companies and the Bell Operating Companies.

At the time of divestiture, restrictions were put on both the RBOCs and AT&T. The RBOCs could not provide long distance service across state lines or across the boundaries of specified calling areas within states; they also could not manufacture telecommunications equipment or provide information services, such as electronic Yellow Pages, cable television, and the like. This restriction on information services was lifted in late 1991; the remaining restrictions were removed in the 1996 Telecommunications Act. AT&T was no longer permitted to provide local telephone service, only long distance. Also, new firms were permitted to compete with AT&T for long distance telephone service. AT&T retained its manufacturing operation (Western Electric) and its research and development subsidiary (Bell

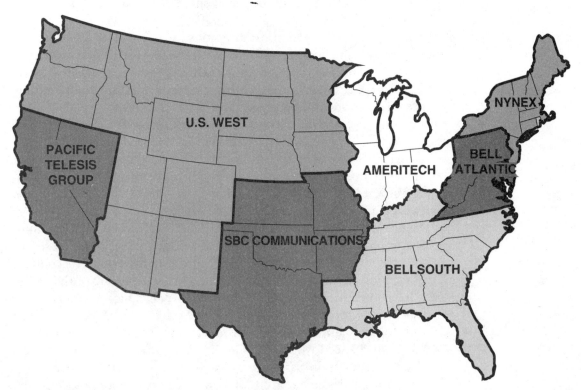

Service area of original seven regional holding companies.

Labs). In addition, the RBOCs established Bellcore to direct their business activities. With over 5800 employees, Bellcore researches business solutions for telecommunications carriers, businesses, and governments worldwide.

Replacing the long-distance telephone service monopoly with competition has had many interesting effects. The average cost of long distance service has dramatically dropped (from about $0.50 per interstate minute in 1982 to about $0.10 in 1997—an 80 percent decrease). Also, interstate computer telephone use has almost tripled since 1984. However, drastic downsizing and reorganization were also results of this movement to a freer market (AT&T, including Lucent Technologies, had over a million employees in the early 1980s; the company has about 400,000 now). Data indicates that employment in the telecommunications industry (as a whole) has remained high. Thus, the breakup of AT&T did not necessarily destroy jobs; rather, many were relocated.

Ameritech

American Information Technologies Corp (Ameritech) serves 11.1 million homes and 1 million businesses with local telephone service. It provides local, long distance, paging, and cellular telephone service, cable TV, security monitoring, electronic commerce, managed services and wireless data communications for much of the U.S. and many parts of Europe. Ameritech subsidiaries (operating companies) include Illinois Bell Telephone Company, Indiana Bell Telephone Company Incorporated, Michigan Bell Telephone Company, the Ohio Bell Telephone Company, and Wisconsin Telephone Company. Internationally, it is involved in Austria, Belgium, China, Croatia, the Czech Republic, Germany, Hungary, Luxembourg, the Netherlands, New Zealand, Norway, Poland, Slovakia, Slovenia, and Switzerland.

Bell Atlantic Corporation

Bell Atlantic's core telephone operations are its local networks that provide advanced voice and data services to 11 million households via 20 million residential and business access lines in six states and Washington, D.C. Bell Atlantic's Advanced Intelligent Network (AIN) is rapidly broadening to include state-of-the-art voice, access, and video services through applications like its Integrated Services Digital Network (ISDN). Bell Atlantic's subsidiaries include New Jersey Bell Telephone Company, the Bell Telephone Company of Pennsylvania, the Chesapeake and Potomac Telephone Company, the Chesapeake and Potomac Telephone Company of Maryland, the Chesapeake and Potomac Telephone Company of Virginia, the Chesapeake and Potomac Telephone Company of West Virginia, and the Diamond State Telephone Company. Internationally speaking, Bell Atlantic is involved in the telecommunications markets of New Zealand, Mexico, Italy, Canada, Australia, Slovakia, and the Czech Republic.

BellSouth

BellSouth Corporation is the largest of the seven holding companies formed at divestiture. It is headquartered in Atlanta, Georgia. With $34 billion in assets, over 20 million access lines, 2.5 million cellular customers worldwide, and 92,000 employees, BellSouth offers local telephone service in 9 southeastern states and provides wireless communications services worldwide. It has recently begun to offer Internet access. BellSouth was the parent company of both Southern Bell Telephone and Telegraph Company and South Central Bell Telephone Company. Today, the regions have been combined, and the entire area is referred to as BellSouth. The BellSouth companies operate throughout the United States and in a number of countries in Europe, Latin America, Asia, and the Pacific.

NYNEX

Nynex took the first four letters of its name from the market it was to serve: New York and New England. The closing "X" represents excellent future prospects. The company is involved in telecommunications, wireless communications, cable television, directory publishing, video entertainment, and information services. Nynex's 67,600 employees and $26.2 billion total assets help support its 13 million telecom customers with 17.6 million domestic access lines and 1.3 million miles of fiber-optic (conductor) lines. Nynex subsidiaries include New England Telephone and Telegraph Company and the New York Telephone Company. Nynex expects to offer long distance service outside its region by midsummer and expects to receive approval for in-region service in New York by the end of the year. Nynex Corporation is a global communications and media company that provides a full range of services in the northeastern United States and select markets around the world, including the United Kingdom, Thailand, Gibraltar, Greece, Indonesia, the Philippines, Poland, Slovakia, and the Czech Republic. Nynex's CableComm, the second largest cable TV and telecommunications operator in the United Kingdom, now passes 1.2 million of 2.7 million homes in 16 franchise areas.

Pacific Telesis Group

One of the nation's largest telecommunications providers, Pacific Telesis Group, has its headquarters in San Francisco, California. With nearly $16 billion in assets, it is among California's largest businesses. In 1995, its revenues totaled $9.04 billion. Pacific Telesis Group owns both the Pacific Telephone and Telegraph Company (PacBell) and the Bell Telephone Company of Neveda (Nevada Bell), which together serve some 20 million customers, providing local and toll telecommunications services, consumer and business broadband networks, data services, and access to long distance and other providers. During the first part of 1997, the company aggressively competed with current long distance companies for customers in California and Nevada, which account for approximately 10 percent of the $75 billion national long distance industry.

SBC Communications

SBC Communications Inc. headquarter's is located in San Antonio, Texas. SBC's largest subsidiary is the Southwestern Bell Telephone Company, which accounted for 63.4 percent of its earnings in 1995. Southwestern Bell serves more than 13 million customers with 14.2 million telephone access lines in its five-state operating region. Southwestern Bell Telephone has ranked first in the telecommunications industry in providing quality service to small business customers. The installation of new network technologies such as ISDN (integrated services digital network) and SONET (synchronous optical network) has expanded Southwestern Bell Telephone's network capacity to include voice, data, and video transmissions. SBC Communications offers local and cellular phone service and now also long distance and cable. Internationally, there are operations in Mexico, France, Chile, South Africa, South Korea, the United Kingdom, and Israel.

U.S. West

In 1988, U.S. West's subsidiaries originally included three Bell operating companies (Mountain States Telephone and Telegraph, Northwestern Bell Telephone Company, and Pacific Northwest Bell Telephone Company), which are now together referred to as the U.S. West Communications Company. The U.S. West Communication Group provides telecommunications services to 25 million customers in 14 western and midwestern states. In February 1996, U.S. West announced it would merge Continental Cablevision into its media group, which would become the nation's third largest cable company. The U.S. West Media Group operates domestic and international cable television and wireless communications networks, publishes directories, and develops and markets interactive multimedia services. Today, the company has interests in cable systems in 10 countries, including the world's largest combined cable and telecommunications system, which is in the United Kingdom. U.S. West has $23 billion in assets and 61,500 employees. U.S. West also has extensive interests in wireless communications services in seven countries. With its Time Warner properties, U.S. West manages 16.2 million domestic cable customers and has access to some 13.9 million homes abroad.

These RBOCs, along with other independent telephone companies, have recently begun to go through some restructuring. On February 8, 1996, the Telecommunications Act of 1996 became law. This act was intended to lift the ban of competition in the telecommunications marketplace within the United States. It was designed to accelerate rapidly the private sector deployment of advanced telecommunications and information technologies and services through a free-market policy. Local exchange carriers (LECs), which formerly provided local service only in rigidly defined geographical areas, could now enter the long distance market. This act requires incumbent local exchange carriers to offer their competitors interconnection and unbundled access to their networks on reasonable terms, as well as the resale of their networks at wholesale rates. The long distance carriers could now operate in the local phone market. Currently, AT&T offers local toll service in 45 states. Multiple system operators (MSOs), who were traditionally in the cable business, were also impacted by this policy. New technologies enabled the MSOs to enter the telephone service market.

MSOs

A cable company that operates in more than one system or strictly defined geographical area determined by the headend (electronic control center of the cable system) is considered a multiple system operator (MSO). The cable industry has had an accelerated growth rate since its beginning in 1952, when it consisted of 70 operating systems and 14,000 subscribers. In 1985, it had 6600 operating systems and 32 million subscribers. In 1995, there were 11,351 operating systems and 56.5 million subscribers, according to the *1995 Cable Television & Cable Factbook*. Nearly 85 percent of these cable operators have a membership with Cable Television Laboratories (CableLabs), established in 1988 as a research and development consortium. The National Cable Television Organization (NCTA) is also a membership organization that makes recommendations to the Federal Communications Commission (FCC). Six of the largest MSOs are discussed hereafter.

Cablevision

Cablevision Systems Corporation is the nation's sixth largest operator of cable television systems. Headquartered in Woodbury, New York, the company serves 2.8 million cable television customers in 19 states. Cablevision owns American Movie Classics, Bravo, SportsChannel, and the Independent Film Channel. Cablevision is now turning its talent for innovation to new areas of electronic communications. In December 1996, the company launched Optimum Online, a high-speed multimedia communications service, to 15,000 homes in the northern part of the township of Oyster Bay, New York. During 1997, Cablevision expects to make this Internet- and online-access service available to more than 150,000 Long Island homes as well as to Cablevision customers in Connecticut.

Comcast Corporation

Today, Comcast provides cable to over 3.4 million subscribers in 18 states and offers cellular communications in markets with a population in excess of 7.5 million in the tristate area of Delaware, New Jersey, and Pennsylvania. Comcast also provides cable and telephone services in the United Kingdom. Comcast Corporation is principally engaged in the development, management, and operation of wired telecommunications, including cable television, telephone services, and telecommunications (cellular, personal communication services, and direct-to-home satellite television).

Continental Cablevision

In 1964, Continental Cablevision was developed in Tiffin, Ohio. Today, it is the nation's third largest MSO, serving over 4 million subscribers. It played a major role in the development of several leading programming services, including helping found the Cable-Satellite Public Affairs Network (C-SPAN). Continental is also an owner of Turner Broadcasting System (CNN, TNT, Cartoon Network), Entertainment Television, and Viewers Choice, the nation's leading pay-per-view service. Continental is also a partner in PrimeStar, the nation's first direct broadcast satellite (DBS) service.

Cox Communications, Inc.

Cox Communications, Inc. (NYSE: COX) is among the nation's largest multiple system operators, serving 3.2 million customers. Acquiring Times Mirror Cable Television increased the number of customers served from 1.9 million to 3.2 million. To reflect the increasing concentration on new businesses such as data services and telephony, the company dropped the word *cable* from its name, becoming Cox Communications, Inc. Then it formed a telecommunications venture with Sprint, TCI, and Comcast and won licenses to deliver PCS wireless communications in 31 major metropolitan areas. As a full-service provider of telecommunications services, Cox has interests in wired telecommunications, including cable television and telephone services; wireless telecommunications, including personal communications services (PCS); and direct-to-home (DTH) satellite television and programming networks.

TCI Group

Tele-Communications, Inc. was formed in 1968, when a small cable company, Community Television, Inc., and a small common carrier microwave company, Western Microwave, Inc., merged. The two wholly owned TCI subsidiaries were renamed Community Tele-Communications, Inc. (CTCI) and Western Tele-Communications, Inc. (WTCI), respectively. TCI's original name was American Tele-Communications, Inc., but was changed in late 1968 to avoid conflict with other competitors. TCI became the nation's largest cable operator in 1982, when it passed the two-million-subscribers milestone. In 1993, a joint venture between TCI, Sega of America, and Time Warner resulted in the Sega Channel. The Discovery Channel, Black Entertainment Television, and CNN are also owned by TCI. Today TCI distributes cable TV to 14 million customers through more than 30,000 employees. It invests in technology development and delivers digital television, telephone, and Internet services via its broadband network. ATCI is a leading operator of cable television, telephone networks, and programming interests in Europe, Latin America, and Asia.

Time Warner

Time Warner is the nation's second-largest MSO with over 11.5 million subscribers. During 1994 and early 1995, Time Warner Cable and Time Warner Inc. entered into agreements to increase the number of subscribers under management from 7.5 million to 11.5 million; also a merger between Time Warner Inc. and Cablevision Industries, added about 1.3 million subscribers. This entertainment giant owns Warner Music Group, Time, Inc., Warner Brothers, and HBO. In the last few years, Time Warner continued its aggressive rollout of fiber-optic cable and multifaceted entry into the telephone business, launching the Full Service Network™. The FSN (as it is called) debuted on December 14, 1994, in Orlando, offering true video-on-demand with full VCR functionality (the ability to fast forward, rewind, and pause), a variety of interactive home-shopping services, video games, and an interactive program guide.

Current Trends

The Telecommunications Act of 1996 recognized that the cable industry faces increased competition from alternative video providers, including wireless multichannel/multipoint distribution system (MMDS) cable, direct broadcast satellites (DBS), telcos, and broadcasters. In order to raise the necessary capital to compete with these powerful competitors, Congress deregulated cable rates under certain terms and conditions. This law also created greater flexibility for acquisitions, mergers, and trading between and among telcos and MSOs. Within two weeks of the law, U.S. West made a $10.8 billion bid for Continental Cablevision. A month later, SBC Communications announced its plan to buy Pacific Telesis, which would form the first union between two Baby Bells. Then, in late April, Bell Atlantic and Nynex ended two years of discussions and announced the largest merger in telecommunications history, a $23 billion deal creating a company second in size only to AT&T. In January 1997, Cox Communications, Inc., and TCI Communications, Inc. announced the completion of a trade of cable television systems representing approximately 600,000 customers. However, the factors involved in these mergers, trades, and acquisitions can be as volatile as the discussions themselves. Since the passage of this law, many discussions have taken place between and among telcos and MSOs. While some telephone companies are looking into cable television, others are choosing direct-broadcast satellite TV. Most are looking for ways to offer high-speed access to the Internet.

Odyssey's homefront study indicated that 37 percent of American households own a personal computer and 46 percent of these (or 17 percent of total U.S. households) are online (March 1997). These figures have been steadily increasing since the Internet became so popular. The exponential growth rate of the Internet astounded most of the telecommunications industry. In the early 1970s, ARPANET (Internet parent) was introduced. The 1980s brought about the linking of the public and commercial networks. Last year, the Internet as a whole doubled in size, as it has done every year since 1988. (Robert Hobbes). The World Wide Web (WWW) had a 342 percent annual growth rate in 1993 since its release (by CERN) in 1991. In 1993, the Internet emerged as more than just a way to send an Email or download an occasional file. It became a place to visit, full of people and ideas. It became *cyberspace*. Its impact quickly spread as everyone wanted to experience this "virtual community." The Internet frenzy is a worldwide phenomena. Europe, Asia, and all other developing countries are expanding and building the Internet infrastructure at a faster pace than the telco. Recently I went to Nazareth-Israel unexpectedly, and, to my surprise, the newest generation of my family in this little biblical town was talking about nothing but being wired to the Internet and its services and market potential.

Customers interested in Internet access have a choice of using an Internet service provider (an ISP such as Netcom, AT&T/WorldNet, MCI, and PSI/Pipeline) or a commercial online service (AOL, CompuServe, Prodigy, MSN, WOW). Internet service providers have a direct connection to the WWW, access to usenet newsgroups, and Email, and they tend to be cheaper and faster. Commercial online services usually provide other services, some of which cost extra to activate. According to a survey done at Odyssey in San Francisco, 48 percent of the online households were using an ISP, while 35 percent were using commercial online services as of July 1996. This is a switch, considering these figures were flipped around only six months prior to this study, when more households were using commercial online services. However, the exponential growth of the Internet was not accurately predicted and therefore nobody—not the telephone companies, the ISPs, the commercial online services, or the MSOs—was prepared to accommodate these new customers. Busy signals and periodic shutdowns created irate customers, which in turn caused lawsuits. States ended up suing Internet providers, which in turn sued telcos. This chaos created a high-demand market for high-speed access to the Internet.

The telephone companies (telcos) intend to provide high-speed access to the Internet with basically two different products: ISDN (integrated services digital network) and ADSL (asymmetric digital subscriber line). ISDN is offered by most of the telcos; it has four times the bandwidth of a standard 28.8 modem. It can run 6–10 times faster than a 28.8 modem. Initial installation of an ISDN line currently runs at approximately $400; then there is a monthly charge of approximately $50 for unlimited access. ADSL has a speed almost 500 times that of an ISDN line. The cable modem is expected to be in the same class as ADSL. There are several cable operators today also competing for these same customers, the ones interested in high-speed Internet access. The technical details of the high-speed cable modem will be found in the following chapters.

On February 8, 1997, more than 60 member countries of the World Trade Organization endorsed a landmark agreement in Geneva to open up their telecommunications markets to all rivals. Countries agreed to end protections for satellite and telephone monopolies, in most cases by the year 2000. More than half of the countries agreed to adopt legally binding regulatory principles that force traditional telephone monopolies to let new rivals connect with their networks at cost-based prices. This will dramatically change the lives of consumers in countries where monopolies control the telecommunications industry and provide poor service at inflated prices. The United States refused to participate in this negotiation last year to hold out for more concessions and a freer trade agreement, which were eventually obtained. The United States is lighting the way for the telecommunications global community.

An abundance of technological alternatives in the telecommunications industry, combined with domestic and international deregulation, has restructured its economic growth potential. In order to survive the stiff competition resulting from the move to a freer market, many telecommunications companies have discussed and decided in favor of acquisitions, merges, joint ventures, and trades. These business discussions between former competitors illustrate just how innovative these industries can be when necessity spurs creativity. The popularity of the Internet has created an astounding and quickly growing new market for high-speed Internet connectivity, an arena in which many different powerful players are competing. The high-demand market for digital, data, voice, image, video, and sound transmission, all at an accelerated speed, has created the necessity for the latest and greatest in the telecommunications world—ADSL for the telco network and the high-speed cable modem for the MSOs. This book describes the details of the high-speed cable modem.

HFC Architecture and Reference Model

1.1 Overview

In this chapter, the cable network infrastructure will be described. The technical and economic challenges to develop and deploy a cable modem in such an environment is very complex, both technically and economically. The cable network architecture was developed for service-specific application and hence was optimized to carry broadcast video services efficiently to the home. Today, users are becoming very sophisticated in terms of their perception of what the network services must provide to satisfy bandwidth-hungry applications. These applications are real today, and entrepreneurs are busy developing even more demanding applications, burdening the network services further. Cable modem will play the major role in providing this bidirectional interactive communication.

The U.S. governmental agencies, with encouragement from the president and vice president, are also championing the effort to provide information access to every citizen in the United States. It is felt that this information superhighway must become part of our daily lives if we, in the U.S., want to maintain our competitive edge in this global economy.

Today's cable network model cannot cope nor was designed to provide these new interactive services. This picture gets even more complex because of the unpredictability of future applications. One might conclude that cable network operators might just forgo this phase of turmoil and "wait and see," as they have often done in the past. Such a notion is also disappearing fast—the Telecommunication (reform) Acts of 1996 will give the telco operators access to the video market. The RBOCs have a deep pocket, and alternate technology is being developed to provide these demanding video services. The digital video broadcasting faction is also entering the market strongly, and its presence is taking the attentions of the large cable companies, if not all of them. It is beyond the scope of this book to describe such a network, but, simply stated, switched digital video and digital TV technology is mature enough and can just as easily take on this task in the near future. Broadband with ATM (asynchronous transfer mode) switching fabric can provide digital video distribution and switch digital video.

If the cable operators took the defensive strategy, then they would rapidly lose their revenue stream to others. They could find themselves in a dilemma and unable to recover fast enough to reclaim their market share. Taking the offensive strategy by providing these demanding network services will be a safer route to take. Such strategy also makes economic sense, since these new applications are considered new market opportunities to them. The challenge, of course, is how the MSOs should approach such a challenge without going broke. CATV networks of today need a major overhaul in order to cope with the applications requiring integrated bidirectional communication.

This chapter is divided into two parts:

1. The cable network in general and the upgrading strategy to hybrid fiber coax (HFC) that is being deployed today

2. The rationale of HFC modernization and an economic model

1.2 History of the Cable Network

The cable network was originally deployed to perform a very simple task. Reception of a TV signal was very poor, especially in suburban areas where the middle class began moving in the 60s. The idea of providing CATV (community antenna television) to improve TV signal reception took off fast, and hence a CATV market emerged. This CATV network used coax shielded cable to deliver strong and equal-in-strength TV signals to the home. A good quality antenna tower received TV channels (6 MHz) from the airwaves and mapped them in the cable spectrum. In North America, the bandwidth 50–550 MHz is reserved for NTSC analog cable-TV broadcasts, as shown in Fig. 1.1.

The TV signals in the cable coax are replicas of the signal broadcasted through the airwaves, so no modifications were needed to the modern television set. Analog amplifiers were installed to compensate for the attenuation of the long coax cable lines. CATV brought about yet another advantage: the ability to provide more channels, with signals equal in strength, to the end user than the usual regular reception through the

Figure 1.1
CATV cable spectrum.

Upstream Channel   Analog Broadcast Channels (6 MHz)   Downstream Digital

0 5   42 50 MHz   550

airwaves. TV signals lose power more readily in the air, and TV receivers and tuners cannot cope with the interference of more powerful adjacent signals. Introducing more and more TV channels to the cable network attracted the general public regardless of the geographic area, not just to those homes with poor TV reception.

1.3 The Road to Regulation

Cable operators became more powerful as more and more homes subscribed to the service. Cable operators became more like broadcasters than TV retransmitters. TV signals from metropolitan areas replaced local channels to make CATV more attractive to subscribers. At this time, the FCC began to regulate the industry and dictate what operator and which channels must be carried to serve the local community.

More regulation followed to deal with monopolies and price control. The Cable Communications Policy Act of 1984 took effect to ease price control due to competition from broadcasters and to encourage growth. By that time, cable operators had become content providers as well.

In 1992, the U.S. congress passed the Cable Television Consumer Protection and Competition Act and reenacted price control regulation with some exceptions.

With the Telecommunication (reform) Act of 1996, congress deregulated the industry with the proviso that the telco industry can now compete in video services and, of course, that the cable operators can also enter the local telephone service market.

1.4 Traditional Cable Network

The technologies of the 60s and 70s were readily available to provide CATV broadcasting services. The networks were built independently to serve particular communities, so the economic model, more or less, dictated a simple and somewhat organized topology of a branch-and-tree architecture. A point-to-point approach was economically prohibitive and did not offer an advantage over a shared medium access, especially for broadcasting applications. Figure 1.2 illustrates a traditional cable network. IEEE 802.14 identified the following functional elements:

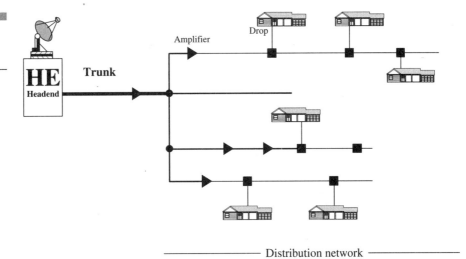

Figure 1.2
Cable system topology.

Distribution network

1. Cable TV headend
2. Long haul trunks
3. Amplifiers
4. Feeders
5. Drops

1.4.1 Headend

The cable TV headend is mainly responsible for the reception of TV channels gathered from various sources: broadcast television, satellite, local community programming, local signal insertion, and so on. These 6-MHz TV analog channels are modulated using the frequency division multiplexing technique and are placed into the cable spectrum as shown in Fig. 1.1.

Depending on a particular economic model, this central control headend can serve thousands of customers by a simple distribution scheme. To achieve geographical coverage of the community, the cables emanating from the headend are split into multiple cables. When the cable is physically split, part of the signal power is split off and sent down the branch. The content of the signal, however, stays intact. In that way, the same set of TV signals reach every subscriber in the community. The network thus follows a logical bus architecture, as shown in Fig. 1.2.

1.4.2 Trunks

High-quality coax cables are used as trunks to deliver the signals into the distribution network and finally to the intended destination. The trunk can be as long as 15 miles. Lower-quality coax are used in the distribution and drop portions of the plant.

1.4.3 Amplifiers

TV signals attenuate as they travel several miles through the cable network to the subscribers' homes. Therefore, amplifiers have to be deployed throughout the plant to restore the signal power. The more times the cable is split and the longer the cable, the more amplifiers are needed in the plant. An excessive cascade of amplifiers in the network creates signal distortion. Amplifiers are also located in the distribution network (sometimes referred to as the last mile). These amplifiers, used in the traditional cable network, are one-way (amplifying the signal from the headend to the subscriber). This scheme introduces several potential problems when a network needs to be upgraded to provide bidirectional communication. In such a case, these amplifiers need to be replaced with new two-way amplifiers.

Amplifiers in the cable network introduce yet another problem. Power to operate the electronics is needed. AC power (60 Hz) is delivered to the amplifier box via the coax cable or can be picked up from various points in its route from power lines. Reliability then comes into play in several areas. Among the most crucial are:

- The system will be vulnerable to power outage. All users lose their communication service when AC fails.
- When active components in the amplifiers fail, thousand of residents may be affected (depending on where the failure is).

Reliability issues are described in Chap. 12.

1.4.4 Feeders

Feeders are sometimes referred to as the distribution network serving residential areas or small communities. The term *home passed* usually refers to homes that are near the distribution network. The coax cables in the distribution network (branch/tree) are usually short and are in a range of

1–2 miles. Amplifiers, therefore, will still be required. The distribution network feeds the lines connected to the neighborhood residential area via bridge amplifiers.

1.4.5 Drops

Drops are usually located on telephone poles or, more recently, in residential pedestrian areas. A lower-quality coax is used to connect from the drop to the home.

1.5 HFC Network

The HFC access network is based on a hybrid of a fiber part and a coaxial part. The fiber extends from the access node to a neighborhood node. This fiber node serves typically about 500–2000 subscribers via coaxial cable drops. These connected subscribers share the same cable and thus the available capacity and bandwidth of this cable. HFC is the first step needed to provide bidirectional services. Because several subscribers share the same downstream and upstream bandwidth, special requirements like privacy and security measures have to be taken into account. A special medium access control (MAC) scheme is required in the upstream direction to prevent collision of information transmitted from customers to the headend. The HFC network is shown in Fig. 1.3.

Figure 1.3
Hybrid fiber coax
topology.

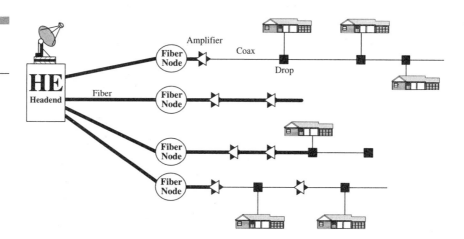

This HFC topology is the choice of many MSOs for new construction or when building or upgrading the older cable system described in Sec. 1.4. It must be emphasized that the distributions part and the drops in the HFC network still remain coax cable. Fiber is just as cheap to install at these locations, but the connector technologies are still prohibitively expensive. Simply put, HFC enhances a bidirectional shared-media system using fiber trunks between the headend and the fiber nodes, and coaxial distribution from the fiber nodes to the customer locations. CableLabs estimates that 12 percent of the cabling is in the trunk portion of the network, 38 percent is in the feeder portion, and the rest is in the drop portion of the network. Here is a summary of the notable changes from the traditional cable network:

- Trunks are deployed in starlike fiber links with a maximum distance between headend and end user of 50 miles (80 km)
- Drops and distribution portion are still coax cables
- Max number of subscribers to fiber node ranges between 500 to 2000 subscribers

1.6 The HFC Advantage

The technical and economic advantages leading the MSOs to embrace the HFC architecture are:

- *Technical.* The fiber truck no longer needs amplifiers. Fiber is less immune to noise, and signal attenuation is practically nonexistent. Amplifiers are still needed at the distribution part of the network, though in limited numbers (3–4 tops). This has the obvious advantages of increasing reliability; hence, an amplifier failure affects only that particular residential area.
- *Economic.* Fiber in the trunk eliminated the cost of maintaining the amplifiers. That in itself would pay (in the long term) for rebuilding the infrastructure. Fiber deployment also means that far more bandwidth/channels will be at the dispense of the cable operator, and thus would be available in the network for the subscriber.

The other advantage of this HFC architecture is its ability to incorporate the communication that is essential to develop a network that can cope with the bidirectional services of contemporary and future applications.

1.7 Bidirectional Communications

Both the traditional cable network and HFC described a unidirectional TV signal distribution from the headend to the subscriber. The fiber and coaxial cable can just as easily accommodate a bidirectional communication and signal from the subscriber to the headend. The directionality of the network from the headend to the subscriber was intentionally designed for one-way TV distribution. Hence, the amplifiers deployed in the network conditioned the TV signals from the headend to the subscriber. Cable systems with bidirectional communication use amplifiers that work in both directions. To accomplish this, back-to-back amplifiers with filters are arranged so that downstream signals (from headend to subscribers) are first filtered and then amplified. Similarly, the upstream signals (from subscribers to headend) will also be filtered and then amplified. The upstream path has an inherent disadvantage because of the branch and tree topology. During amplification of the upstream, the splitter's outputs become its inputs. The splitter simply combines the incoming signals and noise; hence, both are amplified. In the downstream direction, the signals passing through a splitter are attenuated on the splitter outputs, but the noise carried downstream is also attenuated. The upstream-assigned channels (in the frequency range between 5–42 MHz) are also more susceptible to noise. That exasperates the problem even further. See Chap. 7 for more details.

1.8 HFC Capacity

Cable networks were initially designed to support fifty or more analog TV channels (6 MHz each). A study made by Bell Atlantic showed that a fully equipped HFC can deliver:

- Voice telephony
- Up to 37 broadcast analog TV channels
- 188 broadcast digital TV channels
- Up to 464 digital point cast channels (on-demand customer-requested programming)
- High-speed two-way digital data links

In this section we will address the various CATV network topologies and reference models that are emerging in the cable networks.

1.9 CableLabs

Cable Television Laboratories, Inc. (CableLabs) was established in May 1988 as a research and development consortium of cable television system operators. CableLabs is a membership organization; to be a member, a company must be a cable television system operator. Vendors of cable equipment or other telecommunications providers are not eligible. Member companies pay dues based on their subscriber base. CableLabs currently represents more than 85 percent of the cable subscribers in the United States, 70 percent of the subscribers in Canada, and 10 percent of the subscribers in Mexico.

1.9.1 Regional Hub

CableLabs recommended the topology as shown in Fig. 1.4. It was specifically developed for the client MSOs, accommodating various economic models and network needs. IEEE 802.14 recognized such architecture and developed the cable modem standards accommodating this topology.

Figure 1.4
Regional hub HFC
network architecture.

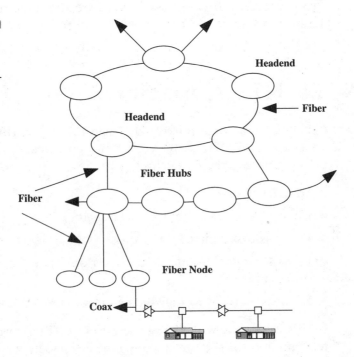

The architecture includes:

■ Centralized regional hub to common equipment among multiple operators within the region
■ Fiber ring connecting hubs and headends providing routing capabilities
■ No amplifiers deployed in the fiber ring or otherwise
■ Maximum distance between farthest end-user node and fiber hub, 50 miles or 80 km

1.10 IEEE 802.14

IEEE (Institute of Electrical and Electronic Engineers) is an international standard body that is ANSI-affiliated. IEEE 802.14 is an official working group under the IEEE umbrella. This IEEE 802.14 working group was chartered to provide for digital communication services (via cable modem) over a CATV branching bus system constructed from fiber optics and/or coaxial cable, as used in the cable TV distribution network. The bidirectional traffic types will be constant bit rate, variable bit rate, and connection- and connectionless-oriented data. See Chap. 11 for more details.

1.10.1 IEEE 802.14 Reference Model

The IEEE 802.14 committee developed a logical reference model. This model, shown in Fig. 1.5, was used as a guide to address the functional requirements of the cable modem in terms of both the MAC (medium access control) and the physical (PHY) layers.

Future cable systems will be limited in size to accommodate various economic models. A region may have ten or twenty different cable systems. IEEE 802.14 is undertaking the task to standardize a cable modem and to perform functions other than one-way TV broadcasting. The other functions as depicted in the reference model include: telephony switches, video on demand, regional video servers, information database access, and telecommuting. This is illustrated in Fig. 1.5 in the "outside world" cloud.

The IEEE 802.14 standard is devoted to communication over the headend. The ability to interoperate efficiently with regional networks was an important consideration in the selection of protocols and design param-

Figure 1.5
IEEE 802.14 network
logical reference
model.

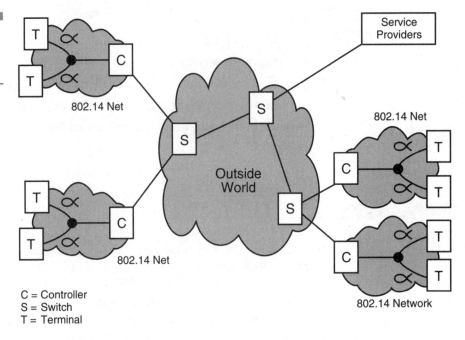

C = Controller
S = Switch
T = Terminal

eters for the branching bus networks and headends. When multiple 802.14
networks are involved, it is necessary to move information from one network to another. The IEEE 802.14 standards did not preclude moving data
from an originating 802.14 network to a destination 802.14 network. Communication between two peers within the same cable system involves
upstream transmission on one branching bus and echoing it back by
downstream transmission from the headend to the subscribers.

1.11 ATM Forum Reference Model

The ATM Forum is an international nonprofit organization formed with
the objective of accelerating the use of ATM (asynchronous transfer
mode) products and services through a rapid convergence of interoperability specifications.

The Residential Broadband (RBB) is a working group of the ATM
Forum. It was formed with the charter to develop access nodes as well as
their associated interfaces. The RBB viewed the HFC in a generic sense
with exclusive emphasis on integrating ATM technologies to the various

network components. Figure 1.6 illustrates the RBB generic model, accommodating the various access architectures including HFC. ATM all the way to the home was one of the RBB's goals when it was officially formed. For example, ATM is to be deployed as a transport as well as service platform. The RBB main charter was to define and complete a set of specifications for an end-to-end ATM for all access nodes including HFC. That also meant ATM to the home and inside the home.

The main goal of the RBB group is to define and specify the following interfaces:

- ANI (access network interface)
- User network interface ($UNI_{w\,x\,and\,y}$) where w = HFC, FTTC, FTTH, VDSL, ADSL, and others
- TII (Technology Independent Interface) with the goal of using existing Forum specifications, ANSI (T1E1), and/or other standards bodies

The RBB group is also looking at minimizing the number of UNI_x variations. To that end, they are working very closely with IEEE 802.14 to define the MAC and physical layers for the HFC network.

1.11.1 RBB's HFC Reference Model

The RBB's view on HFC architecture represents the specific application of the reference model when $UNI_w = UNI_{HFC}$. In this case, the cable modem comes into play as shown in Fig. 1.7. Details of all the functional components shown are described in Chap. 3.

Figure 1.6

Residential Broadband reference configuration model.

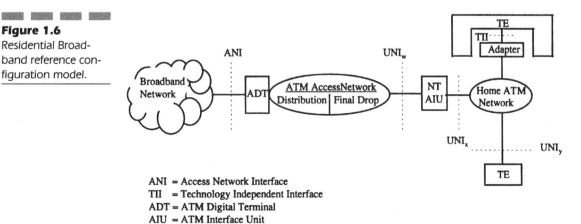

ANI = Access Network Interface
TII = Technology Independent Interface
ADT = ATM Digital Terminal
AIU = ATM Interface Unit

Figure 1.7
ATM Forum reference
model for the HFC
network.

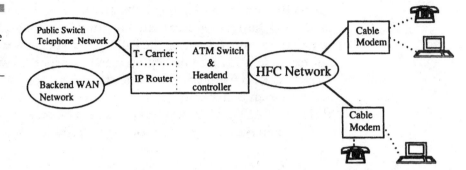

1.12 DAVIC

DAVIC (digital audiovisual council) is a nonprofit international organization started in January 1994 with an initial charter to develop FSN specifications. DAVIC is supported by big conglomerate international companies and is organized into various working groups. Among the related groups are: Set Top Unit Group, Video Server, System & Application, and Delivery Systems.

1.12.1 DAVIC's HFC Reference Model

DAVIC's approach is similar to the ATM forum when defining the access and network nodes. The architecture is generic and accommodates all the various access technologies such as HFC, FTTC, and FTTH. Figure 1.8 illustrates a network architecture including the HFC. Reference points are defined as A1 to A9. A1 to A4 represent the access interfaces and are described below.

Figure 1.8
DAVIC's reference
model.

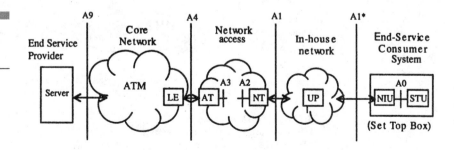

The separation between the core network and the access network is formed between the local exchange (LE) and the access node (AN). The separation between the access network and the in-house network is formed by the network termination (NT). DAVIC considers the access node and the network termination to be part of the access network. A0, A1*, A2, A3, and A4 reference points separate the various system entities.

The end-service consumer system contains a set top box that is divided into two functional components:

1. Set top unit (STU)

2. Network interface unit (NIU)

It is worth noting that the A0 reference point is identical to the TII interface defined in the RBB reference model.

1.12.2 Reference Points Definitions

The reference points defined by DAVIC specification release 1.1 are as follows:

- A1 is defined between the access network and the end consumer in-house network. For a passive NT, A2 and A1 reference points are identical.

- A1 and A1* are identical. In the future, A1* may be a superset of A1, incorporating new in-house network architectures.

- A2 and A3 reference points may differ, and that depends on the optional active components inside the access network.

- The A4 reference point is defined between the access node and the local exchange. The access node takes the signals from the core network. This resembles the ANI interface defined in the RBB group.

1.12.3 DAVIC's HFC View

DAVIC's view on HFC is as shown in Fig. 1.8. The fiber extends from the access node (headend) to a neighborhood node. This neighborhood node serves about 100 to 500 subscribers. These subscribers share the same downstream and upstream bandwidth; hence, special requirements like bandwidth allocations, privacy, and security have to be taken into consideration. A special medium access control (MAC) scheme needs to be devised for the upstream direction to prevent collision of the information that is transmitted from customers to the network.

Fiber extends from the headend (access node) to a neighborhood node. The neighborhood node converts the optical signals to electrical ones transportable across the coaxial network. The coaxial cable serves several subscribers (bus topology) and is fed through a passive NT to the set top box. Figure 1.9 is based on a passive NT; hence, the medium and protocols at reference points A1 and A2 are identical. However, the medium and the lower-layer protocols are different at reference point A3.

1.13 Standards Harmonization Effort

It is clear from the above that several committees are involved in the specifications and standardization effort on HFC. Each of the committees is specialized in a specific area, developing the requirements for HFC. The division of the work can be summarized as follows:

■ IEEE 802.14 is responsible for developing the MAC and the physical layer. It entails a MAC with stringent performance aspect that must meet the services of CBR, VBR, UBR, and ABR.

■ The ATM Forum is responsible for the definition and development of the access network interface (ANI) and the interface to and inside the home at the UNI level. The RBB Group of the ATM Forum is also responsible for all security aspects associated with HFC. The ATM Forum is working very closely with IEEE 802.14 to ensure compatibility, specifically with regard to the MAC and PHY layers. A security and signaling architecture has been passed on to IEEE 802.14 for integrating them into the MAC.

Figure 1.9
DAVIC view on the
HFC access network.

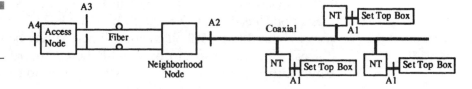

■ DAVIC is also developing specifications addressing the HFC architecture. The mode of operation is quite different from a standard body. A call for proposal is issued to the industry, and, based on input received, a solution is specified. DAVIC is waiting on IEEE 802.14 to deliver the MAC and physical layer specifications so it can be adopted in their specifications.

SCTE (Society of Cable Telecommunications Engineers) and MCNS (Multimedia Cable Network System) are also playing a major role in the specification of a cable modem. They are also addressing network management and other issues covering all aspects of the HFC network. Details on SCTE and MCNS are provided in Chap. 11.

A meeting between officers of IEEE 802.14, the ATM Forum, DAVIC, SCTE, and CableLabs took place recently, and harmonization effort was the key subject of discussion. It was agreed that cooperation was essential between all parties to serve the industry well and create an effective standard and therefore develop specifications. A unified architecture model will be the most likely outcome when all the pieces are eventually put together. A healthy relationship developed, and, as a result, these established liaisons strengthen the communication channels between the groups.

1.14 A Case for Modernization Efforts

As stated previously, today's network, be it telephone, cable, or the Internet, was designed and deployed to provide service-specific application. Subscribers today are much more sophisticated and are now demanding integrated services for multimedia application. This service is no longer a luxury but instead a requirement. The Internet's spectacular success story is a case in point.

Service integration is really nothing new. The telco envisioned such a trend in the 70s and supported the effort in ITU and ANSI to standardize the narrow band integrated service digital network (ISDN). ISDN did not fully succeed because of several factors. Among the most crucial are:

■ Bandwidth was still limited (2B + D = 144 Kb/s total) and therefore did not address all integrated applications.

■ It was not ubiquitous.

- End user equipment vendor support was cool at best.
- User-friendly interactive applications were virtually nonexistent.
- The public was not socially or psychologically ready.

Since then, deregulation, social attitude, technology, and political leadership have played a major role in defining the interactive market. Today, it appears that high-speed interactive services are what consumers want and, more important, are willing to pay for. That being the case, the operators are busy planning to modernize the network to provide such high-speed interactive services. With that in mind, let us examine the modernization plan.

1.14.1 Cable Network Modernization Effort

Modernizing the cable network to provide high-speed interactive service claims to have an advantage over the other networks. HFC bandwidth capacity is enviable. Deploying a cable modem over a modernized HFC might be all it takes to be able to provide today's demanding high-speed interactive applications. A well engineered cable modem (per IEEE 802.14 standard) can provide all the TV cable analog channels, high-speed Internet access, voice, and high-quality interactive video.

The larger cable operators may opt to invest and deploy digital switches (adjunct to the headends) to handle voice traffic and switching. This will elevate cable operators to become telephone service providers, threatening to bypass the RBOC's voice network. Some large cable operators have already tested such a scenario and are deploying it today. Smaller cable operators can simply trunk voice traffic via a T1 for further processing to the RBOC's networks. Access to the Internet can be trunked via an IP router, which may be part or adjunct to the headend.

Some cable operators feel that the lack of switching infrastructure in a cable network is a blessing because of the opportunities it presents. While the RBOCs struggle to reengineer their legacy network, a process that could take many years, the cable industry, on the other hand, will consider its blank drawing board as an asset. It is easier to build a new switching infrastructure from the ground up than it is to upgrade a network that has been in place for years.

HFC ACCESS SHORTFALLS.   It appears that access via an HFC is evolutionary and can be accommodated in a stepwise approach. However, access over HFC can also introduce a host of technically challenging

.problems. Cable operators must address the service effecting problems first and foremost when introducing integrated services. The most crucial ones are:

- Reliability
- Security/privacy
- Operation and maintenance

HFC RELIABILITY. The HFC architecture—one to many, and many to one—although cost effective, is not ideal when it come to service reliability. The following are the critical components that affect service:

- Component failure in an amplifier in the distribution network can render the entire neighborhood out of service. This can be solved by bypassing the failed amplifiers upon failure or providing the more expensive solution of a redundant amplifier (hot standby).
- AC power failure (powering the amplifiers) is a more serious problem that must be resolved. AC power outage can render the entire serving area out of service. Providing AC backup to power the amplifiers must be provided so customers can still make voice calls.
- Because of the shared medium topology, the action of a malicious user can affect the operation and communication of all those connected users in the branch or tree in both directions. A failed cable modem may have the same effect (of disrupting the shared bus), but it is expected that cable modems will be designed to isolate such failure if economically feasible.
- Upstream transmission path is prone to noise of all kinds. The entire cable network must be well maintained to ensure that ingress noise is not leaking into the system and causing failure to users who are on the bus.

HFC SECURITY/PRIVACY PROBLEMS. The HFC topology of a shared medium environment creates several problems facing the cable operators when delivering services. Among these are service theft and eavesdropping on private conversation. In the upstream direction, a malicious user can pretend to be another user and access the network resources. Clever malicious users have successfully done this and faked the ID access security in a cellular network.

Modern cryptosystem technology is being specified in the ATM Forum security working group. It is designed to solve the sorts of privacy and security problems that crop up in cable networks. The ATM Forum is considering the technique of the public key cryptography, which will be effective when applied to the cable network. IEEE 802.14 had adopted this technique and will be partially included in the first release standard. This technology is also being considered for security on the Internet network so it can be used for banking transactions and the like. More details on security are provided in Chap. 10.

HFC OPERATION AND MAINTENANCE PROBLEMS. All network operators, sooner or later, experience network failures. The way service providers respond to these problems will have a major impact on the loyalty a subscriber will have to the service providers. Two of the reasons cable companies are so strongly avoided are:

1. Poor reputation in network quality/reliability
2. Poor customer service

These are the top two attributes that telecommunications managers consider when selecting a telecommunications service provider.

In the media, we hear many stories about the terrible customer service the cable industry has. This reputation must change if cable operators hope to survive the competition resulting from the convergence of telephony services, data services, and broadcast entertainment services. Customer service calls should take no longer than two to three minutes in total. With today's available technologies, there is no justification for a call lasting longer than that. Technologies such as computer telephone integration (CTI), automatic number identification (ANI), and voice response units (VRUs) enable a customer's account information to appear on the screen before the customer service representative even answers the call. These technologies do more than satisfy customers—they also reduce operation costs.

The MSOs are drafting new support infrastructures tailored to accommodate the multiservice platform. If that support infrastructure is built correctly, the cable industry stands a good chance in the struggle to survive the fiercely competitive war between multiple services.

1.14.2 The Telecommunication Network Modernization Effort

Unlike the cable operators, the RBOCs are planning to deploy a broadband network to provide high-speed multimedia interactive services. The access portion for the network was chosen by most RBOCs to be asymmetric digital subscriber line (ADSL) technology. The most important feature of ADSL is that it can provide high-speed digital services on the existing twisted-pair copper network in overlay without interfering with the traditional analog telephone service (plain old telephone service: POTS). ADSL thus allows subscribers to retain the (analog) services to which they have already subscribed. Due to its highly efficient line-coding technique, ADSL supports new broadband services on a single twisted pair. As a result, new services such as high-speed Internet access, on-line access, telecommuting (working at home), VOD, and others, can be offered to every residential telephone subscriber. The technology is also somewhat independent of the characteristics of the twisted pair on which it is used, thereby avoiding cumbersome pair selection and enabling it to be applied universally.

The asymmetric bandwidth characteristics offered by the ADSL technology (64 to 640 kbit/s upstream + 500 kbit/s to 8 Mbit/s downstream depending on the line quality and distance) fit in with the requirements of client–server applications such as WWW access, remote LAN access, and VOD, where the client typically receives much more information from the server than he or she is able to generate. A minimum bandwidth of 64–640 kbit/s upstream guarantees excellent end-to-end performance.

THE ADSL/TELCO NETWORK ADVANTAGE. The telco network is one of the most reliable telephone networks in the world. This was not an accident—both the RBOCs and Bellcore played a major role in specifying stringent requirements before a switch vendor can gain entry into that market. Subscribers expect that such reliability will be maintained. Building redundancies in a telco switch is a very complex subject demanding complex development both in the hardware and software. Data vendors are just beginning to comprehend the implementation complexities when developing redundant systems with terms such as hot standby, N + 1 protection, software and hardware extensions, and so on. These features are the reasons we do not have more congressional hearings on why the network went dead. Yes, Congress considers this, understandably, a national security question. If the airport in Baltimore cannot communicate with neighboring New Jersey, this is a big deal. In this case, public safety is in jeopardy. In business, total outage can cost some companies million of dollars in lost revenue. A kid playing an online Nintendo game or surfing the Web should not prevent his neighbor from acquiring dial tone to reach 911 (due to bad traffic engineering of a shared medium). The point is that future networks must maintain unquestionable service quality to serve us properly and be viable in the marketplace. To do less will be proof that these future interactive services are to be no more than an intellectual exercise.

1.14.3 Rationalizing the Business Case

The rationale for modernizing an HFC network to provide integrated services is going to be rather convoluted. Planning and deploying new products for the interactive services will, no doubt, depend on several factors, such as geographic area, ease of access, existing network capacity, and so on. The question facing planners will also include the need to integrate network management resources, IN (intelligent network) connectivity, and platform integration. Some operators may very well opt for building an overlay network with high-speed fabric. The ultimate decision on how certain networks will be upgraded will depend on that particular economic model.

1.14.4 Potential Revenue

The potential revenue is very attractive to all operators (revenue estimates are based on averaging the numbers extracted from various published reports). At the minimum, the following market opportunities are subject to competition:

■ The access market for POTS (plain old telephone services), including long-distance access revenue, is about $70 billion. That is about $475 per household.

■ CATV broadcast services included charges for premium channels is about $390 dollars per year.

■ Internet access charges are fluctuating, but the best estimates today are about $160 per year minimum. Online services are also experiencing phenomenal growth. Internet users numbered 30 million in 1994, a projected 100 million by the end of 1995, and an estimated 300 million by 1998. In 1994, the World Wide Web grew over 1700 percent. There are currently over 200,000 Web sites, with the population doubling every 53 days. One estimate shows that a WWW page is created every 10 seconds. Recent studies show that 33 percent of all U.S. households now own a PC, and sales grew 21 percent in 1994. PC sales now exceed TVs. The majority of homes in the U.S. are expected to have one or more PCs in the next few years. These statistics will, most likely, be invalidated by the time this book is published. However, if one assumes 100 million users are devoted Internet subscribers, then the revenue will be about $16 billion per year.

- The video-on-demand (VOD) market has been puzzling, and the prediction of its demise has been exaggerated. The fact remains that subscribers will use such service if it is part of a package but will not invest in it economically or emotionally. So, one concludes that the market numbers are real and account for around $200 per year per household.

There are numerous other new services (and potential new revenue) broadband network promises to deliver that will be appealing to consumers, such as online video games, video telephony, video poker, virtual malls (home shopping), electronic banking, and work at home. The list seems endless. Some of these services will no doubt contribute billions of dollars more to the revenue stream. But for the purpose of this exercise, we will refer to the four services shown above: local access telephony, CATV, the Internet, and VOD.

Figure 1.10 shows a simplified model of the revenue stream an operator will use to build a rational business case. Future services will add more to this projection, but the minimum shown should give an operator an indication of the sort of market opportunity and hence evaluate the sort of capital investment needed to show a credible margin.

Figure 1.10
Yearly revenue stream for the residential market.

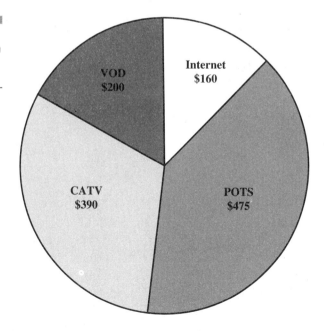

1.14.5 Cost of Modernization

Cable operators must invest and upgrade the cable network to HFC so it can support high-speed Internet access as well as voice telephony and the other interactive services. Economic analyses made by David P. Reed showed that building the HFC from scratch does not cost that much more than building the traditional cable network. Reed estimated that building the HFC system will total about $450 per home passed, based on 60 percent penetration rate and 500 home nodes. Installing fibers instead of coax in the trunk without the need for special amplifiers accounts for the low cost differential. Major cable operators recognize this and are basing their strategy on building HFC networks without even having to deliver new interactive services. In the future, the network will be two-way-communication ready.

Upgrading the traditional cable network is a more complex economic model. The cost of upgrading depends on several factors and how much rebuilding needs to be done. In order to fully upgrade a traditional cable network to an HFC two-way transmission, the following upgrades will be needed:

- At the minimum, the trunk cable (from the headend to the distribution network) must be replaced with fiber.
- The amplifiers in the distribution network must be added to or replaced, enabling them to transmit in both directions.
- The headend must be rebuilt.

The cost range of rebuilding the system was estimated in various studies to be between $100 to $240 per home passed. This is based on 60 percent penetration with 500 subscribers per node.

1.14.6 Economic Model

The economic model to build a business case will be a complicated task for some small cable operators and a challenging one to larger MSOs. There are over 11,000 cable systems in the United States. Large cable operators claim to have revamped their systems, and a good percentage of it is HFC-ready. The numbers are disputed by many experts. Table 1.1 is Merrill Lynch's estimation of HFC modernization effort.

TABLE 1.1

HFC Moderniza-
tion/Penetration

Year	% of homes passed by upgrades
1995	18%
1997	50%
1999	77%
2001	85%

OTHER FACTORS.   Technically, the economic model is going to be a bit more complicated. The HFC topology does not lend itself to dynamic reconfigurations or scaleability. The HFC systems of today seem to be hardwired to provide that specific service. For example, if more upstream or downstream channels are needed to accommodate the increase in traffic or the introduction of new applications, the analog amplifiers in the HFC distribution part of the network will have to manually be modified or may need to be replaced. If video telephony becomes a popular service, then increasing the upstream bandwidth will be required (allocating more channels in the RF spectrum). Reducing the homes passed per node would be a costly alternative. Cable modems and possibly also the head-ends will need to be modified to handle tuning to these new channels. IEEE 802.14 addresses this, and the cable modem can be developed with frequency-agile features to deal with such future uncertainty, but the pressure of competition will force many manufacturers to develop a bare-bones cable modem to meet the minimum requirements.

Going digital all the way is another parameter in the equation that must be evaluated. A 6-MHz analog channel with QAM modulation (see Chap. 5) can deliver over 30 Mb/s of digital bandwidth and possibly more in the future. This is quite a remarkable increase in bandwidth. Not only will a better-quality video be delivered, but there would be as much as a tenfold increase in the amount of TV-channel carrying capacity to the subscriber. A good-quality video picture requires 4 to 6 Mb/s. Near video on demand (NVOD) has been one of the applications that is well suited for this. The cable operators are facing increasing competitive pressure to deliver digital channels, not only from the video digital video broadcasting (DVB) market, but also because of the recent agreement reached for standardizing HDTV (high-definition TV). Rebuilding the HFC to all-digital is a formidable engineering task. Fiber may have to penetrate the distribution network, and more expensive but flexible digital repeaters must replace the analog amplifiers in the network.

Competition from other large cable companies is an unlikely scenario. According to the FCC (CS Docket #95-61), cable television is available to 95 percent of the American households, and only 1.5 percent of American homes are passed by multiple systems. Therefore, the competition will exclusively be from other network operators, mainly the telco or the newcomers to the marketplace.

The uncertainty of the use of new technology and present and future service offerings are fundamental difficulties facing the cable operators. Most of the upgrades that are needed are referred to as sunk cost (unrecoverable). The cable operators cannot afford to make a big strategic mistake. Such mistakes can be very costly, and going out of business will not be an unlikely scenario. We can go into details of microexamining every aspect of an economic model: from how many channels the HFC needs to support to the sort of spectrum allocation required to the all-digital upgrade. Such exercises and other extensive studies have been made in the past. What is clear from all these studies comes back to the basic equation that is well known in a competitive environment. Operators must be players in that market and, therefore, the issues are:

1. Time to market
2. Service differentiation

TIME TO MARKET. Time to market is an obvious strategy the cable operators must use. Deploy cable modems now or simply lose out to ADSL or other technologies. It is really as simple as that. Bringing back a customer was found to be much more difficult than signing on new ones.

Internet access is a very demanding service today, and consumers are eager to use high-speed access to satisfy their bandwidth-hungry applications. Forester Research Inc., per Table 1.2, forecast the estimated number of homes equipped with cable modems rather optimistically. Paul Kagan Associates, per Table 1.3, estimated a bit more pessimistically. These num-

TABLE 1.2

Optimistic Cable
Modem Forecast

Year	Number of cable modems
1996	90,000
1997	550,00
1998	2 million
2000	7 million

Source: Forester Research Inc.

TABLE 1.3

Pessimistic Cable
Modem Forecast

Year	Cable modems shipped in millions
1996	0.1
1997	0.4
1998	1.4
1999	2.2
2000	4.0

Source: Paul Kagan Associates

bers, to some, are obviously very impressive. But it is also obvious that MSO will be modernizing the cable network on a case-by-case basis.

SERVICE DIFFERENTIATION.   Service differentiation is another tool the cable operators must use. Service bundling is one possible approach the cable operators can use effectively. Service bundling is packaging multiple services offered to subscribers at an attractive price. Bundling multiple services, including TV distribution and premium channels, might be an attractive pricing strategy. A cable operator may be able to attract those customers who want to buy several services together, even if the competitor can attract others by offering individual services at a lower price.

Broadband Services in Cable Environment

2.1 Overview

ITU, in Recommendation I.211, fully describes all aspects of broadband services. It is a rather generic model of all the services one might anticipate. The text of Recommendation I.211 is reprinted in Sec. 2.2 (with permission granted from ITU). I.211 lays the foundations of all the broadband services that the industry has since it began exploring.

The business and entertainment aspects of broadband services have been fluctuating in the last two years. HFC makes it more interesting because of its inherent architectural capability to include entertainment TV. In this chapter, we focus on describing the services as defined in the standard bodies and today's hot applications that can justify crafting a rational business case. HFC and, hence, cable modem requirements are described in detail. In this chapter, we also describe services and associated applications as defined by ITU. Then the service concept is described in the context of service categories and QOS (quality of service) as applied to ATM. The application scenarios envisioned by NII (National Information Infrastructure) are then described, followed by the applications that are most popular and most likely to be implemented first. Finally, what that all means is spelled out in terms of requirements for standardizing the cable modem.

2.1.1 Services/Application Definition

We are all aware of the services and applications the entertainment industry has been defining. It is very important that we distinguish the terms *service* and *application*. ITU, the ATM Forum, and IEEE 802.14 have defined services very eloquently. Services are classified as constant bit rate, variable bit rate, available bit rate, and so on. *Application*, on the other hand, is defined as a session used by a subscriber to perform a task. With that in mind, services and applications can be generically defined as:

Service.   Has to do with what the network must provide to the end user so that the application can be run successfully and correctly.

Application.   A session running on a PC or other home electronic appliance: a video conference call, a voice call, downloading a video clip, voice over the Internet, and so on.

A network supporting constant bit rate services will handle voice application successfully. Handling voice application over the Internet cannot be performed successfully because the Internet network today is not equipped in terms of service to handle constant bit rate services.

2.2 ITU Services Aspects

ITU defined the fundamentals of all the services expected from a broadband network. Section 2.2 is a reprint (with permission from the ITU) of selected portions of Recommendation ITU-T I.211 (March, 1993).

2.2.1 Introduction

This recommendation should be interpreted as a guideline to the objective of providing detailed recommendations on specific standardized services to be supported by a B-ISDN (Broadband Integrated Services Digital Network). The purpose is:

1. To provide a classification of such services
2. To provide some considerations on the means to describe services based on the description method as defined in Recommendation I.130
3. To give a basis for the definition of the network capabilities required by B-ISDN

The service concepts considered in this Recommendation are in accordance with I.210. The Recommendation takes into account some of the known and relevant aspects of the B-ISDN, including:

- Capabilities to increase flexibility to both user and network operator including independent call and connection control
- The quality of service implications of information being structured and transported in cells
- Capabilities for flexible bandwidth allocation
- Capabilities for the provision of service timing information
- The overall interface capabilities

The Recommendation also gives guidance on video coding aspects taking into account the characteristics of the ATM-based network and recommends a common approach to video coding for all visual services including both interactive and distribution-type services.

2.2.2 Classification of Broadband ISDN Services

GENERAL.   This subclass describes the classification of broadband services, the definition of those service classes, and gives examples of services in each service class proposed to be supported by the B-ISDN.

This classification does not take into account the location or the implementation of the function either in the network or in the terminals. This classification is primarily from the point of view of the network and not the user point of view.

Depending on their communication functions and applications, the services to be supported by the B-ISDN may be internationally standardized and offered by the Administration as bearer services or teleservices.

SERVICE CLASSES.   Depending on the different forms of the future broadband communication and their applications, two main service categories have been identified: interactive services and distribution services. The interactive services are subdivided into three classes of services—the conversational services, the messaging services, and the retrieval services. The distribution services are represented by the class of distribution services without user individual presentation control and the class of distribution services with user individual presentation control (see Fig. 2.1).

DEFINITION OF SERVICE CLASSES.   *Conversational services* in general provide the means for bidirectional communication with real-time (no store-and-forward), end-to-end information transfer from user to user or between user and host (e.g., for data processing). The flow of the user information may be bidirectional symmetric, bidirectional asymmetric, and, in some specific cases (e.g., such as video surveillance), the flow of information may be unidirectional. The information is generated by the sending user or users, and is dedicated to one or more of the communication partners at the receiving site.

Examples of broadband conversational services are videotelephony, video conference, and high-speed data transmission.

Messaging services offer user-to-user communication between individual users via storage units with store-and-forward, mailbox, and/or message handling (e.g., information editing, processing, and conversion) functions.

Examples of broadband messaging services are message handling services and mail services for moving pictures (films), high-resolution images, and audio information.

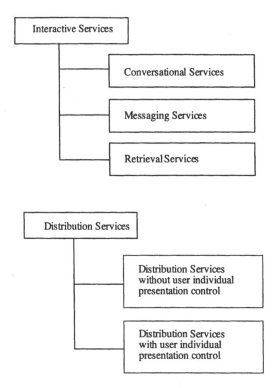

Figure 2.1
Classification of
broadband services.

Retrieval services. The user can retrieve information stored in information centers provided for public use. This information will be sent to the user on demand only. The information can be retrieved on an individual basis. Moreover, the time at which an information sequence is to start is under the control of the user.

Examples are broadband retrieval services for film, high-resolution image audio information, and archival information.

Distribution services without user individual presentation control. These services include broadcast services. They provide a continuous flow of information which is distributed from a central source to an unlimited number of authorized receivers connected to the network. The user can access this flow of information without the ability to determine at which instant the distribution of a string of information will be started. The user cannot control the start and order of the presentation of the broadcasted information. Depending on the point of time of the user's access, the information will not be presented from the beginning.

Examples are broadcast services for television and audio programmes.

Distribution services with user individual presentation control. Services of this class also distribute information from a central source to a large number of users. However, the information is provided as a sequence of information entities (e.g., frames) with cyclical repetition. So, the user has the ability of individual access to the cyclical distributed information and can control start and order of presentation. Due to the cyclical repetition, the information entities selected by the user will always be presented from the beginning.

One example of such a service is full-channel broadcast videography.

IDENTIFICATION OF POSSIBLE BROADBAND SERVICES. Table 2.1 contains examples of possible services, their applications and some possible attribute values describing the main characteristics of the services.

The identification and full specification of specific services for standardization can be completed only after a thorough examination of the needs of users by, for example, market research. The full specification of such services should be based on the application of appropriate description methodology.

2.2.3 General Network Aspects of Broadband Services

The purpose of this subclause is to give guidance concerning some of the important aspects which need to be taken into account when supporting and developing services for the B-ISDN. See Table 2.1.

MULTIMEDIA ASPECTS. Broadband services may involve more than one information type. These services are termed *multimedia services* (see Recommendation I.374). For example, video telephony will include audio, video, and possibly some form of data. Other information types may be text and graphics, for example. A structured approach to the development of multimedia services is recommended to ensure:

- Flexibility for the user
- Simplicity for the network operator
- Control of interworking situations
- Commonality of terminal and network components

The B-ISDN provides independent call and connection control facilities which should be exploited to help achieve the above objectives. The B-ISDN will make it possible, within a single call associated with a specific service, to establish a number of connections which may each be

TABLE 2.1
Possible Broadband Services in ISDN[a]

Service classes	Type of information	Examples of broadband services	Applications	Some possible attribute values[g]
Conversational services	Moving pictures (video) and sound	Broadband video telephony[b,c]	Communication for the transfer of voice (sound), moving pictures, and video scanned still images and documents between two locations (person to person)[c] • Tele-education • tele-shopping • tele-advertising	• Demand/reserved/ permanent • Point-to-point/ multipoint • Bidirectional symmetric/ bidirectional asymmetric • (Value for information transfer rate is under study)
		Broadband videoconferencing[b,c]	Multipoint communication for the transfer of voice (sound), moving pictures, and video scanned still images and documents between two or more locations (person to group, group to group)[c] • Tele-education • Tele-shopping • Tele-advertising	• Demand/reserved/ permanent • Point-to-point/ multipoint • Bidirectional symmetric/ bidirectional asymmetric
		Video-surveillance	• Building security • Traffic monitoring	• Demand/reserved/ permanent • Point-to-point/ multipoint • Bidirectional symmetric/ unidirectional
		Video/audio information transmission service	• TV signal transfer • Video/audio dialogue • Contribution of information	• Demand/reserved/ permanent • Point-to-point/ multipoint • Bidirectional symmetric/ bidirectional asymmetric

TABLE 2.1 CONTINUED

Possible Broadband Services in ISDN

Service classes	Type of information	Examples of broadband services	Applications	Some possible attribute values
	Sound	Multiple sound-programme signals	• Multilingual commentary channels • Multiple programme transfers	• Demand/reserved/permanent • Point-to-point/multipoint • Bidirectional symmetric/bidirectional asymmetric
	Data	High-speed unrestricted digital information transmission service	• High-speed data transfer • LAN (local area network) interconnection • MAN (metropolitan area network) interconnection • Computer-computer interconnection • Transfer of video and other information • Still image transfer • Multisite interactive CAD/CAM	• Demand/reserved/permanent • Point-to-point/multipoint • Bidirectional symmetric/bidirectional asymmetric • Connection-oriented connectionless
		High-volume file transfer service	• Data file transfer	• Demand • Point-to-point/multipoint • Bidirectional symmetric/bidirectional asymmetric
		High-speed teleaction	• Real-time control • Telemetry • Alarms	
	Document	High-speed telefax	User-to-user transfer of text, images, drawings, etc.	• Demand • Point-to-point/multipoint • Bidirectional symmetric/bidirectional asymmetric

Service category	Information type	Service	Examples	Attributes
Message services		High-resolution image communication service	• Professional images • Medical images • Remote games and game networks	• Demand • Point-to-point/multipoint • Bidirectional symmetric/bidirectional asymmetric
		Document communication service	User-to-user transfer of mixed documents[d]	• Demand • Point-to-point/multipoint • Bidirectional symmetric/unidirectional (for further study)
	Moving pictures (video) and sound	Video mail service	Electronic mailbox service for the transfer of moving pictures and accompanying sound	• Demand • Point-to-point/multipoint • Bidirectional symmetric/unidirectional (for further study)
	Document	Document mail service	Electronic mailbox service for mixed documents[d]	• Demand • Point-to-point/multipoint • Bidirectional symmetric/unidirectional (for further study)
Retrieval services	Text, data, graphics, sound, still images, moving pictures	Broadband videotex	• Videotex including moving pictures • Remote education and training • Tele-software • Tele-shopping • Tele-advertising	• Demand • Point-to-point • Bidirectional asymmetric
		Video retrieval service	• Entertainment purposes • Remote education and training	• Demand/reserved • Point-to-point/multipoint[f] • Bidirectional asymmetric

TABLE 2.1 CONTINUED

Possible Broadband Services in ISDN

Service classes	Type of information	Examples of broadband services	Applications	Some possible attribute values
Distribution service without user individual presentation control	Video	High-resolution image retrieval service	• Entertainment purposes • Remote education and training • Professional image communications • Medical image communications	• Demand/reserved • Point-to-point/multipoint[f] • Bidirectional asymmetric
		Document retrieval service	"Mixed documents" retrieval from information centers, archives, etc.[d,e]	• Demand • Point-to-point/multipoint[f] • Bidirectional asymmetric
		Data retrieval service	Telesoftware	
		Existing quality TV distribution service (PAL, SECAM, NTSC)	TV programme distribution	• Demand (selection)/permanent • Broadcast • Bidirectional asymmetric/unidirectional
		Extended quality TV distribution service • Enhanced definition TV distribution service • High-quality TV	TV programme distribution	• Demand (selection)/permanent • Broadcast • Bidirectional asymmetric/unidirectional
		High-definition TV distribution service	TV programme distribution	• Demand (selection)/permanent • Broadcast • Bidirectional asymmetric/unidirectional

38

	Examples of broadband services[b]	Applications	Possible attribute values[c]
Distribution services with user individual presentation	Full-channel broadcast videography	• Remote education and training • Tele-advertising • News retrieval • Telesoftware	• Permanent • Broadcast • Unidirectional
Moving pictures	Video information, distribution service, and sound	• Distribution of video/audio signals	• Permanent • Broadcast • Unidirectional
Data	High-speed unrestricted digital information distribution service	• Distribution of unrestricted data	• Permanent • Broadcast • Unidirectional
Text, graphics, still images	Document distribution service	• Electronic newspaper • Electronic publishing	• Demand (selection)/permanent • Broadcast/multipoint[f] • Bidirectional asymmetric/unidirectional
	Pay TV (pay-per-view, pay-per-channel)	TV programme distribution	• Demand (selection)/permanent • Broadcast/multipoint • Bidirectional asymmetric/unidirectional

[a] In this table, only those broadband services are considered which may require higher transfer capacity than that of the H capacity. Services for sound retrieval, main sound applications, and visual services with reduced or highly reduced resolutions are not listed.

[b] This terminology indicates that a redefinition regarding existing terms has taken place. The new terms may or may not exist for a transition period.

[c] The realization of the different applications may require the definition of different quality classes.

[d] A *mixed document* may contain text, graphic, still, and moving picture information as well as voice annotation.

[e] Special high-layer functions are necessary if postprocessing after retrieval is required.

[f] Further study is required to indicate whether the point-to-multipoint connection represents in this case a main application.

[g] For the moment this column merely highlights some possible attribute values to give a general indication of the characteristics of these services. The full specification of these services will require a listing of all attribute values which will be defined for broadband services in recommendations of the I.200 series.

In addition, Recommendations I.362 and I.363 describe ATM adaptation layer functions for B-ISDN services identified on the basis of timing relation (between source and destination), bit rate (constant or variable), and connection mode (connection-oriented or connectionless).

associated with a specific information type. The B-ISDN will enable the addition and/or deletion of optional information types during a call.

It is recommended, therefore, that the development of multimedia services proceeds on the basis of the following principles:

- That a limited set of standardized information types be developed
- That the association of services and standardized information types be controlled but in a flexible manner

2.3 ATM Service Architecture

The ATM service architecture applies to ATM high-speed packet-switching technology that is used to deliver the broadband services. The broadband services described by ITU in Sec. 2.2 are not necessarily ATM-specific. However, ITU selected ATM as the target solution for broadband ISDN services. Hence, we will describe services in the context of ATM being the underlining technology. Chapter 3 describes ATM in detail. IEEE 802.14 also used these service categories for standardizing the cable modem.

2.3.1 Service Categories

There are four service categories defined by the ITU and adopted by IEEE 802.14. They are:

1. Constant bit rate (CBR)
2. Non-real-time variable bit rate (nrt-VBR)
3. Unspecified bit rate (UBR)
4. Available bit rate (ABR)

The ATM Forum also defined real-time VBR (rt-VBR) as another service category.

These services demand certain QOS and traffic characteristics from the network. The network performs services such as routing, admission control, and resource allocation for each of the service categories shown above. For real-time traffic, there are two categories, CBR and rt-VBR. The spacing between ATM cells must be controlled and minimized by the network in order to provide real-time service. Non-real-time (nrt-VBR), UBR, and ABR traffic behaves in a non-real-time fashion, and it can vary depending on the category the user requests from the network.

DEFINITIONS FOR SERVICE CATEGORIES. A service category is built from parameters the network must provide. A set of QOS parameters shown below defines a service category. ITU, and later the ATM Forum, defined them as:

1. Peak-to-peak cell delay variation (CDV)
2. Maximum cell transfer delay (Max CTD)
3. Mean cell transfer delay (Mean CTD)
4. Cell loss ratio (CLR)

CONSTANT BIT RATE (CBR) SERVICE DEFINITION. The CBR service is intended for real-time applications. This service is applicable for voice and video applications. The ATM cells carrying voice or video information must maintain very short and bounded jitter (cell delay variation) by the network. Cells that are delayed will reduce the value to the application. Example applications for CBR are:

- Interactive video (videoconferencing)
- Interactive audio (telephone)
- Video distribution (television, distributed classroom)
- Audio distribution (radio, audio feed)
- Video retrieval (video on demand)
- Audio retrieval (audio library)

REAL-TIME VARIABLE BIT RATE (rt-VBR) SERVICE DEFINITION. This service category was defined by the ATM Forum. Real-time VBR service is similar to the CBR service and intended for real-time applications, or those requiring tight delay and delay variation. This service is also applicable for voice and video applications. Unlike CBR service, the sources are expected to transmit at a bursty rate varying with time. Statistical multiplexing gain is assumed for this application. Cells that are delayed by the network beyond the specified parameter will reduce the value to the application. Examples of real-time VBR are:

- Any CBR application for which the end system can benefit from statistical multiplexing by sending at a variable rate, and can tolerate or recover from a small but nonzero random loss ratio (VBR voice is a possible application)
- Any CBR application for which variable rate transmission allows more efficient use of network resources

NON-REAL-TIME VARIABLE BIT RATE (nrt-VBR) SERVICE DEFINITION. This non-real-time VBR service is intended for non-real-time applications that have bursty traffic behavior. The parameters that define this service are peak cell rate and sustainable cell rate. This service may be applicable to large file transfer applications. Cells that do not conform to the specified traffic will reduce the value to the application and experience cell loss ratio. Example applications for non-real-time VBR are:

■ Response-time-critical transaction processing

■ Airline reservations

■ Banking transactions

■ Process monitoring

■ Frame relay interworking

UNSPECIFIED BIT RATE (UBR) SERVICE DEFINITION. The UBR service is intended for non-real-time applications (i.e., very bursty traffic not requiring tightly bounded delay and delay variation). Examples of such applications are computer communications applications such as file transfer and Email messages. UBR sources are expected to be very bursty; therefore, the service supports a statistical multiplexing gain among sources. Like Email applications today, UBR service does not specify service guarantees. No numerical commitments are made with respect to the cell loss ratio experienced by a UBR connection. Example applications for UBR:

1. Interactive text/data/image transfer

   Banking transaction or credit card verification

2. Text/data/image messaging

   Email, telex, or fax transaction

3. Text/data/image distribution

   News feed, weather satellite pictures

4. Text/data/image retrieval

   File transfer, library browsing

5. Aggregate LAN

   LAN interconnection or emulation

6. Remote terminal

   Telecommuting, telnet

AVAILABLE BIT RATE (ABR) SERVICE DEFINITION. The ABR service was defined by the ATM Forum, and IEEE 802.14 adopted it for the cable modem standard. Depending on the network traffic resources, the service characteristics, provided by the network, were made to change during the life of the connection. A flow control mechanism is specified to support several types of feedback to control the source rate of information flow in response to changing network available resources. An end system that adapts its traffic to the feedback (i.e., for reducing its traffic) will encounter a low cell loss ratio and also obtain a fair share of the available bandwidth network resources. ABR service is well suited for support of the Web-browsing applications. An ABR connection is established by the end system and specifies to the network the maximum required bandwidth (in peak cell rate) and a minimum usable bandwidth (in minimum cell rate). The minimum cell rate may be specified as zero. The bandwidth available from the network may vary and will be based on information gathered by a remote maintenance cell scouting the network from end to end. More details on ABR are described in Chap. 3 and Chap. 8 (the latter for cable modem applications). Examples of applications for ABR are:

- Any UBR application for which the end system requires a guaranteed QOS
- Critical data transfer (e.g., defense information)
- Supercomputer applications
- Data communications applications requiring better delay behavior, such as remote procedure call, distributed file service, or computer process swapping/paging

2.4 Applications Promoted by National Information Infrastructure

Cross Industry Working Team (XIWT) was inspired by President Clinton and Vice President Al Gore's vision of a national information infrastructure (NII). XIWT crafted a functional services framework model and a reference architecture model. The functional services model describes the NII's building blocks and components, particularly an *enabling services layer* that must exist to facilitate the rapid development and deployment of applications and the integration of all components. These enabling services include common capabilities needed by most of the applications and specialized services for specific application domains such as health care, finance, or manufacturing.

2.4.1 Applications per XIWT and NII

Ten applications were defined by XIWT. Here is a brief summary of each:

1. *"Homing" from work.*   This multimedia application illustrates a fully interactive dialog with home, office, and school.

2. *Businesses unit.*   This multimedia application describes an interactive video Email and video communication for business applications.

3. *Home entertainment, information, and shopping.*   This application describes a user surfing the network for leisure video activities (VOD) and shopping via the virtual mall.

4. *Intelligent transportation systems.*   This application describes an intelligent highway system with video surveillance cameras and roadway sensors to collect information on how to reach a destination using satellite positioning services and personal portable information devices with video display capabilities.

5. *Senior citizen use.*   This application illustrates a video visit scenario via a large-screen, high-resolution computer display and features other kinds of access where users become information providers and show each other prerecorded video information.

6. *Starting a business.*   This application shows how small business can be established via videoconferencing and accessing government databases. These provide both information and services on an immediate basis and allow online interstate registrations, online legal services, and worldwide protection mechanisms for intellectual property.

7. *Middle school from home.*   This application illustrates a computerized data network in schools, with high-speed connection into the NII, audiovisual capabilities, and networked "virtual classrooms" composed of students and teachers at different locations.

8. *Telemedicine.*   This application utilizes high-speed networking for videoconferencing, large file transfers, real-time computer visualization of data, and networked supercomputing resources for medical and educational applications.

9. *Government services.*   This application illustrates how government electronic service entities provide personal services using fast data-

bases. These governmental information databases are or will be available to the public with easy-to-use navigational capabilities.

10. *Law enforcement.* This application shows how a wireless network, both terrestrial cellular in populated areas and by low-earth-orbiting satellite with emergency video and image transmission, will be used by law enforcement to apprehend suspects.

Some of the ten applications shown above exist today or most likely will emerge in the near future. They are intended to illustrate how people, with different backgrounds, education, and personal needs, could benefit from the NII. These scenarios are just a few of the applications targeted for use in the NII. There is a vastly broader range of possibilities, most of which are difficult to imagine today. The people of the United States, and of the world to which they are linked, will imagine and try out ideas for economic and human development, collaboration, and creativity that would seem wildly improbable if they could be articulated today.

2.5 Key Applications

When developing and standardizing a cable modem, all aspects of the services and applications described in the above sections were considered. IEEE 802.14 has taken an active role in defining requirements that deal with all aspects of the services. These requirements are described later in the chapter. A more realistic approach, however, is developing services that makes business sense. The key to survival in a multiservice, multiprovider market is differentiation. From today's viewpoint, it appears that all players/operators, either alone or in partnerships, will eventually offer the same services. The "most wanted" list of applications is:

- High-speed Internet access
- Digital VOD or NVOD
- Digital broadcasting
- Telephony over cable (wireline and wireless)
- Work at home

There are, of course, plenty of applications that will bring in more revenues when the infrastructure is built. Among the most popular are:

- Telemedicine
- Distant learning

- Video telephony
- Home shopping/virtual mall
- Digital audio/CD-ROM on demand
- Local information services
- Advertising and direct marketing
- Electronic games
- Telemetry

2.5.1 High-Speed Internet Access

High-speed Internet access is the key application that operators are banking on to build a rational business case. If there were a killer app, high-speed Internet access would fit the definition adequately. The phenomenal growth of the Internet has been documented in every journal and news media; suffice to say, the Internet has the attention of almost every business executive including those in the White House. The success of the Internet should not be all that surprising. The *ubiquity* of the service was the key—that, coupled with changing social attitude and the PC revolution in homes, was the catalyst that made it happen. People were tired of using their PCs to play primitive games and store recipes or other useless information. Being wired to the net is the thing of the 90s.

THE INTERNET AND QOS DILEMMA.   The Internet, like cable TV and the telephone network, was designed for service-specific applications, mainly connectionless services such as the UBR service described above. Sophisticated browsers and ease of use elevated the Internet to a state where some referred to it as the information superhighway. This image of the superhighway is different from the vision promoted earlier by the standards and telephone companies. The real superhighway integrates all communications: voice, data, and video. Some developed applications to integrate the services, but the quality will remain poor unless major changes, mainly the concept of QOS, are developed in the Internet infrastructure. Otherwise, voice over the Internet will remain no more than ham-radio quality.

The Internet is now plagued with access delay, and it is slow when downloading files or video clips. This slow-speed access is turning off a lot of potential users to a point that may threaten its growth. ADSL and cable modems will play a major and positive role to remedy this frustrating experience. Assuming the servers are also equipped with this high-speed access, the deployment of ADSL and cable modems will be like changing the highway speed limit from 1 mph to 250 mph or even 1500 mph. This will defiantly abate the access speed problem on the Internet. Unfortunately, this is not the whole story. The concept of QOS in the Internet network is practically nonexistent. High-speed access alone will not solve the multiservice problems. QOS is analogous to a modern highway discipline: the road must be well maintained, give priority or a special lane to passing cars, have traffic yield to an emergency vehicle, make trucks take a different route, or construct special lanes designed for HOV. And, of course, a highway patrol is needed to monitor traffic. This sort of discipline is necessary if the Internet is to behave as the information superhighway. A car cannot travel 250 mph on a highway if trucks are blocking the road or there is no passing lane. The scenario described is a classic one of goods or services versus application. The ATM Forum and ITU developed all the QOS needed to engineer a well-behaved network, which accommodates the highway scenario above. The Internet Engineering Task Force (IETF) is now developing QOS features onto the Internet. IEEE 802.14, in standardizing the cable modem, recognized this Internet (IP protocol) and ATM direction and left IP hooks in the architecture.

2.5.2 Digital VOD or NVOD

Video on demand and near video on demand were originally thought to be killer applications. A VOD application is defined as the ability to check out any movie electronically, watch it at home, and have full VCR control capability. NVOD application is similar to VOD without VCR control capability. In NVOD, only selected movies are played and repeated periodically (every 5 minutes or less) to accommodate user timetables. Both DAVIC and the ATM Forum developed a complete and credible set of specifications addressing the VOD/NVOD applications. Most marketing studies also alluded that the VOD market and its periphery applications would be embraced on a large scale. Several factors are responsible for delaying the introduction of VOD to the market: priorities in developing strategies within the network operators because of the Telecommunication Act of 1996, access technology choices, the Internet frenzy, RBOC's occupation with the long distance market, and the risk involved in building such a huge and expensive infrastructure.

When ADSL and cable modems are deployed in the network for Internet access, the VOD/NVOD will surface again and may take a central role. The revenue associated with the VOD market is unquestionable in terms of justifying a business case.

2.5.3 Digital Broadcasting

Digital broadcasting will be of particular interest to the cable operators. A digitally compressed video signal can increase the HFC network capacity tenfold. This allows cable operators the extra capacity to deliver NVOD movies and even provide broadcast premium programming on demand—HBO, for example.

2.5.4 Voice Telephony over Cable

This new revenue application is of particular interest to the cable operators. Telephone service does not require a lot of bandwidth from cable networks, but it does impose a very short delay and delay variation requirements on the cable system. If long delay is encountered, then an echo will result, disrupting the conversation. An expensive echo-cancelling technique can correct this problem. The issue of telephony over cable is not a big problem to integrate, either in the cable modem or otherwise. The problem the cable operators face is one of reliability and privacy. A "teen line" can be marketed effectively by cable operators as part of a bundling service package. Voice over HFC is described in more details in Chap. 9.

2.5.5 Other Applications

Other applications that are on the wish list with a good revenue potential are shown below. The list is by no means extensive, but it reflects the direction of the market as seen today.

TELEMEDICINE. Telemedicine is the ability to provide high-quality video telephony service interactively for the purpose of transmitting digitized medical data, X-ray images, and others. The lines of communication will primarily be established between hospitals, labs, pharmacies, and so on. Other applications may be a routine yearly physical, the ability to consult with medical experts, and home video care.

HOME OFFICE/TELECOMMUTING.   *Home office* is linking the worker with the computer at the place of business and providing the user with processing speed and processing power. It creates the environment of a virtual office. Telecommuting is growing at 12 percent CAGR.

DISTANT LEARNING.   Distant learning is the ability to bring educational experiences of the campus environment into the home interactively.

VIDEO TELEPHONY.   Video telephony is the use of a TV or PC screen to see the person or other images on the screen while talking. Video telephony (the death of the blind date) can be further segregated into:

- Communication with relatives, "video visit"
- Alarm security monitoring
- Merchandising product

HOME SHOPPING/VIRTUAL MALL.   The virtual mall is where companies can market their products and services effectively. The application varies and would include interactive communication between a buyer and a seller. High-quality video may be required.

DIGITAL AUDIO/CD-ROM ON DEMAND.   Users currently pay from $30 to $100 to purchase a CD-ROM title. Network providers can deploy CD-ROM servers to allow users to access high-quality stereo titles on demand. There will be no need for a local CD-ROM drive, and users could sample titles before purchasing them. Software developers would enjoy lower distribution and packaging costs, along with increased exposure and automated updating (especially valuable with reference-based titles).

LOCAL INFORMATION SERVICES.   Multimedia-based classified ads, yellow and white pages, restaurant guides, auto traders, local library, matchmaking services, and so on.

ADVERTISING/DIRECT MARKETING.   Advertising will become a viable source of revenue. The Internet is evolving into an interactive multimedia platform. This will, no doubt, nurture and accelerate direct marketing to the home.

ELECTRONIC GAMES.   Some cable operators already provide electronic game machines or deliver software games through downloads to game machines. With directional capabilities, games can become interactive and play between different users on the network. This application applies to adults as well, who are eager to find a partner for chess or popular card games.

TELEMETRY.   Telemetry applications will play a major role in a cable network. There are several potent subapplications. The most popular are:

- Remote sensing and remote monitoring of gas, electric, and water utility
- Allowing utility companies to monitor usage frequently and to initiate and disconnect service during peak and off-peak hours via remote control at a cost saving to customers
- Providing burglar and fire alarm services and video monitoring

In the previous sections, we described the requirements of the network services that are required to run applications and examined some of the hot and not-so-hot applications the industry will most likely pursue or is pursuing. In the next section, we will examine the impact of the above on the cable modem and the service requirements to deliver such applications.

2.6 HFC Service Requirements

In an HFC environment, several critical issues must be addressed for a shared media access network to meet the service requirements. The cable modem is the key element in the cable network that will be responsible for conforming to the service requirement. The cable modem must also operate in the HFC environment—that means shared medium. In a shared medium environment, a medium access control (MAC) is needed to police the traffic and maintain the QOS and traffic parameters that are characteristic of the services as described in the section titled "Definitions for Service Categories." The function of a medium access control protocol is to allow users to obtain resources of the upstream channel (user-to-headend direction) when they need it and the downstream channel (from headend to user) when present. There are a number of approaches to develop a MAC protocol. MAC is described in more detail in Chap. 5. Three categories of MAC protocols have been discussed in IEEE 802.14 or developed in first-generation cable modem products:

1. Contention-free MAC protocols with static scheduling (e.g., time division multiple access)
2. Purely contention-based protocols (e.g., CDMA and slotted ALOHA)
3. Reservation-based MAC protocols

2.6.1 Traffic Characteristics and Performance Parameters over HFC

The requirements of the traffic characteristics of an HFC user application is provided, as an illustrative set of performance parameters was used to measure the performance of the cable modem. This set of performance parameters was used for general network performance measurements and also for performance and capacity analysis for supporting various services. Average delay (or mean packet delay) and channel throughput are two of the parameters used to evaluate the overall network performance. The average delay and throughput of an access protocol is highly sensitive to traffic load on the network. Special considerations were given to CBR traffic, because variable transmission delays would otherwise severely compromise the quality of CBR service. Some applications have a plurality of protocol data unit (PDU) types (e.g., high and low priorities) with different delay and loss of tolerance. Many such service requirements must be considered to evaluate the overall performance of a MAC protocol. The tables below illustrate the traffic characteristics and performance requirements for a variety of application types that will be supported by IEEE 802.14 MAC protocol.

- CBR traffic performance requirements. Table 2.2 is for voice, video telephony, and video on demand (VOD) services.
- Table 2.3 is for video games, data, and multimedia services. Included are requirements such as video games involving fast button pushes or mouse clicking, work-at-home, home shopping, and remote library access.

The term *PDU* in the tables refers to a protocol data unit, which may be, for example, an ATM cell when the protocol uses ATM cells to package user data in the upstream and downstream. The term *take rate* refers to the ratio of the number of lines actually subscribed to the number of houses passed. Also, the forward and reverse bandwidth values are on a per-user basis.

TABLE 2.2

ATM-Based CBR Traffic Characteristics and Performance Parameters

	Voice (end-to-end access)	Video telephony (end-to-end access)	MPEG-2 VOD (end-to-end access)
Number of homes passed	2000	2000	2000
Take rate	125%	10%	40%
Calls/busy hour/ subscriber line	3	1	0.1
Call holding time	3 min	9 min	120 min
Call blocking problems	<.001	<.01	<.01
Forward bandwidth	64 kbps	384 kbps	5–8 Mbps
Reverse bandwidth	64 kbps	384 kbps	(non-CBR)
Call control and management: messages per call	28 (Q931/2931)	34 (Q931/2931)	NA (Q931/2931)
Call control and management: cells per call	70 (Q931)	76 (Q931)	NA (Q931 and VCR remote control)
Call set-up delay	NA	NA	NA
Dial tone delay	<300 ms	<300 ms	<300 ms
Round-trip delay (UNI-UNI)	TBD	TBD	TBD
Call control and management: cell delay (99.99%)	<20 ms	<20 ms	<20 ms
Bearer information cell delay (99.99% end-to-end)	<150 ms	<150 ms	<150 ms
Information cell loss rate, lower limit	$<10^{-3}$	$<10^{-3}$	$\leq 10^{-6}$
Cell delay variation	<2–3 ms	<2–3 ms	<0.3–3 ms

TABLE 2.3

Video Games,
Data, and Multi-
media Services

	Video games	Data (Work at home)	Multimedia (Home shopping)
Number homes passed	2000	2000	2000
Take rate	20%	80%	50%
Calls/busy hour	1	0.8	0.5
Call holding time	30 min	10–120 min	5–60 min
0.001	≤.01	≤.01	≤.01
Forward bandwidth	0.5–4 Mbps	64 kbps–10 Mbps	64 kb/s–20 Mbps
Reverse bandwidth	30 PDU/s peak 3 PDU/s avg. (sporadic)	10 Mbps peak 10 kbps–500 kbps avg.	64 kb/s–500 kbps
Call control/management: messages per call	85 (Q931)	145 (Q931)	85 (Q931)
Call control/management: PDUs per call	127 (Q931)	187 (Q931)	127 (Q931)
Call control/management: PDU delay (99.99%, upstream)	≤20 ms	≤20 ms	≤20 ms
Upstream information PDU delay (99.99%)	≤20 ms	≤20 ms	≤20 ms
Downstream information PDU delay (99.99%)	≤40 ms	≤100 ms	≤100 ms
Information PDU loss for low priority	$\leq 10^{-3}$	$\leq 10^{-3}$	$\leq 10^{-3}$
Information PDU loss for high priority	$\leq 10^{-8}$	$\leq 10^{-8}$	$\leq 10^{-8}$

DATA TRAFFIC PERFORMANCE REQUIREMENTS.   Traffic characteristics and performance requirements for data services include applications such as Email, interactive transactions, bulk data transfer, work-at-home, and so on. The parameter values shown in the tables are based on an illustration reflecting ANSI X3.102 and ITU Recommendation I.350.

2.6.2 General Comments

The following comments are worth noting. They complement and clarify some aspects of the services and traffic characteristics.

DATA TRAFFIC. Delay tolerance characteristics of data traffic can be further classified as:

- Delay-sensitive
- Moderately delay-tolerant
- Highly delay-tolerant

For example, interactive data services can be considered as delay-sensitive and should be classified under CBR service. Online file transfers and remote library access can be considered as moderately delay-tolerant data traffic (this can be classified as ABR service). While Email and bulk data transfers (off-line) can be considered as highly delay-tolerant data traffic, in this case the service classification is considered as UBR traffic.

VIDEO TRAFFIC. Video traffic has stringent end-to-end delay requirements and must have a low probability of PDU/packet loss and bounded delay jitter. When transmitting video traffic packets over HFC networks, the MAC and PHY layer protocol must guarantee a high quality of service for various classes of video compression standards such as JPEG, H.261, and MPEG I and II. Video traffic is particularly sensitive to PDU delay jitter. The protocol must keep the jitter (CDV) as low as possible within the HFC access network.

MULTIMEDIA TRAFFIC. For supporting multimedia traffic, synchronization issues between the transmitter and receiver must be considered when evaluating the protocol. Synchronization is categorized into two types:

1. Continuous
2. Point synchronization

Video telephony is an example of continuous synchronization, where audio and video signals are created at a remote site and transmitted over the network; then the two signals are synchronized continuously at the receiver site for playback. SRTS is one method used in ATM to perform this type of synchronization. A slide show with blocks of audio allotted to each slide is an example of point synchronization, where certain transition time points of one media block coincide with corresponding points of another media block. In a two-way audiovisual communication, when voice and video information are carried in one single stream (or virtual channel), delays on the order of 200 ms can be tolerated. But when the voice and video information are carried as different streams, the need to achieve synchronization will reduce the delay tolerance for either stream to about 150 ms.

CHAPTER **3**

ATM over HFC

3.1 Overview

In this chapter, ATM (asynchronous transfer mode) will be described in the BISDN and HFC environment. ATM is still a controversial subject among some cable operators (MSOs). It is perceived to be immature market-wise and/or lacking the necessary standards. Operationally, ATM is perceived to be too complex to implement and much beyond what is needed to accommodate initial deployment for Internet application or other interactive data or video services. These views were reflected in the ATM Forum general meeting in August of 1993 when Dr. Richard Green (CEO/Cable-Labs), as the keynote speaker, had the opportunity to address the ATM Forum members on the subject of ATM. Dr. Green was supportive of this new technology, and he articulated his views on how one must position ATM in the cable network environment.

Interesting key points and observations are:

■ The cable TV industry is accelerating the growth and interest in deploying ATM as a backbone technology to the video distribution network (it is beyond the first step). There are those (MOSs), however, who are not fully sold on ATM yet. Therefore, ATM must prove itself in the performance/cost ratio.

■ Cable operators are enjoying the attention of the telco industry. People are realizing that ATM is here today.

■ Market research is leading CATV operators to jump into interactive video (home video) and provide VOD, MM services, electronic video stores, and video games.

■ CATV offers a viable solution and alternative to the emerging PCS market.

■ ATM figures heavily in the Time Warner Orlando project in support of VDT (video dial tone).

The cable industry view on ATM has not changed much since Dr. R. Green addressed the ATM Forum. One possible exception would be an initial offering of services. Internet, for example, is now playing a major role as *the* market pull and taking the lead as the initial service to be offered over high-speed cable modem.

This chapter will also examine the structure of ATM inside the home, which is sometimes referred to as *ATM end-to-end*. We will also explore the impact of ATM over HFC and highlight all the debate that will most likely stay with us for a while. The first part of the chapter will describe ATM in general, and then ATM over HFC will be examined in particular.

3.1.1. Organization of This Chapter

This chapter is organized into two parts. The first part describes ATM and broadband, and the second part describes ATM over HFC. The intent is to give the reader the background needed to understand ATM and how it relates to the cable network environment and the cable modem in particular. The topics covered include:

Part 1
- Broadband ISDN and the rationale of deploying ATM as the underlying switching fabric
  Lower layers architecture
  Higher layers architecture
  Network aspects
  Service categories and traffic control

Part 2
- Residential broadband
- ATM over HFC

This chapter is not intended to give a comprehensive view on B-ISDN or the intricate detail mechanics of ATM. One needs more than a chapter to do justice to the work done in the last 10 years or so in ANSI, ITU, and recently in the ATM Forum. Instead, this chapter will focus on broadband and ATM issues that are particularly focused and of interest to the ATM over HFC. In collapsing ATM and B-ISDN into one chapter, I took the liberty of not using any disciplined approach when describing the B-ISDN and ATM subject. Rather, the topics covered will be described to give the reader the needed background materials to assimilate the overall picture of ATM over HFC.

3.2 Broadband ISDN

In this section, the concept of broadband ISDN and ATM will be described. Enough details will be included to understand better the rationale of why ATM will eventually play the major role in HFC and other nodes in the public and private networks

3.2.1 Broadband vs. ATM

Although there are a variety of definitions of what is broadband and how ATM fits into the picture, my favorite definition is the following: broadband is defined as a network element capable of delivering BB services to the user access interface beyond 1.5 Mbits/s or 2 Mbits/s.

ATM, on the other hand, is the high-speed packet-switching technology that can be used to deliver broadband services. In this context, broadband services can be delivered to the user in today's circuit switch fabric, or synchronous transfer mode (STM). In fact, existing networks deployed today can deliver such service via DS3 links. ITU, however, in 1988, selected ATM as the target solution for broadband ISDN. The rationale for choosing ATM over STM should become obvious by the end of this chapter.

There are few in the industry who are still skeptical and are questioning the complexities and maturity of ATM as the technology that can meet the market demand, especially in the near term. The skeptics want to see ATM prove itself in the marketplace first before investing heavily in it.

3.2.2 Technology Push, Market Pull

The advent of high-speed networking has always been part of our daily reality, especially in business. High-speed equipment has been responsible for productivity, cost reduction, and therefore the improvement of our daily lives. High-speed memory and personal computers are just a few examples we are all aware of, and we expect such trends to continue. Such progress is, no doubt, putting pressure on the network to match that industry and accommodate the high-speed data and video market. Intelligent processing power of peripheral equipment is also contributing to the demand of deploying high-speed networks to facilitate remote transaction, efficiency, and therefore cost savings.

Throughout this high-speed evolution of user equipment, the public network, in particular, has stayed more or less stagnant. As the demand for higher data bit rate increased worldwide, new services, including video, became more and more attainable, attractive, and therefore a market reality. As a result, the market for the broadband network is becoming evident, and realistic talk about broadband services and deployment started in the late 80s.

The Internet phenomenon, a late entry into the broadband service portfolio, fueled the discussion of the market need to deploy high-speed network nodes. Hence, operators today are responding by deploying ATM switches for the Internet backbone and access networks.

3.2.3 History of ATM

ATM (asynchronous transfer mode) was first introduced in ITU, previously CCITT, in 1984. The story is that the CNET conceived the idea and sketched it on a napkin during lunch in Geneva. Alcatel Antwerp was one of the first companies to develop ATM, originally ATD (asynchronous time division), in 1985.

In 1987, ITU selected ATM as the switching technique for broadband ISDN. Finally, in 1990, ITU released a set of recommendations on B-ISDN using the ATM techniques. ANSI, ITU, and the ATM Forum are still in the midst of developing and completing all aspects of broadband ISDN services and standardization for both public and private networks. Ongoing work in ANSI (T1S1) is defining the U.S. position on signaling capability set 3 (multimedia) and its interworking in the intelligent network (IN). ITU recognizes ANSI as a credible contributor, reflecting realistic U.S. public network views.

The ATM Forum was established in October 1991 and made remarkable progress in developing ATM specifications while promoting interoperability. The success of the ATM Forum was phenomenal and reflective of the industry's intention to deploy and embrace ATM in both the public and private networks. I share the view of many that the success of the ATM Forum was due to the refreshing idea of giving the data vendor community equal voice in developing specifications to meet the overall market requirements. Developing broadband specifications demands the talent of all the industry, and the ATM Forum was there to serve and accommodate the need. After combating the cultural shock of the data and telco community, one must say that the data community made several remarkable and positive contributions to the success of ATM. One of their flagship developments was the private NNI (network–network/node interface), which accelerated ATM deployment of the enterprise network. Another development was the successful specification of the available bit rate (ABR) traffic control service. This service accommodates the market needs for developing generic services requiring flexible traffic management control.

3.2.4 What Is ATM?

ATM is a fast packet-switching technology transferring digital information in the form of cells. Cells are constant in length and convey the information in an asynchronous manner, occupying no identified position in time (hence asynchronous). This is unlike circuit switch technique in which the information transmitted constantly relies on time and space.

ATM is also differentiated from the X.25 packet switch data network in that ATM operates at very high speed and is connection-oriented with fixed cell size and no retransmission at the network layer. The concept of quality of services is also embedded in the ATM layering structure.

The ATM cell, shown in Fig. 3.1, can deliver voice, data, and video services. The cell size is 53 bytes and is divided into two functional components:

1. A 5-byte header
2. A 48-byte payload or information field

The header functions:

- Identification of up to 2^{28} logical addresses
- Management information
- Error protection/correction, congestion information, and others

The 48 bytes may contain signaling, management, or data information such as digitized speech, images, and so on.

3.2.5 Why ATM?

Before describing the advantages of ATM and why ITU selected ATM as the target solution for B-ISDN, it would be worthwhile to compare it with the existing circuit switching fabric, namely STM. This comparison

Figure 3.1
ATM cell.

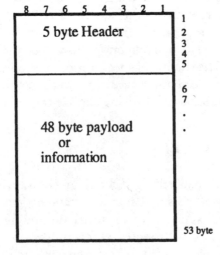

should be viewed in the context of integrating services of voice, data, and video. Such service integration is not far-fetched and evident in the services the Internet is providing today. It would be more difficult to justify ATM as the switching or transport platform for a specific application only. For example, ATM is not likely to be efficient if used as the switching fabric for data applications exclusively. The frame relay technology is likely to be more cost-effective than ATM for such applications. Some, however, are devising clever solutions to do so, especially for voice. The feeling is that if ATM can be justified to transport voice, then future deployment of it in the network will be secured.

3.2.6 ATM vs. STM

Today's STM switches in the telco network worldwide were developed and deployed in the 70s and 80s. The switch was optimized to handle voice traffic only. The network is synchronous (time-dependent), hence the name *synchronous transfer mode* (STM). The time relationship between users was continuously maintained to ensure real-time voice communications. The voice channel was assigned a symmetrical and fixed slot in time in the connection and delivered to its peer every 125 μsec (traditional sampling rate of an 8-kHz clock). Moreover, this time slot remains allocated to the connection throughout the life of the call, even during the silent period, or during the inactive interval when the channel is used as a data link (via the modem).

Data traffic behavior is bursty in nature and very likely asymmetrical. Using a symmetrical fixed time slot in an STM network is obviously not an efficient way of communicating. A data user may need a megabit of bandwidth for a short period of time but not nearly as much or none at all in the return path or otherwise. Today, one can easily experience this stricken principle as an Internet user connected to the telco access line via a modem. Another noticeable paradox is that it seems that the STM network is slaved to the end user. The network is wasting its resources waiting on a user for input. It would, obviously, be ideal if this dead time (resource wise) is allocated to other users who are on the network. ATM solved that problem, using the asynchronous transfer method. The network processes an incoming ATM cell when it receives one (independent of the time). The connection is identified using a label at every ATM cell received. The label belonging to that ATM cell can then be translated by the network and dispatched accordingly. In fact, the ATM cell is oblivious to its content (voice, data, video). As far as the network is concerned, they all look alike.

3.2.7 Advantages of ATM

There are several reasons for selecting ATM as the target solution for B-ISDN. Among the most crucial are:

1. Service integration as briefly described above
2. Service transparency
3. Multiplexing gain

SERVICE INTEGRATION. Service integration is an important feature that makes ATM attractive to implement. ATM transfers information in cells that could contain voice, data, or video. Hence, with a single physical line, a single network could deliver a variety of services to the user. Figure 3.2 is a typical configuration in a home today: a connection for the telephone, another one for Internet access (assuming a second POTS line is used), and a third connection to the cable TV.

In a broadband network using ATM, as shown in Fig. 3.3, a single physical line will be sufficient and can provide all three services mentioned.

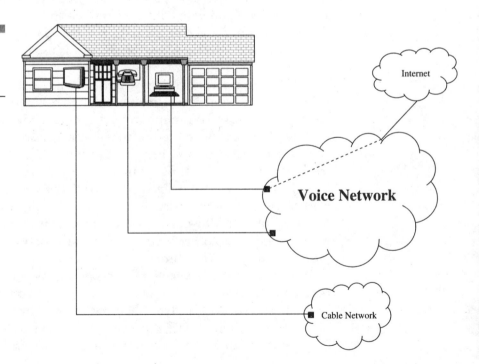

Figure 3.2
Typical access line connections to a home.

Figure 3.3
Typical UNI$_{FTTC}$ connection to a home.

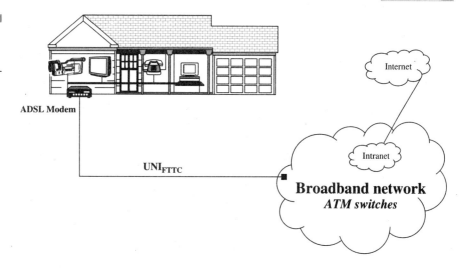

The ATM Forum defined that interface to be a UNI$_{FTTC}$ (user network interface for fiber to the curb). This UNI$_{FTTC}$ represents the model of an ATM broadband network in the telco environment. UNI$_{FTTC}$, in this case, is most likely to be an ADSL line. The ATM Forum also defined an interface inside the home called the technology independent interface (TII). TII connects the various devices from the network terminal (NT) unit.

When the service is provided by a cable operator, the interface as defined by the ATM Forum is UNI$_{HFC}$. The terminating unit in this model will contain a high-speed cable modem embedded in a set-top box or other home network equipment. It is as shown in Fig. 3.4.

Figure 3.4
Typical UNI$_{HFC}$ connection to a home.

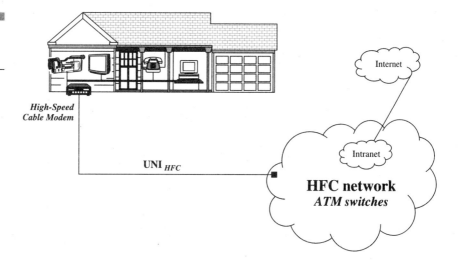

SERVICE TRANSPARENCY.   Service transparency is one fundamental advantage ATM brings when integrating services. In the conventional network, the network operator must provide the hardware interface at the central office every time a user requests new service. With ATM, it will be possible to provide additional services to a user simply by negotiating the needed bandwidth and quality of service (QOS) with the network. No additional hardware will be needed at the central office or at the home to provide this new service.

Assume, for example, a user as shown in Fig. 3.4 wishes to add a video camera to his or her home network. The idea is to buy the video camera with a standard generic interface much like it is done today when buying a telephone with an RJ-11 connector. Then, simply connect the camera to the home network. To go online, one simply negotiates with the network bandwidth needed and the sort of quality picture the user is willing to pay for when communicating with the outside world. A stereo system can also be installed in the same manner. Other services are also contemplated for home appliances to control various devices, monitor home security, utility meter reading, and the like.

MULTIPLEXING GAINS.   ATM inherently has multiplexing capabilities. One fundamental benefit of ATM over circuit switch techniques is the reliance on time domain when information is transferred. Unlike circuit switch, ATM has no predefined structure. This means that communication can be multiplexed regardless of the data rate. The statistical nature allows integration of services and also provides flexibility for transporting the communication channels. This can best be illustrated by the example shown in Fig. 3.5.

Let us assume a scenario in which a user is connected to the Internet through the traditional PC modem. If called, the line will be busy and remain so for the duration of the Internet connection. In this simplified model, if one assumes that the connection is ATM, having the same data rate, the multiplexing nature of ATM and its independence on time domain will allow the user to communicate with the Internet while talking on the phone using the same physical line. ATM cells can deliver voice samples, and, during the silent period, data could then be transferred transparently. Data cells simply stay in queue and are transferred in a multiplexed manner. Management of the queue is a function of the quality of service needed for the connection. Voice cells are time-critical and will have priority when extracted from the queue, while data cells are more immune to delay but, unlike voice, less so to cell loss.

Figure 3.5
ATM cell multiplexing.

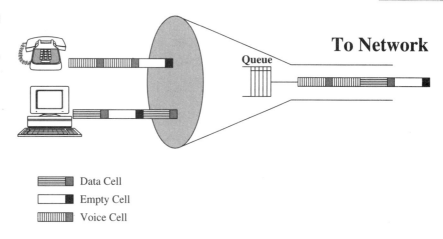

Data Cell
Empty Cell
Voice Cell

3.3 Network Architecture

Broadband ISDN (B-ISDN) was a term defined by ITU for public networks, while the ATM Forum developed specifications for the private networks. With the release and acceptance of the P-NNI, both the private and public networks are deviating architecturally in the areas of routing, addressing, and network connectivity.

Distinctions between the two networks are likely to disappear in the future. Presently, however, private networks still lack several features the public operators require before fully embracing it for public use. These are features such as billing, interfacing physical reach, network reliability and availability, networking to IN, and others. Figure 3.6 illustrates a broadband network. These interfaces are defined by ITU and/or the ATM Forum.

3.3.1 Public UNI

The public UNI is the interface connecting an ATM user directly to the public network. This interface was developed by ANSI and ITU standards and fully conforms to the B-ISDN reference model as will be described in more detail later in this chapter. The ATM Forum adopted it in the UNI 3.x series specifications.

Figure 3.6
Interfaces of the
broadband network
architecture.

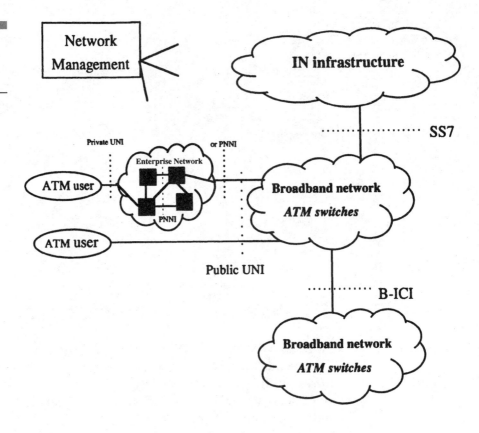

3.3.2 Private UNI

The ATM Forum developed the concept and specifications of the private UNI. The private UNI is the interface connecting an ATM user to the corporate or enterprise network. Unlike the public UNI, it is optimized to operate in a campus environment with limited physical reach connecting to customer premises equipment (CPE).

3.3.3 B-ICI

The B-ICI interface was developed by ANSI and completed by the ATM Forum. It is the interface between two different public network providers or carriers when crossing the demarcation point. The specification also includes service-specific functions needed when intercarrier crosses the LATA boundaries.

3.3.4 P-NNI

The P-NNI interface was developed exclusively by the ATM Forum. It is the interface between private ATM switches (B-ICI-like). P-NNI contains two protocols: one protocol is defined for distributing topology and routing based on the link-state routing technique; the second protocol defines signaling to establish connections across the private networks.

3.3.5 Signaling System Number 7 (SS#7)

SS#7 is a signaling link used as the out-of-band channel to exchange signaling information between public switches. The SS#7 interface was developed in ANSI and ITU in the 1970s. It is presently used in the public networks to provide 1-800 and other enhanced services. SS#7 is also used as the link to provide management information from/to the telco IN (intelligent network) infrastructure so it can extend enhanced supplementary services to users. ITU and ANSI are presently working on upgrading the protocols and call models to accommodate broadband services in the IN environment.

3.4 B-ISDN Protocol Reference Model (PRM)

The protocol reference model is the blueprint one needs to comprehend the fundamentals of ATM at all levels. The PRM, as shown in Fig. 3.7, was first developed in ITU.

The PRM contains three planes:

1. *User plane.* Mainly used to transport user information
2. *Control plane.* Used for signaling
3. *Management plane.* Maintains the network and operational functions

Before describing functions of the three planes, the lower layers will be described first. They are the foundations upon which these planes are constructed.

Figure 3.7
B-ISDN protocol ref-
erence model (PRM).

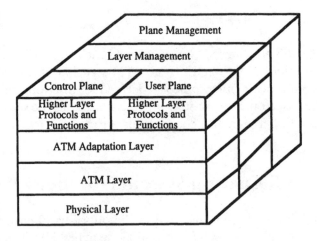

3.4.1 B-ISDN Lower Layers

The lower layers of the protocol reference model follow the OSI discipline and regiment in all respects. Each layer functions independently of the others and is designed to perform services that are needed by the layer above or below. The three layers are:

1. Physical layer

2. ATM layer

3. ATM adaptation layer

PHYSICAL LAYER. The physical layer, as illustrated in Fig. 3.8, contains two sublayers: the physical medium dependent (PMD) sublayer and the transmission convergence (TC) sublayer. These sublayers take on the personality of the attached physical link to perform the needed functions. A generic description of these sublayers are shown below.

PMD SUBLAYER. The PMD sublayer addresses aspects that are dependent on that particular transmission medium. The functions of this sublayer are:

■ Correct transmission and reception of bits

■ Ensure bit timing construction/reconstruction

■ Line coding at the transmitter and receiver

TRANSMISSION CONVERGENCE (TC) SUBLAYER. At the transmission convergence sublayer, the bits are assembled to fit into the frame of that par-

Figure 3.8
Sublayers of the physical layer.

Header Error Check (HEC) Cell Scrambling/Descrambling Cell Delineation (HEC) Path Signal Identification Frequency Justification Scrambling/ Descrambling Transmission frame Generation/ Recovery	**Transmission Convergence** ***TC***	**Physical Layer**
Bit timing Line coding Physical Medium	**Physical Media Dependent** ***PMD***	

ticular physical media. ITU defined this sublayer in I.442. The functions performed by the TC are:

- *HEC (header error check).*   A one-byte code generated using a specific polynomial. It is generated by the transmitter and inspected at the receiver to check the sanity of the received ATM cell. Two modes of operation are defined: correction mode and detection mode. In correction mode, only a single-bit error can be corrected, while detection mode provides for multiple-bit error detection. Errored cells are discarded. HEC code is the last octet of the ATM cell header.

- *Cell delineation.*   HEC is also responsible for identifying cell boundary. A number of consecutive and correct HECs received marks the boundary of the cells that are in a frame. A detailed description is provided in the section on header error control.

- *Cell scrambling/descrambling.*   Performed at the receiver and transmitter side. Scrambling avoids continuous nonvariable bit patterns and improves the efficiency of the cell delineation algorithm. Scrambling also prevents any malicious operation on the user information.

- *ATM cell mapping.*   Performed in order to align by row the byte structure of every cell with the byte structure of the SONET STS-3c payload capacity (synchronous payload envelope). This is a specific function when SONET is used as the physical layer.

- *Insertion of idle cells.*   Performed at the transmitter to accommodate the rate of the transmission system. If the physical rate is at DS3 and the data transmitted on that link is less, then idle/unassigned cells are inserted to maintain the DS3 clock rate.

PHYSICAL LINK.   Both the PMD and TC are the building blocks needed to assemble the physical layer. There are many interfaces defined for the public and private interfaces. These specifications were defined in ANSI

(T1S1 and T1E1) and ITU. The ATM Forum developed several other interfaces for the private network. A partial list is shown below. More interfaces are still being specified in the ATM Forum and T1E1 to address specific markets for residential broadband or to further reduce component cost. The list now includes:

- SONET 155.52 (SM and MM fiber)
- DS3 (PLCP and direct mapping)
- ADSL
- FDDI (100 Mb/s)
- UTP 5—155 Mbps
- UTP 3—25.6 and 52 Mbps
- DS1/E1
- E3
- E4
- Utopia
- Serial Utopia (NIRVANA)
- 6.3 Mbps
- 622 Mbps
- Cell-based TC
- 155 Mbps over UTP3
- 155 Mbps over 120 copper link segment

The physical layer for the HFC (cable modem) is fully described in Chap. 5.

DS3 Physical Layer Interface. DS3 physical layer interface is one example described below on how an interface, in this case DS3, is mapped using PMD and TC sublayers. This mapping of ATM cells over DS3 is very much unlike the mapping of ATM cells over cable modem physical layer. In this case, the mapping is based on a point-to-point, fully digital link. The mapping of all the other interfaces shown above are similar in principle and can be obtained from the ATM Forum office.

DS3 PLCP mapping was initially developed in ANSI standard (IEEE 802.6 completed it for DQDB). The functions of the DS3 physical layer are depicted in Fig. 3.9 and seen by the physical media dependent (PMD) sublayer and the transmission convergence (TC) sublayer.

Figure 3.9
DS3 physical layer
(PMD + TC sublayers).

Transmission Convergence **Sublayer**	HEC generation/verification
	PLCP Framing and Cell Delineation
	Path Overhead utilization
	PLCP Timing (125 μsec clock recovery)
	Nibble Stuffing
Physical Media Dependent **Sublayer**	Bit timing, Line coding
	Physical medium

Format at 44.736 Mbps. To carry ATM traffic over existing DS3, the Physical Layer Convergence Protocol (PLCP) defines the mapping of ATM cells by inserting the 53-byte ATM cells into the DS3 PLCP (Fig. 3.10). The PLCP is then mapped into the DS3 information payload. Extraction of ATM cells from the DS3 payload operates in the analogous reverse procedure (i.e., by framing on the PLCP and then simply extracting the ATM cells directly).

PLCP format. The DS3 PLCP is in a 125-μs frame within a standard DS3 payload. There is no fixed relationship between the DS3 PLCP frame and the DS3. The DS3 PLCP frame, as shown in Fig. 3.10, consists of 12 rows of ATM cells, each preceded by 4 octets of overhead. Although the DS3 PLCP is not aligned to the DS3 framing bits, the octets in the DS3 PLCP frame are nibble-aligned to the DS3 payload envelope. Nibble stuffing is required after the twelfth cell to frequency-justify the 125-μs PLCP frame.

Figure 3.10
DS3 PLCP Frame
(125 μs).

PLCP Framing		POI	POH	PLCP Payload	
A1	A2	P11	Z6	First ATM Cell	
A1	A2	P10	Z5	ATM Cell	
A1	A2	P9	Z4	ATM Cell	
A1	A2	P8	Z3	ATM Cell	
A1	A2	P7	Z2	ATM Cell	
A1	A2	P6	Z1	ATM Cell	
A1	A2	P5	X	ATM Cell	
A1	A2	P4	B1	ATM Cell	
A1	A2	P3	G1	ATM Cell	
A1	A2	P2	X	ATM Cell	
A1	A2	P1	X	ATM Cell	
A1	A2	P0	C1	Twelfth ATM Cell	Trailer
1 Octet	1 Octet	1 Octet	1 Octet	53 Octets	13 or 14 Nibbles

PLCP overhead utilization. The following PLCP overhead bytes/nibbles are required to support the coding/functions across the UNI:

- A1—Frame alignment
- A2—Frame alignment
- B1—Bit interleaved parity
- C1—Cycle/stuff counter
- G1—PLCP path status
- Px—Path overhead identifier
- Zx—Growth octets
- Trailer nibbles

Framing octets (A1, A2). The PLCP framing octets are used to identify framing. The A1 and A2 octets framing have the same pattern used in SONET and SDH. They are used for ATM cell delineation without using the HEC mechanism (described later in the chapter).

Bit interleaved parity-8 (B1). The bit interleaved parity-8 (BIP-8) field supports path error monitoring.

Cycle/stuff counter (C1). The cycle/stuff counter provides a nibble-stuffing opportunity cycle and length indicator for the PLCP frame. A stuffing opportunity occurs every third frame of a 3-frame (375-μs) stuffing cycle. The value of the C1 code is used as an indication of the phase of the 375-μs stuffing opportunity cycle (see Fig. 3.11a).

PLCP path status (G1). The PLCP path status is as illustrated in Fig. 3.11b. The G1 octet subfields are: 4-bit far end block error (FEBE), a 1-bit RAI (yellow), and 3 X bits (X bits are ignored).

FEBE provides a count of 0 to 8 BIP-8 errors received in the previous frame (i.e., G1 (FEBE) = 0000 through G1 (FEBE) = 1000). If not imple-

Figure 3.11
(a) Cycle/stuff counter (C1); (b) PLCP path status (G1) definition.

C1 Code	Frame Phase of Cycle	Trailer Length
11111111	1	13
00000000	2	14
01100110	3 (no stuff)	13
10011001	3 (stuff)	14

(a)

Far End Block Error (FEBE)	RAI (Yellow)	X-X-X
4 Bits	1 Bit	3 Bits

(b)

mented, G1 (FEBE) shall be set to 1111. Any other value of G1 would be caused by other errors and is interpreted as 0 errors.

RAI (yellow) alerts the transmitting PLCP that a received failure indication has been declared along the path. When an incoming failure condition is detected that persists for a "soaking period" (typically 2–10 seconds), an RAI shall be sent to the far end by setting G1 (RAI) = 1. The RAI will be detected when G1 (RAI) = 1 for 10 consecutive PLCP frames. The indication is cleared by setting G1 (RAI) = 0 when the incoming failure has ceased for 15 ± 5 seconds. At the receiving end, removal of the RAI signal is recognized by detecting G1 (RAI) = 0 for 10 consecutive PLCP frames.

Path Overhead Identifier (P0–P11). The path overhead identifier (POI) indexes the adjacent Path Overhead (POH) octet of the DS3 PLCP. Figure 3.12 shows the coding for each of the POI octets.

The above section addresses mapping of ATM over the DS3 physical layer. Each interface has specific personality and hence is documented separately. The specifications for a particular interface can be obtained from various sources. In the following section, the ATM layer will be described along with its associated services.

ATM LAYER. The ATM layer will now be described. It is the layer above the physical layer, providing services to the layer above it.

The ATM layer is independent of the physical layer. This is very important—being independent means it can be common to all the physical interfaces described previously. That means designers will not have to develop different ATM modules for different physical layers. The ATM layer is made up of four sublayers as shown in Fig. 3.13.

Figure 3.12
POI coding.

POI	POI Code	Associated POH
P11	00101100	Z6
P10	00101001	Z5
P9	00100101	Z4
P8	00100000	Z3
P7	00011100	Z2
P6	00011001	Z1
P5	00010101	X
P4	00010000	B1
P3	00001101	G1
P2	00001000	X
P1	00000100	X
P0	00000001	C1

Figure 3.13
The four sublayers for
the ATM layer.

Generic Flow Control Cell VPI/ VCI Cell Multiplexing/ Cell Rate	*ATM LAYER*

SUBLAYERS OF THE ATM LAYER. The four sublayers of the ATM layer are GFC (generic flow control), VPI/VCI translation, cell multiplexing/demultiplexing, and cell header generation.

Functions of the sublayers are:

Generic Flow Control (GFC). The GFC is mainly used as a flow control mechanism on customer equipment and therefore is applicable only at the user network interface (UNI). As such, GFC information is communicated between the user and the network in every ATM cell. Idle cells will not pass the GFC information to the ATM layer. The unassigned cells can then be used instead to convey flow control information to the ATM layer.

VPI/VCI Translation. The incoming VPI/VCI (virtual path/connection identifier) is translated to the address of an outgoing VP/VCI link. For a VP cross-connect, only VPIs are translated. On a VC switch, both VPI and VCI are translated.

Cell Multiplexing/Demultiplexing. The multiplexing and demultiplexing of ATM cells of different connections are concatenated into a single cell stream and passed on to the physical layer. Demultiplexing is done in reverse manner.

Cell Header Generation. The header is generated and/or extracted when received from or delivered to the adaptation layer. The adaptation layer is the layer above the ATM and will be described later.

ATM CELL STRUCTURE. The ATM cell contains a 5-byte header and 48-byte payload. Two ATM header cell structures were defined by ITU:

1. ATM header for the UNI. This is shown in Fig. 3.14.

2. ATM header for the NNI (network network interface). See Fig. 3.15.

With the exception of the GFC field, the two structures are functionally similar and therefore only the UNI header will be described. The four GFC bits are added to the VPI field for the NNI.

The structure of the ATM cell at the UNI, shown in Fig. 3.14, contains the following fields:

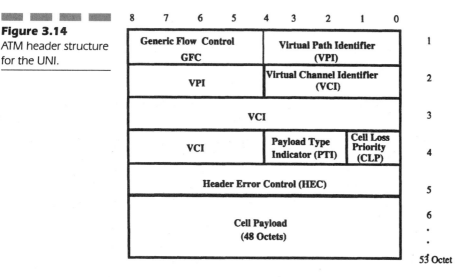

Figure 3.14
ATM header structure for the UNI.

Generic Flow Control (GFC). This field has local significance only and can be used to provide standardized local functions (e.g., flow control) on the customer site. The value encoded in the GFC is not carried end to end and will be overwritten by the ATM switches. In the uncontrolled mode, the GFC will be set to a zero value.

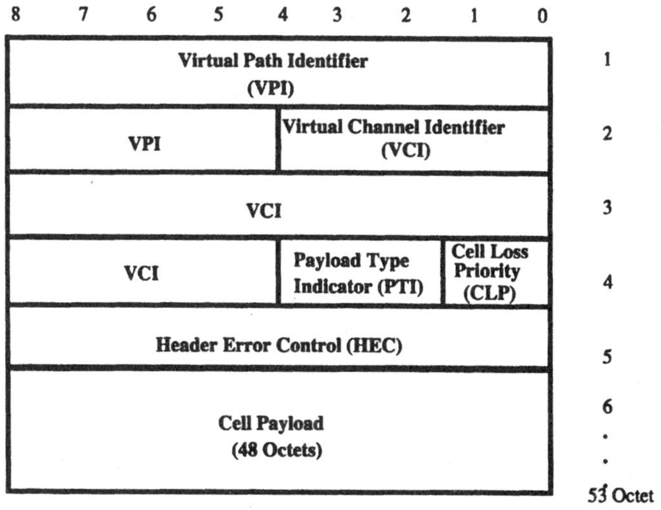

Figure 3.15
ATM header structure for the NNI.

Virtual Path/Virtual Channel (VPI/VCI) Identifier.   The number of routing bits in the VPI (8 bits) and VCI (16 bits) subfields are used for routing. The VPI/VCI value is allocated by the network during call negotiation between the user and the network.

Preassigned VPI/VCI Values.   ITU and the ATM Forum defined a set of preassigned VPI/VCI values for specific use. These values are unique and cannot be used for any other function. Table 3.1 shows the values and associated functions.

Metasignaling cells are used by the metasignaling protocol when establishing and/or releasing signaling connections.

General broadcast signaling cells are used by the ATM network to broadcast signaling information independent of service profiles.

TABLE 3.1

Predefined VPI/VCI
Header Values

Use	Value[1,2,3,4] Octet 1	Octet 2	Octet 3	Octet 4
Unassigned cell indication	00000000	00000000	00000000	0000xxx0
Metasignaling (default)[5,7]	00000000	00000000	00000000	00010a0c
Metasignaling[6,7]	0000yyyy	yyyy0000	00000000	00010a0c
General broadcast signaling[5]	00000000	00000000	00000000	00100aac
General broadcast signaling[6]	0000yyyy	yyyy0000	00000000	00100aac
Point-to-point signaling (default)[5]	00000000	00000000	00000000	01010aac
Point-to-point signaling[6]	0000yyyy	yyyy0000	00000000	01010aac
Invalid pattern	xxxx0000	00000000	00000000	0000xxx1
Segment OAM F4 flow cell[7]	0000aaaa	aaaa0000	00000000	00110a0a
End-to-end OAM F4 flow cell[7]	0000aaaa	aaaa0000	00000000	01000a0a

1. "a" indicates that the bit is available for use by the appropriate ATM layer function.
2. "x" indicates "don't care" bits.
3. "y" indicates any VPI value other than 00000000.
4. "c" indicates that the originating signaling entity shall set the CLP bit to 0. The network may change the value of the CLP bit.
5. Reserved for user signaling with the local exchange.
6. Reserved for signaling with other signaling entities (e.g., other users or remote networks).
7. The transmitting ATM entity shall set bit 2 of octet 4 to zero. The receiving ATM entity shall ignore bit 2 of octet 4.

Segment OAM F4 flow cells have the same VPI value as the user data cell transported by the VPC but are identified by two unique preassigned virtual connections.

Payload Type (PT). The PT is a 3-bit field used to indicate whether the cell contains user information or layer management, network congestion state or network resource management. The detailed coding of this 3-bit field is as shown in Table 3.2.

Code points 0 to 3 are used to indicate user cell information. Bit 2 indicates the congestion condition of the connection referred to as EFCI (explicit forward congestion indicator).

Code points 4 and 5 are used for VC level management functions. Code point 4 is used for identifying OAM cells, while the PTI value of 5 is used for identifying end-to-end OAM cells, bit 3 of octet 4.

Cell Loss Priority (CLP). This 1-bit field in the ATM cell header allows the user or the network to indicate the explicit loss priority of the cell. The user may want to set priority on ATM cells that could be subject for discard if the network is congested.

Header Error Control (HEC). The HEC field is used by the physical layer for detection/correction of bit errors in the cell header. HEC is also used to perform delineation of cell boundary in the physical layer. Figure 3.16 shows the HEC mechanism used in a state machine to perform cell delineation.

When a framing error occurs, the HEC sequence will invariably be corrupted. The state machine goes into a hunt state looking for a correct

TABLE 3.2

PTI Codings

PTI coding	Description
000	User data cell, congestion not experienced, SDU type = 0
001	User data cell, congestion not experienced, SDU type = 1
010	User data cell, congestion experienced, SDU type = 0
011	User data cell, congestion experienced, SDU type = 1
100	Segment OAM F5 flow-related cell
101	End-to-end OAM F5 flow-related cell
110	Reserved for future traffic control and resource management
111	Reserved for future functions

Figure 3.16
HEC state machine transition.

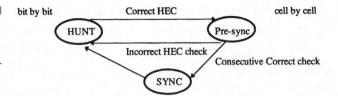

HEC. It will slide the window by one bit and recalculate HEC until a correct one is found. The calculation is HEC multiplied by 8 and divided by the polynomial $X^8 + X^2 + X + 1$. Once the correct HEC is found, the state machine goes into a presync state. In the presync state, the correct HEC must be derived n number of times before proceeding into a sync state (to ensure validity). Otherwise, it goes back to the hunt state. In the sync state, m incorrect consecutive HECs will bring the system into hunt state. The numbers for n and m are defined by ITU to be 6 and 7 for the SONET-based physical interface.

The associated services of the ATM layer will be described later in the chapter.

ATM ADAPTATION LAYER (AAL).   The AAL is the layer above the ATM layer. Again, this layer is independent of the layers above or below it. The function of AAL is to enhance the service provided by the ATM layer and support the specific adaptation of services above it. The upper layers provide control, management, and user services. AAL has two sublayers, as shown in Fig. 3.17:

1. Segmentation and reassembly (SAR) sublayer
2. Convergence sublayer (CS)

SEGMENTATION AND REASSEMBLY (SAR).   The SAR sublayer on the transmit side will segment the information from a higher layer into a size to fit the ATM payload (48 bytes without overhead). On the receiving end, the SAR will assemble the data into packets for delivery to the higher layer.

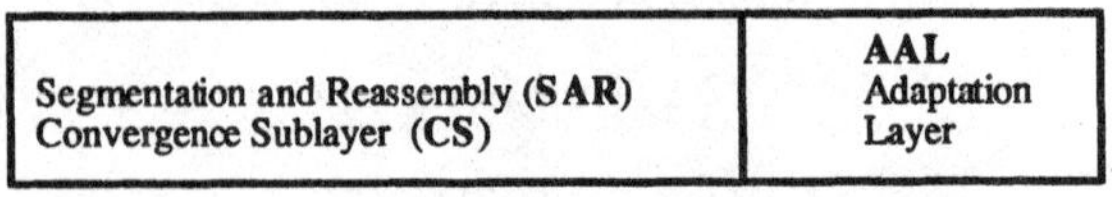

Figure 3.17
SAR and CS: sublayers of AAL.

Segmentation and Reassembly (SAR) Convergence Sublayer (CS)	AAL Adaptation Layer

CONVERGENCE SUBLAYER (CS). The CS sublayer performs several functions. It is service-dependent. Services performed over the ATM layer are grouped into four classes. The four classes are:

1. Class A
2. Class B
3. Class C
4. Class D

Three basic parameters are used to classify each of the services. The parameters are: time relation between source and destination, bit rate, and connection mode.

Time relation between source and destination. Voice and video services usually require time relationships between end users. Data services, on the other hand, need no time relation. Data can wait in buffers before being dispatched to its destination.

Bit rate. This parameter differentiates between a constant bit rate and variable bit rate.

Connection mode. The connection parameter can be either a connection-oriented or connectionless service.

ITU, in Recommendation I.362, defined the four classes of services and mapped these associated parameters as shown in Fig. 3.18.

Class A Service. Class A service is classified as a connection requiring time relation between end users, connection oriented, and constant bit rate (CBR). This service is sometimes referred to as circuit emulation (CE). Transporting a DS1 between end users (PBX trunk) is a typical example of the service. MPEG-2 video may also use class A service.

Class B. Class B service is classified as a connection requiring time relation between source and destination operating in a connection-oriented mode with a variable bit rate (VBR).

Figure 3.18
Adaptation service classes.

	Class A	Class B	Class C	Class D
Timing Relation	Required		Not Required	
Bit Rate	Constant	Variable		
Connection Mode	Connection Oriented			Connectionless

Class C.   Class C service is classified as a connection requiring no time relation between end users and operating in a connection-oriented model with a variable bit rate. Typical service over class C is the transfer of medium or large files.

Class D.   Class D service is classified as a connection requiring no time relation between end users and operating in connectionless-oriented mode with a variable bit rate. SMDS (switched multimegabit data services) is one typical application of this service.

There are three AALs that have been well defined in ANSI, ITU, and the ATM Forum. Another AAL-CU (AAL-composite user) is in its final stage of definition. Recently, ITU gave it the name AAL2. The three AALs that are well defined are AAL1, AAL3/4, and AAL5. AAL1, AAL5, and AAL-CU (AAL2) will play a major role in the HFC and cable modem development. Therefore, a description of these AALs follows.

AAL1.   AAL1 is defined under class A service. It is connection-oriented, requiring time relation between the source and destination and constant bit rate.

AAL1 Sublayers.   The SAR sublayer as shown in Fig. 3.19 adds the control header byte to the 47 bytes it receives from the AAL1 CS. So, a total of 48 bytes is sent to the ATM layer. This header/control field definition is as follows:

CSI.   The CSI (convergence sublayer indicator) is a single bit used as a peer-to-peer communication channel conveying time stamp information. Time stamp information of an ATM user is the difference between the receiver's local clock and network clock. This network clock must be common to both end users and can be easily derived from a SONET physical layer. This information is crucial when delivering CBR services. Hence, it is used for clock recovery and synchronization between end users. This delta information (time stamp), when received by the peer CS, will be used to control its local clock rate via a phased lock loop. This is done to ensure that the buffer, containing the transported voice information, will not overflow or underflow. This mechanism is referred to as the SRTS (synchronous residual time stamp) mechanism and is defined in ITU.

Figure 3.19
AAL1 SAR structure.

CSI SN SNP	47 bytes Data

1 bit 3 bits 4 bits

This SRTS mechanism should take care of the problem of drift/wander of the local clock on the user equipment. Low stability oscillators are used in home devices to keep them cost-effective. The buffer mechanism and SRTS will also solve the problem (to a certain extent) of cell delay variation (CDV) when cells are switched in an ATM network. CDV is the jitter of the successive ATM cells in a connection.

In the absence of a network/common clock to both sender and receiver, an adaptive technique was devised to use in place of the SRTS mechanism whereby the local clock of the user is controlled by a phase lock loop that is rigged by the fill of the buffer pointer.

Sequence number (SN). The SN is a 3-bit sequence number (modulo 8) incremented and tagged onto each cell. The peer CS validates that sequence to determine if a cell was lost or a foreign cell was inadvertently injected by the network. The error will be reported to the management plane. The management plane maintains various corrective actions. In a video or voice session, the corrective action may use the previous frame or previous ATM cell instead, or freeze action for the duration of the errored video frame, and so on.

Sequence number protection (SNP). This 4-bit field is the sequence number protection, mainly used to detect a multiple-bit error and/or correct a single-bit error of the information field.

AAL5. Data equipment manufacturers first started AAL5 development in ANSI (T1S1). It was then called SEAL (simple and efficient AAL). The ATM Forum and ITU completed this AAL5 specification. The objective of AAL5 was to reduce the overhead in the SAR/CS. The other objective was to devise a better error-detection mechanism for data services. ITU selected AAL5 for transporting signaling information for the control plane.

SAR Sublayer of AAL5. The AAL5 SAR sublayer accepts the protocol data unit from the CS, which is an integral multiple of 48 octets. The PTI value in the ATM layer defines the data boundaries as described in the section on payload type. A value 1 indicates the end of the SAR data unit. A value 0 means the beginning or continuation of an SAR data unit. Using this technique, one can efficiently queue the data stream until the last cell is received with a PTI = 0X1. CRC is then calculated and passed on to the layer above.

The Convergence Sublayer for AAL5. The common part of the convergence sublayer (CPCS) is illustrated in Fig. 3.20. The definitions of the fields are as follows:

Figure 3.20
CPCS format for
AAL5.

User Data (0-65535 bytes)	PAD	UU	CPI	Length	CRC-32

Padding (PAD) field. Varies from 0–47 unused bytes. This *n* number of bytes is used to align the protocol data unit into an integer number of 48 bytes.

User-to-user (UU) field. A one-byte user-to-user piece of information.

Common part indicator (CPI) field. A one-byte field for future use. Presently it is set to zero.

Length field. A two-byte field containing the length of the user data field.

CRC-32. A 4-byte field that contains the calculation of the entire data unit including the CPI, UU, PAD, and length field. The CRC-32 polynomial is: $x^{32} + x^{26} + x^{23} + x^{22} + x^{16} + x^{12} + x^{11} + x^{10} + x^8 + x^7 + x^5 + x^4 + x^2 + x + 1$.

AAL-CU (AAL2). ITU and ANSI (T1S1) have long discussed voice over ATM. Voice samples carried over the ATM payload is very inefficient, since voice samples are very short. Filling up the ATM cell with voice samples will cause echo due to the delay encountered in transmission and cell reassembly. A partially filled cell was another approach discussed but did not go far because of its obvious inefficiency of transporting voice. The composite user (CU) method is a way of multiplexing several user voice samples into minicells and embodying them into a single ATM cell payload as shown in Fig. 3.21.

Figure 3.21
AAL-CU minicells
multiplexed into an
ATM cell.

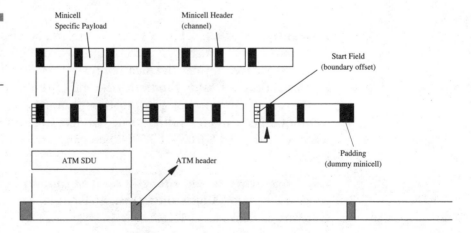

In summary, the two fundamental benefits of AAL-CU are:

1. *Efficient bandwidth utilization.* Multiplexing users over a single ATM VC allows for efficiency in bandwidth.

2. *Low delay.* Multiplex of short packets (minicell) in the payload of a single ATM cell can be reassembled with much less delay.

ITU gave it the preliminary name of AAL2. This is unfortunate, since it would invalidate several ATM published books and other literature in the market. AAL2 was originally set aside by ITU for the (incomplete) variable bit-rate service. That AAL cell structure is radically different from the one recently agreed to in *ITU for AAL-CU.* The new AAL2 accommodates voice service adaptation of mobile voice applications as well as landline trunking. There are some who are still advocating the renaming of AAL-CU to AAL6 instead of AAL2.

Structure of AAL-2 Sublayers. The AAL2 layer includes two sublayers. They are the CPS and SSCS sublayers:

Core/common part sublayer (CPS). Performs the basic functions of assembly and disassembly of user data (SDU) into minicell PDUs. The minicells are in turn multiplexed and demultiplexed into the ATM VC connection. The main function of this sublayer is to perform the core functions of minicell identification, multiplexing, and delivery.

Service-specific convergence sublayer (SSCS). Performs service-specific functions such as segmentation and reassembly of long user SDUs into shorter SDUs. This sublayer can also support protection of the SDU by a CRC, length indicator, and attached sequence number.

The format of CPS of AAL2 is as shown in Fig. 3.22. Three types of SSCS service-specific convergence sublayers are also proposed: SSCS/basic (simple mapping), SSCS/CRC (CRC and sequence number), and SSCS/long-data.

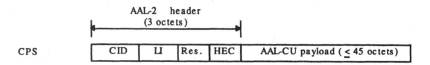

Figure 3.22
Structure and format of AAL2.

CID. Circuit identification (8 bits or more)

LI. Length indication (6 bits)

Res. Reserved (5 bits or less)

HEC. AAL-CU header error check (5 bits)

HEC is capable of either correcting one bit error or detecting more than two bit errors. AAL-CU PDU can be self-delineated in a sense that:

- A PDU boundary can be identified by using LI (unlike the fixed 53 octets of an ATM cell delineation).
- All other mechanisms of an ATM cell delineation will apply (e.g., once delineation is lost, a receiver goes to the hunting state of octet-based operations).

The length of AAL-CU payload is limited to 45 octets, so that one AAL-CU PDU can be conveyed (and possibly switched) by a payload of an ATM cell. Reserved bits may be utilized to indicate payload type (e.g., OAM flow or F7) or binary indication (e.g., an SSCS/long-data) when needed.

Similarities of MAC/AAL2 Cell Transport. AAL2 is expected to play a major role in the future development of the cable modem. The minislot concept adopted by IEEE 802.14 (see Chap. 5) for the upstream channel is a close relative of AAL2. Adaptation of this AAL2 to voice as well as MPEG video over HFC simplifies the delay requirements and bandwidth efficiency in the transport. This will become more evident when the new real-time-specific services are developed in future releases of the cable modem.

This ends the brief descriptions of the foundation layers (low layers). Next, the higher layers of B-ISDN are briefly described.

3.4.2 B-ISDN Higher Layers

The higher layers, as mentioned in the protocol reference model, are:

- User plane
- Control plane
- Management plane

USER PLANE. The user plane contains user information and service-specific protocols that ride on top of ATM. IP over ATM or frame relay over ATM are a few examples in which such services are delivered end to end. The associated service protocol is encapsulated using the ATM data unit and transported via the appropriate AAL. In most cases, these services have their own protocol but are preempted by the B-ISDN protocol layers.

CONTROL PLANE OR SIGNALING. This control plane provides signaling message transport and call control/connection control capabilities.

SIGNALING DEFINITION. Signaling is a mean of dynamically establishing, clearing, and/or modifying a call or connection. A call may have several connections. Permanent connections between users is practiced today. The operator simply establishes the call (via a network management element), assigning a VPI/VCI connection for the requested users. Tearing down that connection will also require operator intervention to clear the connection manually. This is obviously a clumsy and expensive task if the call duration is short. Signaling, therefore, is required when the average holding time of a call does not exceed a certain period of time. Four to six minutes is the average holding time of a voice call today.

ITU released several recommendations dealing with signaling protocol, procedures, and interworking. Recommendations Q2931 for user signaling and Q2761 for NNI (B-ISUP) signaling were based on a remnant of the signaling recommendations developed for narrow-band ISDN (NB-ISDN). Together, these two recommendations comprise the foundation needed to develop signaling for broadband private and public networks. Unlike NB-ISDN, broadband-added capabilities include features such as modifying a live connection (e.g., changing bandwidth), modifying the connection's QOS, interworking with narrow-band signaling, multimedia signaling capabilities, and so on.

Both the ITU Q2931 and Q2761 use out-of-band signaling, meaning that the signaling messages have an independent communication channel. VPI = 0 and VCI = 5 virtual channels were assigned exclusively (as mentioned in the section on payload type). This VPI/VCI = 5 is used for signaling communication between the user and the network.

Figure 3.23 illustrates the set of the Q-series recommendations (mapped into a broadband network) developed or under development by ITU to fully implement signaling in the network. It is beyond the scope of this book to describe each one of the Q-series recommendations, but we will briefly describe a few that will have immediate impact on HFC/cable modem development:

Figure 3.23
Q-series recommen-
dations mapped into
a broadband net-
work.

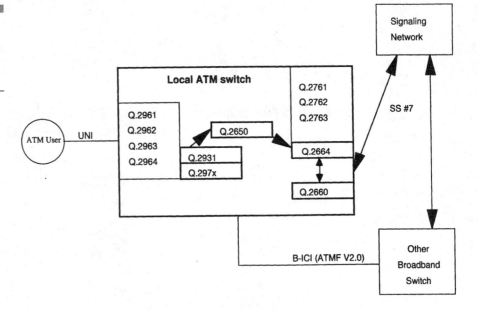

SIGNALING AT THE UNI. UNI 4.0, released in July 1996 from the ATM Forum, is aligned with ITU signaling recommendation. It includes the following capabilities:

- Point-to-point calls
- Point-to-multipoint calls
- Signaling of individual QOS parameters
- Leaf-initiated join
- ATM anycast
- ABR signaling for point-to-point calls
- Generic identifier transport
- Virtual UNIs
- Switched virtual path (VP) service
- Proxy signaling
- Frame discard
- Traffic parameter negotiation
- Supplementary services
- Direct dialing in (DDI)
- Multiple subscriber number (MSN)

- Calling line identification presentation (CLIP)
- Calling line identification restriction (CLIR)
- Connected line identification presentation (COLP)
- Connected line identification restriction (COLR)
- Subaddressing (SUB)
- User-user signaling (UUS)

IEEE 802.14 started to specify the signaling portion for the cable modem. It will be based on UNI 4 with some of the capabilities shown above. Below is a simple scenario on how to set a point-to-point connection. The message flow for setting up a call is as shown in Fig. 3.24.

MULTIMEDIA IN RESIDENTIAL BROADBAND. More demanding features such as multimedia are important aspects for residential broadband. ITU has extended the call model of an ATM call such that a call can include multiple connections or zero connections. Hence, a call between two users becomes an association that describes a set of connections between them. This concept was extended to three or more users to support multiparty/multiconnection calls. There can be multiple calls between two

Figure 3.24
An example of call setup.

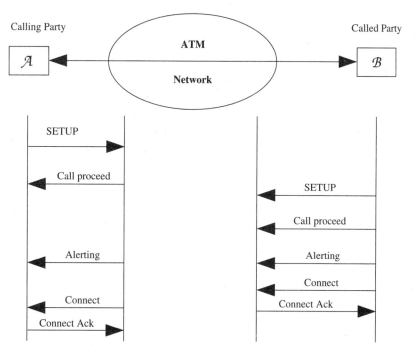

users. By extending the call between two users to support multiple connections, multimedia applications can be supported more effectively. Each media type of a multimedia communication application has its own bandwidth and QOS requirement. Separate VC connections became feasible and satisfied the multimedia communication requirements. The concept of a call with multiparty is very important for residential broadband service, because many applications running on the set-top box could require communication with various servers in the headend or at the network.

A call without a connection is also useful. When making a call, a user can probe the network without actually setting up the connection, wasting valuable network resources. If a connection is established first, network resources could be held up temporarily only to learn that the call cannot be completed because the end user is busy or the terminal capabilities do not match.

Connection Types.　ITU defined five types of connections covering all possible aspects needed for connecting users, especially for residential broadband. It is beyond the scope of this book to describe the details of these connections and their associated performance requirements. A general and brief definition is shown below. The five connection types are:

Type-1 connection.　A point-to-point configuration with bidirectional and asymmetric communication. See Fig. 3.25.

The bandwidth is independently specified in each direction; hence, it is an asymmetric connection. The type-1 connection is unidirectional and can have zero bandwidth in one direction. It can also specify that the physical route taken by the connection in each direction must be the same. Type 1 must be supported for residential broadband services.

Type-2 connection.　Point-to-multipoint unidirectional configuration, unidirectional communication with a root and leaf control. See Fig. 3.26.

Figure 3.25
Type-1 connection.

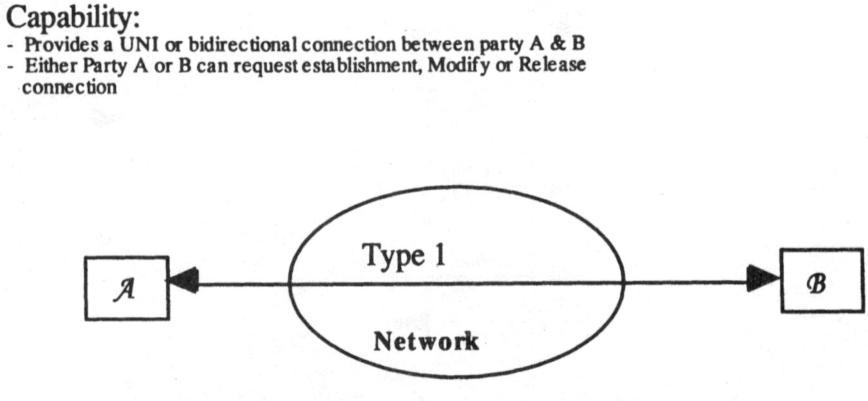

Capability:
- Provides a UNI or bidirectional connection between party A & B
- Either Party A or B can request establishment, Modify or Release connection

Figure 3.26
Type-2 connection.

Capability:
- Provide a unidirectional connection between party A, B & C.
  Originating party A terminating to B, C or more
- Provides multicasting capabilities by the network
- B&C are unable to talk back to A party
- Either party may request the connection (establish, modify, release) and may
  be allowed to deny the connection

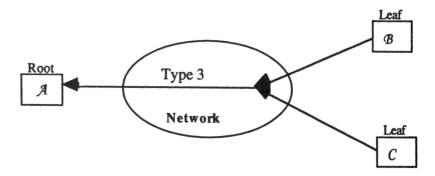

The type-2 connection involves three or more parties with the calling party being the root. This is a unidirectional connection where the root is the only source, multicasting its information over the type-2 connection to all the other parties of the connection (called leaves). The root can add and drop parties. The leaf party can also drop from such a connection. It is up to the root to specify during the connection setup time whether it allows independent add from leaf parties (known as leaf-initiated join). The type-2 connection is required to support multicast services (e.g., broadcast TV). The leaf-initiated join mechanism allows the viewer to join a particular TV channel.

Type-3 connection. A multipoint-to-point unidirectional configuration with unidirectional communication, leaf and single root control. See Fig. 3.27.

Figure 3.27
Type-3 connection.

Capability:
- Same as type 2 configuration except B& C party can talk back to A party.

This type-3 connection has the same logical configuration as type-2 connections, except the information flow is reversed, from the leaves back to the root. A typical application for a type-3 connection is advertisement insertion. This requires merging of multiple video clips from different servers. It is important to have a multipoint-to-point connection from all these servers to each. The reason is that, at any one instant, there is only one ad playing from the viewer's point of view. The type-3 connection is the most efficient connection type for these applications. Otherwise, if a separate point-to-point connection is established from each server to the viewers, only one of the connections is active at any one time, and the rest are wasted.

Type-4 connection. A multipoint-to-multipoint bidirectional configuration with bidirectional and full party control. See Fig. 3.28.

This connection is the bidirectional multipoint-to-multipoint connection that allows communications between multiple parties. The information sent by any party will be received by all other parties to the connection. The type-4 connection is important for supporting group communication such as off-campus education, multiuser games, and multiparty video calls.

Type-5 connection. A point-to-multipoint bidirectional connection with bidirectional communication and root and leaf control. See Fig. 3.29.

This type is similar to a type-4 connection, but it has features as shown in the capability section.

MANAGEMENT PLANE. The management plane is composed of two layers:

Figure 3.28
Type-4 connection.

Capability:
- Provide a bidirectional connection between party A, B & C.
- Provides multicasting capabilities by the network
- Either party may request the connection (establish, modify, release) and may be allowed to deny the connection

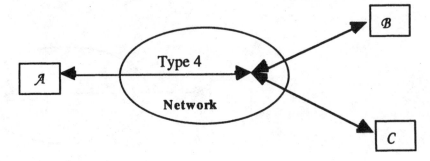

Figure 3.29
Type-5 connection.

Capability:
- Provide a bidirectional connection between party A, B & C.
- Provides multicasting capabilities by the network
- Root party A can send information to Leaf party B or C but the Leaf party can only communicate with party A
- Either party may request the connection (establish, modify, release). & may be allowed agree or deny the connection

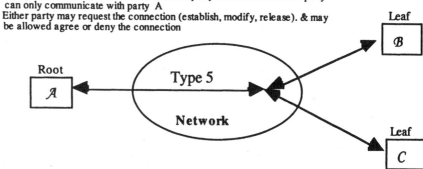

1. Plane management
2. Layer management

PLANE MANAGEMENT. Plane management is responsible for overall functions of the system end-to-end. Its main role is to report, extract, and coordinate all management information between all the planes shown in the broadband reference model.

LAYER MANAGEMENT. Layer management performs management-related functions such as performance, operation, administration, resource management, and parameters associated with each of the user planes. ITU recommendation I.610 defined basic OAM principles that provide the status, testing performance, monitoring, and so on. The basic functions of OAM defined by ITU are:

- Performance monitoring
- Defect and failure detection
- System protection
- Performance information
- Fault localization

Performance Monitoring. Performance monitoring periodically scouts and controls the internal operation of the layers. The data is stored and collected for later evaluation of the system performance.

Defect and Failure Detection. Defect and failure detection periodically checks for failures and, when detected, it will localize them and attempt to isolate them from the network.

System Protection. The main function of system protection is to isolate failed entities and exclude them from operation. This protects the system from excessive failures.

Performance Protection. Performance protection is used to inform all neighboring entities of a detected failure so the failure will be avoided by other layers within the system. The status information is also sent to the networks, so they can avoid the failed entity.

Fault Localization. Fault localization detects failures by internal or external means (test modules). It determines the identity of the failed component and enters into a protection state that will take it out of service. ITU defined five levels of OAM to identify the faulty equipment. Figure 3.30 shows the five (F1 to F5) functions that can be invoked to determine the exact fault. The physical layer is composed of the lower three levels: F1, F2, and F3. The ATM layer consists of the remaining two levels.

For a physical layer based on SONET/SDH or PDH, flows of F1, F2, and F3 are carried in synchronous channels in the overhead of the physical layer. For the ATM layer itself, the F4 flow is carried in cells distinguished by preassigned VCIs. F5 flow is carried in cells distinguished by special PTI codes in the virtual circuit. A summary of the F1 to F5 OAM follows.

Virtual channel (F5). Both endpoints perform VCI termination functions for a B-ISDN connection. Such a connection is composed of several virtual paths. The OAM functions are performed on a VCI level and may give input to any of the five phases described above.

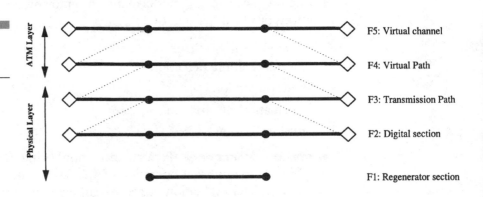

Figure 3.30
OAM hierarchical levels.

Virtual path (F4). Both endpoints perform VPI termination functions for a B-ISDN connection. Such a connection is composed of several transmission paths. Again, one of the five phases described above may be involved in the virtual path maintenance.

Transmission path (F3). Both endpoints perform the assembly/disassembly of the payload and the OAM functions of a transmission system. Since cells must be recognized at a transmission path to extract OAM cells, cell delineation and HEC functions are required at the termination points of each transmission path. A transmission path is composed of several digital sections.

Digital section (F2). Both endpoints are section termination points. A digital section comprises a maintenance entity. It is capable of transporting OAM information from adjacent digital sections.

Regenerator section (F1). This is the smallest recognizable physical entity for OAM, and is located between the repeaters.

3.4.3. Other Aspects of B-ISDN

In this section, the following will be described:

1. User-network interface
2. Traffic control
3. Service categories
4. Routing

USER-NETWORK INTERFACE. ITU Recommendation I.413 defines the user network interface for B-ISDN as shown in Fig. 3.31. The reference configuration is generic and covers all aspects of the user interface configurations and connectivity with various private-to-public operator switches.

There are four reference points: R, S, T, and U. Definition and functions of each are as follows.

B-TE1 and B-TE2. Broadband terminal equipment 1 and 2 terminating all protocols from the low and higher layers.

Figure 3.31
B-ISDN reference configuration of customer equipment.

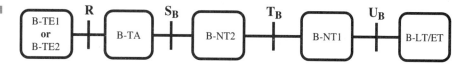

B-TA. A terminal adapter when needed.

B-NT2. The broadband network termination 2. It could be a LAN, PBX, or smart cable modem (like the set-top box). It performs signaling protocol, adaptation layer handling, policing, cell multiplexing, traffic control, and so on.

B-NT1. The broadband network terminal 1. It performs line termination and OAM functions.

B-LT. Broadband line termination at the end office.

Possible variations of this configuration are the star and star-shared-medium configuration. Functionally, however, they all behave in a similar manner.

TRAFFIC CONTROL. Traffic control is the central nervous system responsible for the overall operation of the broadband network. The primary role of the parameters and procedures used for traffic control and congestion control is to protect the network and the end system in order to achieve network performance objectives. The network must operate under the worst-case condition and must also match today's network reliability and availability that the user has enjoyed for decades. To do less will undermine the deployment of broadband in general and ATM in particular.

The previous chapters describe how a user can have all these multimedia ATM features with associated QOS on demand and so on. These features are unique and extremely valuable in making multimedia work in the way that it is envisioned. The industry has little practical experience in the area of complicated traffic control coupled with maintaining a QOS. The issues that are partially addressed are:

- How to deal with all these traffic variables
- How to keep the network sane under fault condition and avoid congestion collapse
- How to protect the network by maintaining the performance objectives of cell loss, CDV, and cell transfer delay
- How to maintain objective performance in the presence of unpredictable statistical traffic fluctuation (Mother's Day syndrome)
- How best to engineer the network in such an environment while optimizing network resources to achieve practical network efficiency

In summary, traffic management objectives must exhibit:

- Robustness
- Flexibility

■ Simplicity (e.g., they must be operator-friendly to avoid network congestion collapse or total traffic outage)

ANSI, ITU (I.371/I.311 Recommendation), and the ATM Forum TM v.04 addressed most of these issues. They are described below. The topics that will be covered are:

■ Connection admission control (CAC)

■ Usage parameter control/network parameter control (UPC/NPC)

■ Priority control (PC)

■ Traffic shaping (TS)

■ Network resource management (NRM)

■ ABR traffic control

Figure 3.32 illustrates the positioning of these functions in the network.

CONNECTION ADMISSION CONTROL (CAC).   CAC contains a set of actions performed by the network during initial call setup with an ATM end user. CAC accepts the requested connection if sufficient resources are available in the whole network to accommodate such QOS and bandwidth requests. The call is accepted if and only if the system can maintain and preserve the QOS of all other connections already established. If resources are not available, the network will simply deny the connection or may renegotiate the parameters with the end user. The parameters that are negotiated are:

■ Limit on the volume of the traffic (per traffic descriptor); e.g., average/peak bit rate, burstiness, and peak duration

Figure 3.32
Traffic control network functions.

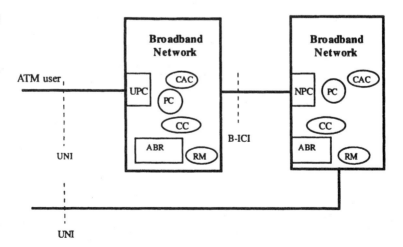

- QOS class requested:
  Cell loss ratio (CLR)
  Cell transfer delay (CTD)
  Cell delay variation (CDV)

CAC traffic contract may change during the life of the call. An ATM user may want to add, delete, increase, or decrease the QOS or bandwidth that was originally negotiated during the initial call setup time, in which case CAC performs the same functions as described above, as if a new call were established. CAC is operator-implementation-dependent. An operator may even choose to control his or her network resources to accommodate his best-paying customers first.

USAGE PARAMETER CONTROL (UPC). In an ATM environment, unlike in a synchronous transfer mode, the ATM user may at any time opt to send information up to his or her physical access rate. Therefore, throughput must be policed at the UNI interface by the usage parameter control (UPC) function in the network. This ensures that the contract originally negotiated during the setup phase with the user is respected and fair to the other connections. The set of actions taken by UPC are:

1. Monitoring
2. Enforcing

In the monitor mode, the UPC validates routing of the VPI/VCI connections that are being used, monitors every connection, and validates its traffic characteristics and volume per the negotiated contract. As the traffic cop, the UPC enforces the negotiated traffic contract of any and all misbehaving users (or due to equipment failure). The minimal actions taken by UPC at the cell level are:

- Passing the cell when the connection is respecting its contract.
- Discarding the ATM cell if the connection is violating its negotiated contract and the network is reaching its resource limit. This ensures that adequate resources are available to the conforming users.
- Tagging the violating cell by setting its CLP bit to 1. CLPs set to 0 are considered by the network as conforming cells. Setting the CLP to 1 will be discarded first by any node in the network that is experiencing congestion. This guarantees that service-quality-compliant users are not affected.

The UPC may opt to release the connection if it concludes that fault equipment is the cause or malicious user behavior is assumed. ITU I.371 Annex B describes in details the algorithm and behavior of the UPC when monitoring and/or enforcing the traffic contract.

The network parameter control (NPC) at the B-ICI (interface between two public broadband switches) functions in the same manner.

PRIORITY CONTROL (PC).   The priority control is a feature the user may elect to use by setting priority on his or her traffic. He may choose to set the CLP bit to 1 (low-priority traffic) so the network, if congested, will discard them first as opposed to the cell with CLP = 0. This network behavior ensures that performance is preserved for those of high-priority cells in case of traffic congestion or equipment failures. This also guarantees a priority service for the user who inadvertently submits traffic beyond his or her negotiated contract.

TRAFFIC SHAPING.   Traffic shaping partially alters the characteristic of the traffic and the effects of CDV on the peak cell rate of the ATM connection. An example of a traffic-shaping mechanism is to throttle and therefore limit burstiness of the peak cell rate traffic so it may comply with the negotiated peak cell rate contract. This mechanism also allows the user to prioritize his or her burtsy traffic by setting the CLP bit accordingly. At the ATM switch, traffic shaping alters the traffic characteristics cell flow so it can reduce the network jitter or cell delay variation (CDV). It is done by suitably spacing cells in time.

NETWORK RESOURCE MANAGEMENT.   The main function of the network resource management is to allow an operator to segregate his or her traffic characteristic based on the differing QOS. This mechanism simplifies CAC and reserves capacity in virtual path connections separated from other traffic. This mechanism may reduce operational control cost vs. increased capacity cost.

ABR TRAFFIC CONTROL.   ABR is defined under the umbrella of service categories. However, its mechanism is discussed here because of its contribution to the network behavior in abating or reducing the chances of network congestion and so on. In the HFC environment, ABR will play a major role in determining the traffic behavior between the headend and cable modem. The ABR described below is by no means complete. It is intended to illustrate the basic mechanism as it may be applied for correct cable modem operation.

Connection Establishment for ABR. Upon the establishment of an ABR connection, the user specifies to the network both a maximum required bandwidth and a minimum usable bandwidth. These parameters are defined as the peak cell rate (PCR) and the minimum cell rate (MCR). The MCR may be specified as zero. The bandwidth available from the network may vary, but it should not become less than the negotiated MCR.

Data application traffic behavior. Traditionally, most data applications have the ability to regulate and moderate the information transfer rate as a function of the network's loading status. They can reduce the information transfer rate if the network requires them to do so; likewise, they may wish to increase their throughput if there is extra bandwidth available in the network. The traffic generated by data applications is highly unpredictable and bursty in nature. This is to say that a fixed (rigid) traffic contract as defined for connections using the CBR or VBR service category is not ideally suited for this type of data application.

Telco application traffic behavior. The telecommunication applications rely on fixed bandwidth resource reservations. Moreover, a telco application, such as voice, can tolerate cell loss but not cell delay. Data applications, on the other hand, are rather sensitive to cell loss but can tolerate substantial variations in delay. Hence, there is a need for the network to inform the user via feedback of an impending exhaustion of resources that could result in cell loss. Therefore, additional ATM layer traffic management facilities are necessary in order to transport traffic effectively from such sources in ATM networks so that network operators can make the most of any unused link capacity without affecting the performance of the CBR and VBR connections.

ABR service definition. ABR is an ATM layer service category where the throughput, provided by the network, may change subsequent to connection establishment. It is expected that a user who adapts his or her traffic to the changing ATM layer transfer characteristics, as indicated by feedback control information received from the network, will experience a low cell loss ratio. Cell transfer delay and cell delay variation are not controlled by the ABR connections. Therefore, ABR is not intended to support real-time applications.

ABR connection establishment. On the establishment of an ABR connection, the originating party or calling party of the connection, denoted as the source end system (SES) by the ATM Forum, will specify a maximum bandwidth, designated as the peak cell rate (PCR), and a minimum usable bandwidth, denoted as the minimum cell rate (MCR). The PCR is negotiated between the end systems and the networks. The MCR is also specified on a per-connection basis, but it may be specified as zero. The bandwidth available at a particular instant somewhere in the network may become as small as the MCR. Hence, the available bandwidth of an ABR connection varies between the MCR and the PCR during the connection lifetime. This was designed to accommodate the following scenarios:

- Bandwidth becomes free again when CBR and VBR connections are released
- Bandwidth advertised as available to other ABR connections sharing that same link possibly remains unused
- Bandwidth resources in the network are being reserved for CBR and VBR connections that are set up

ABR service requirements. The ATM Forum and the ITU define the following criteria of any networking mechanism of the ATM layer ABR service:

- *Fairness.* The available resources may be shared according to a network-operator-defined policy amongst all ABR connections set up on a particular link. No set of ABR connections should be arbitrarily or systematically chosen over another set.
- *Robustness.* The operation of the ATM network should not rely upon the correct cooperation of the end users to comply with the provided control information (i.e., the network should protect itself from possible misbehaving ABR users).
- *ABR may apply to a virtual channel connection (VCC) or a virtual path connection (VPC).* ABR VCCs within a VPC share the capacity of the corresponding VPC. This VPC may either be allocated with a static bandwidth or the VPC bandwidth itself may be changed dynamically through ABR mechanisms operating at the virtual path level.
- *Any ABR connection has two components.* First is a minimum usable cell rate MCR, which may be zero, but for which there is a quantitative commitment made by the network. Second is an elastic component on top of MCR, for which the commitment made by the network is only a relative assurance that bandwidth will be shared via a defined network allocation policy amongst the competing ABR connections. Once MCR commitments are met, the network policy for fairly allotting the remaining available bandwidth amongst the

ABR connections is not subject to standardization. However, it is expected to be consistent and predictable. Yet, this network allocation policy cannot be observed in isolation by an ABR user.

- *A reference behavior for both the source and destination will be specified.* ATM-layer QOS performance commitments regarding a network-specific CLR will apply only to those ABR connections whose end systems follow the defined reference behavior.

- *During an ABR connection establishment for a connection with MCR = 0, the CAC function will not block the connection because of bandwidth allocated to other connections.* Yet, blocking due to CAC for other reasons (e.g., lack of VP/VC identifiers) is not precluded.

- *The ABR service category should have the goal of providing access as rapidly as possible to any unused bandwidth.* Up to the ABR connection's negotiated PCR, whenever the network bandwidth is available, this should be the goal.

ABR traffic control mechanism. The ATM Forum adopted a rate-based, as opposed to a credit-based, approach with buffer allocation. This ABR traffic control mechanism will enable the source to dynamically adjust its cell-sending rate based on feedback control information received from the network, indicating the availability status of bandwidth resources.

End-to-end feedback control. To obtain network feedback information, the ABR connection's source injects resource management (RM) cells into the data cell stream on a regular basis. The network will then look for available bandwidth resources (i.e., the RM cells inserted by the source will actually fulfill the role of network bandwidth scout). The RM cell uses the same VPI/VCI as the user data cells and is standardized by the ITU. It is identified in the ATM header payload type identifier (PTI) code point equal to 110 (see Table 3.2).

The source end system (SES) RM cells are then reflected by the destination end system (DES), thereby closing the information control loop back toward the SES (see Fig. 3.33). Thus, with respect to the regular user data cell flow from the SES to the DES, there are two RM cell flows: one in the forward direction from SES to DES, and one in the backward direction. The direction of flow to which an RM cell belongs will be indicated by means of a one-bit direction indicator. This DIR is a bit in the RM cell information field. DIR = 0 indicates that the RM cell flows in the forward direction, while DIR = 1 identifies a backward RM cell flow.

RM feedback. RM cells with DIR = 0 and the congestion indication bit in the RM cell set to zero (CI = 0) is generated by the SES and forwarded

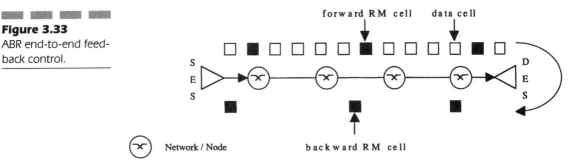

to the network. This represents future opportunities for a cell-rate increase on this ABR connection. The SES then waits for the network's response. It will increase its cell-sending rate upon the reception of backward RM cells with CI still equal to 0. The absence of backward RM cells due to occasional RM cell loss, for instance, or the presence of backward-received RM cells with CI = 1, indicate to the SES that it has to decrease its cell-sending rate. The decision to deny such rate-increase opportunities for an ABR SES is made independently by each intermediate network possibly or each intermediate switching node on the end-to-end path of the ABR connection. This happens on the basis of its internally defined and monitored congestion status indicators (network-operator-specific) of bandwidth and possibly also buffer resources that need to be protected.

ABR feedback control, described in the above sections, illustrates a simple scenario. This should give the reader operational understanding of ABR service and how it must eventually be adapted in the HFC network and the cable modem in particular. Chapter 8 discusses ABR implementation in a shared medium environment.

SERVICE CATEGORIES. The architecture for services provided at the ATM layer consists of the following five service categories:

1. CBR   Constant bit rate
2. rt-VBR   Real-time variable bit rate
3. nrt-VBR   Non-real-time variable bit rate
4. UBR   Unspecified bit rate
5. ABR   Available bit rate

This section defines the ATM service categories using the following QOS parameters:

- Peak-to-peak cell delay variation (CDV)
- Maximum cell transfer delay (Max CTD)
- Mean cell transfer delay (Mean CTD)
- Cell loss ratio (CLR)

Constant bit rate (CBR) service category definition. The CBR service category is intended for real-time applications, i.e., those requiring tightly constrained delay and delay variation, as would be appropriate for voice and video applications.

Real-time variable bit rate (rt-VBR) service category definition. The real-time VBR service category is intended for real-time applications, i.e., those requiring tightly constrained delay and delay variation, as would be appropriate for voice and video applications.

Non-real-time (nrt-VBR) service category definition. The non-real-time VBR service category is intended for non-real-time applications that have bursty traffic characteristics and that can be characterized in terms of a PCR, SCR, and MBS.

Unspecified bit rate (UBR) service category definition. The unspecified bit rate (UBR) service category is intended for non-real-time applications, i.e., those not requiring tightly constrained delay and delay variation. Examples of such applications are traditional computer communications applications such as file transfer and Email. UBR sources are expected to be bursty. UBR service supports a high degree of statistical multiplexing among sources. UBR service does *not* specify traffic-related service guarantees. Specifically, UBR does not include the notion of a per-connection negotiated bandwidth.

Available bit rate (ABR) service category definition. ABR is an ATM layer service category for which the limiting ATM layer-transfer characteristics provided by the network may change subsequent to connection establishment. ABR service was described in the previous section.

ROUTING. ATM is basically connection-oriented. The ATM header values are assigned to each end of a connection for the duration of the call. The header identifier is translated in a switch from one port to another. Signaling and user information are carried on separate virtual channels. There are two types of connections that are possible: virtual channel connections (VCCs) and virtual path connections (VPCs). A VPC can be considered as an aggregate of VCCs. When cell switching is performed, it must be done based on the VPC, then on the VCC. The five-byte ATM header of an incoming port is translated by the ATM layer and switched to an output port based on the assigned VPI/VCI value. Figure 3.34 illustrates how switching is performed in ATM. In general, all ATM switches behave in the same manner. Input ports are $ports_1$ to $ports_n$ transport ATM cells based on their header information to $output_1$ to $output_m$. Headers of input ports are translated using the translation table to the appropriate output port having a new header value. For example, an ATM header c on input $ports_1$ becomes (per translation table in Fig. 3.34) ATM header w $output_2$.

In this way, ATM cells are switched based on their respective VPI/VCI values. These VPI/VCI values have only local significance. The table is created during call setup, when the switch assigns these VPI/VCI values to the calling party. The table is maintained throughout the life of the call and is disassembled (by removing VPI/VCI association from the table) when the call is terminated.

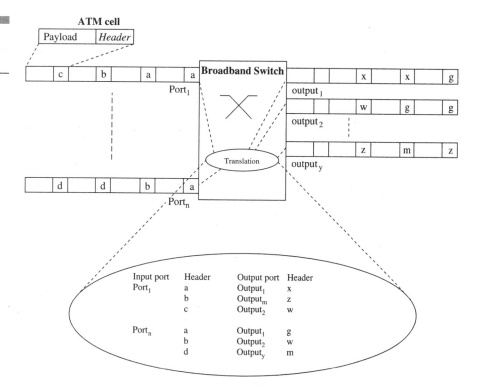

Figure 3.34
ATM switching principle.

3.5 ATM over HFC (Residential Broadband)

The first part of this chapter describes general aspects of broadband and ATM. In this part of the chapter, we will discuss how ATM fits into HFC and hence the cable modem. Therefore, the following will be described briefly:

- ATM access node reference model
- HFC network supporting ATM
- ATM over HFC reference architecture
- Transport protocol model
- MAC layer requirement
- QOS as seen by HFC

The challenges of implementing ATM over HFC are plenty. HFC operates in a shared medium environment and that in itself presents a set of issues that HFC must resolve. Home network distribution also behaves in a shared medium environment and that also is contributing to the challenge by adding another host of issues that must be addressed. ATM behaves best in a point-to-point environment—at least it is well understood and specified in this capacity.

The advantages of bringing ATM to the home based on the sections above should be obvious by now. In summary, an ATM-based network is capable of providing multimedia and all broadband services and applications with specified QOS on demand. Few believe that ATM to the home can be deployed soon. That may very well be true. However, the idea is novel and forward-looking. When ATM reaches the home, a consumer can purchase conforming ATM appliances from any department store (such as video camera, TV, cable modem, and security monitoring devices), hook them to the home network, and go online regardless of which operator one subscribes to. The operator may be your friendly neighborhood RBOC or the cable company.

ATM ACCESS NODE REFERENCE MODEL. ATM in the home was one of the ATM Forum's goal when the Residential Broadband working group was officially formed. The charter of this organization was to define and complete a set of specifications for an end-to-end ATM—that means ATM to the home and inside the home. The description of HFC below comes from the ATM Forum perspective. IEEE 802.14's notions on HFC, the physical layers, and the MAC layers are described in Chaps. 5 and 6. IEEE 802.14 adopted the ATM over HFC concept but also provided the hooks to accommodate other protocols such as IP. The ATM Forum views are more partial to the telco approach, and the description below reflects that sentiment.

Figure 3.35 depicts the reference model for residential broadband as defined by the ATM Forum. The reference model is generically conceived and accommodates the HFC access network as well as fiber-to-the-curb (FTTC), fiber-to-the-home (FTTH), VDSL, ADSL, and others. The ATM Access Network also takes on the personality of its environment. For HFC, the access network becomes the headend. Below is a low-level description of the interfaces.

ACCESS NETWORK INTERFACE (ANI).   The ANI interface is an ATM interface between the ATM digital terminal (ADT) and the access network. The ATM Forum agreed to develop two sets of interfaces for the ANI. The two solutions are:

1. A transparent interface where the access node relies on the network to handle all important aspects of call setup and administration thereof

2. A smart interface in which the access node takes charge, to the extent possible, of handling all aspects of the call, including billing

Europe seems to be partial to solution 1, while the U.S. has taken more of a liking to solution 2.

Figure 3.35
Residential Broadband reference configuration model.

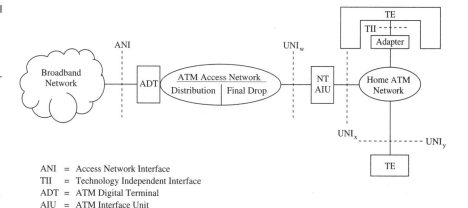

ANI = Access Network Interface
TII = Technology Independent Interface
ADT = ATM Digital Terminal
AIU = ATM Interface Unit

UNI$_x$ AND UNI$_y$. UNI$_x$ and UNI$_y$ are the entry points to the home, and specifications are already released or soon forthcoming.

TECHNOLOGY-INDEPENDENT INTERFACE (TII). The technology-independent interface (TII) is a common interface that operates at the ATM layer and permits user terminals to be independent of any particular home or access network. The TII defines the insulating boundary between network-specific adapters and technology-independent terminal equipment. A TE device can be moved from one network type to another by changing its adapter device.

ATM OVER HFC REFERENCE ARCHITECTURE. ATM over HFC reference architecture is as shown in Fig. 3.36. It is the transformation from the generic model shown in Fig. 3.35. The HFC access network can be viewed as a broadband network that transports radio frequency signals between the ADT and the AIU. Functional components and interfaces are described below.

UNI. The UNI interface is an ATM interface complying with an appropriate subset of UNI x.y described previously. This subset is intended to include all the functionality of UNI x.y but would not include all the physical layer interfaces.

ANI. The ANI interface is the ATM interface between the ADT and an ATM network. This interface will be based on the existing ATM standards, and no differences at the physical or ATM layers are expected. An example of a potential difference is the transport of user signaling messages to an ATM switch within the ATM network.

Figure 3.36
HFC over ATM reference architecture.

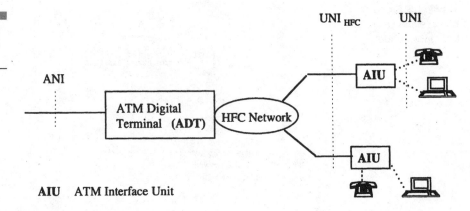

UNI_{HFC}. The UNI$_{HFC}$ interface describes an HFC-specific interface that supports all the protocol elements necessary to allow the ATM interface unit to support a standard UNI and provide connectivity across the HFC to an ATM network over ANI. In the case of HFC, the physical interface associated with UNI$_{HFC}$ will generally also include other services (e.g., NTSC/PAL video) that are multiplexed onto this interface using frequency division multiplexing (FDM).

ATM DIGITAL TERMINAL (ADT). The ATM digital terminal (ADT) interfaces to an ATM network and the HFC. The ADT (called *headend* in a cable environment) is responsible for managing all the HFC-specific aspects of the ATM transport system, and together with the AIU is responsible for providing transparent ATM transport service between the UNIs and the ANI over the HFC.

Figure 3.37 illustrates the functional architecture of an ADT. ADT interfaces with the HFC at a splitter and with the ATM network over an HFC via the ANI.

The ADT consists of four functional modules:

1. ADT RF interface (ADTRI)

2. ADT bandwidth manager (ADTBM)

3. ADT network interface (ADTNI)

4. ADT controller

ADT RF interface. The digital RF Interface (ADTRI) is responsible for interfacing with the HFC and modulating the downstream digital signals onto the HFC. These downstream digital signals contain the ATM cells received from the ADT bandwidth manger (ADTBM) and may contain messaging control information for supporting ATM over an HFC. ADTRI also demodulates the signals received from the HFC,

Figure 3.37
Functional architecture of the ATM digital terminal (ADT).

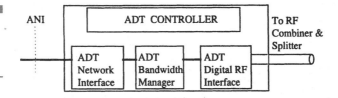

ANI = Access Network Interface
ADT = ATM Digital Terminal

extracts ATM cells and all the control information, and delivers them to the ADTBM. The ADTRI function will also perform FEC (forward error correction) of the ATM cells in order to meet the ATM QOS requirements.

ADT bandwidth manager (ADTBM).   The ADTBM is responsible for implementing an access control protocol for managing the upstream ATM bandwidth on the HFC network. ADTBM must quickly (perhaps in μsec) respond to AIU requests for upstream ATM bandwidth in order to meet the ATM QOS associated with the upstream ATM connections.

ADT network interface (ADTNI).   The ATDNI is responsible for terminating the physical and ATM layers of the ANI. Its functions include ATM operations and ATM traffic management. ADTNI also provides the multiplexing function of the upstream ATM virtual connections from the AIUs onto a single ATM interface.

ADT controller.   The ADT controller is responsible for controlling the activities of the three modules of the ADT. The controller will also be responsible for terminating the ATM UNI signaling protocol and a standard ATM NNI. Both of these capabilities are still under study by both the ATM Forum and IEEE 802.14.

ATM INTERFACE UNIT (AIU).   The ATM Interface Unit (AIU) is as shown in Fig. 3.38. In IEEE, this is referred to as the station or cable modem. AIU is responsible for terminating all the HFC-specific aspects of the ATM transport and for providing the ATM interface within the home. The AIU includes four modules:

1. AIU digital RF interface (AIURI)
2. AIU bandwidth manager (AIUBM)
3. AIU network interface (AIUNI)
4. AIU controller

AIU digital RF interface (AIURI).   The AIU digital RF interface (AIURI) terminates the modulated RF signals and passes ATM cells and any HFC message control information to the AIU's bandwidth manager. Like the ADTRI, it performs FEC.

AIU bandwidth manager (AIUBM).   The AIU bandwidth manager (AIUBM) terminates the MAC protocol and presents the appropriate ATM cells to the AIU network interface (AIUNI). Cells from the AIUNI are scheduled for transmission upstream under the control of the MAC and delivered to the AIURI.

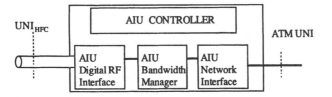

Figure 3.38
Functional architecture of ATM interface unit (AIU).

UNI$_{HFC}$ = HFC - UNI
AIU = ATM Interface Unit

AIU network interface (AIUNI). The AIU network interface (AIUNI) presents a standard ATM Forum UNI and interfaces with the AIUBM. It is the objective that the interface presented to the user will be no different from the ATM Forum UNI 3.x specification.

AIU controller. The controller manages all AIU modules. Signaling, when or if developed at the AIU, will be handled by the controller. Signaling is under study by both the ATM Forum and IEEE 802.14. The AIU controller, however, will likely implement HFC-specific operations functions and perhaps some ATM operations functions.

ATM TRANSPORT PROTOCOL MODEL. The ATM transport protocols at a UNI$_{HFC}$, as illustrated in Fig. 3.39, consist of a physical layer and a MAC. This architecture is only intended to address the transport of ATM.

The physical layer includes the modulation schemes for both the upstream and downstream channels. There can be more that one type physical layer in a single direction.

Figure 3.39
ATM transport protocol model.

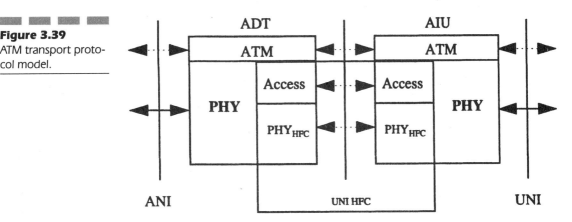

The MAC layer is responsible for managing the distributed access to the upstream HFC resource across the multiple AIUs. This is a key protocol element and will directly affect the resulting ATM QOS. The ATM protocols should see no change in the way they operate over the HFC. Within both the ADT and the AIU, the functions performed at the ATM layer at both an ADT and AIU includes cell relaying.

MAC LAYER REQUIREMENTS. The ATM Forum came up with several important issues that need to be addressed in order to support ATM adequately over shared media access networks. In the HFC environment, several users share an upstream channel; however, relatively few will typically be active at any time. For such applications, a MAC protocol must be designed to allow users to obtain use of the upstream channel when they need it. An efficient MAC protocol, in combination with effective scheduling, permits high utilization of the upstream channel and thus increases the number of subscribers that may be on a segment, improving subscriber satisfaction. There are a number of possible approaches to constructing MAC protocols, each of which has a different trade-off.

Three categories of MAC protocols have been discussed in IEEE 802.14 or fielded in first-generation products:

1. Contention-free MAC protocols with static scheduling (e.g., time division multiple access)

2. Purely contention-based protocols (e.g., CSMA and slotted ALOHA)

3. Reservation-based MAC protocols

The ATM Forum left the work to be completed by IEEE 802.14. Details of the MAC are fully described in Chap. 6. However, it is worthwhile noting some of the observations the ATM Forum sent (directly and indirectly) to IEEE 802.14 to help in the development of the MAC requirements.

Chapter 4

Cable Modem Architecture and Evolution

4.1 Overview

The previous three chapters provided an overall picture of the cable environments and outlined the ground rules for the modernization plan of developing and preparing requirements for the *cable modem*. In this chapter, we further explore the infrastructures and their convergence and focus on the evolution of the cable modem.

Modern technologies, such as ATM, promise high-speed integrated services for all the different applications. With a broadband as the kernel, network convergence will accelerate, and one would no longer be able to distinguish one network from another. The Internet, a late market entry into the equation of the telco and cable modernization plan, is playing a central and confusing role. *Intranet* (with an *a*) is a new term created by marketers to personalize it. The Internet has changed the way the populace, experts, and technologists view the information superhighway. There is little doubt that it is viewed today as *the* information superhighway. That was not its original purpose, although some say that it was under construction for the last thirty years or so. None of the technologies really thought of it becoming the information superhighway, if in fact it is. Even Microsoft only discovered the Internet in 1995. The Internet was considered by standard experts in ITU, ANSI, and others to be no more than an adjunct to the information superhighway that is yet to be built with broadband networks over native ATM services. The epilogue of the Internet story should become clear in the next few years. Issues such as willingness to pay, service quality, and traffic congestion will be major factors in determining the eventual outlook of the real superhighway.

The information superhighway was on the RBOC's future planning strategy since it is the natural evolution of the telecommunication network. NB-ISDN (Narrow-Band Integrated Services Digital Network) has been fully standardized and well understood since the 1980s. ISDN coupled with the intelligent network services platform could have offered similar (if not better) services in terms of service quality than the Internet. Despite that, ISDN did not flourish. The reason why is a long story, but it had mostly to do with market readiness, attracting the peripheral vendors, and lack of long-term strategic planning. It is probably too late for ISDN today to become a competing alternative to other high-speed access devices. NB-ISDN is considered too slow to satisfy the bandwidth-intensive applications that are in demand by today's sophisticated consumer.

The technical challenge of developing a cable modem that will deliver integrated services over cable is significant. Recently IEEE 802.14 provided an IP port from the cable modem to accommodate the Internet network. The state of digital cable technology today is such that we cannot fully predict how systems will be built. The sort of quality of service needed, the cost, and particularly the uncertainty of the market demand for applications and services that do not yet exist are not fully predictable. With all such uncertainties, IEEE 802.14's approach was to develop a standard for cable modems intended to deliver integrated and multiple services with the necessary QOS with flexibility and efficiency.

In the next three chapters, the cable modem will be described in detail. This chapter will focus on the cable modem architectural issues, long-term architecture envisioned by ITU, digital cable technology, and its evolution and impact on cable modem future trends.

The chapter is organized as follows:

- Summary review of the cable television environment
- Exploring the digitizing of the cable network
- Cable modem architecture
- Future trends as envisioned by ITU, and the role of cable modem

4.2 Cable Television Review

In the United States, TV signals are broadcasted and transmitted in 6-MHz channels. In Europe, the TV signal is 8 MHz. Channel allocation is regulated by the Federal Communications Commission (FCC). The FCC also regulates the frequency, location, and signal power used by TV broadcasters. These FCC rules guarantee that stations that use the same TV channel are far enough apart so as not to interfere with one another. That, coupled with TV signal power rapid attenuation in the air, enables cable operators to deliver more channels with equal signal power to homes. Television signals delivered over cable networks have the same signal characteristics as the TV signals sent over the air. This enables modern TVs to operate in the same way whether signals are received over the air or by cable hookup. Cable operators, by virtue of the increased market demand due to TV program variety, flexibility, and premium channel availability, were elevated from CATV to broadcasters, and later they became content providers. During that period, the cable network evolved little if any in terms of upgrading to two-way communication media. Lately, however, HFC modernization plans accelerated not only to set the stage for providing an infrastructure for bidirectional communication, but also to increase channel capacity due to fiber optics installation and system reliability and reduce operating and maintenance costs. With those as prerequisites, the cable modem became practical to develop because:

- Technically the infrastructure was conducive to bidirectional communication.
- The interactive market was in demand.
- The 1996 Telecommunication (Reform) Act was a reality.

CableLabs worked hard to keep up with this accelerating development and issued an RFI (Request For Information) to the industry, soliciting proposals for the Integrated Multiple Service Communication Network. Since then, several organizations have become involved in providing accelerated procedures to provide specifications not only for the cable modem but also to the overall management aspect of this HFC network. These specifications also focused on defining the standard interfaces to the outside world, as well as management, security, and administration of HFC. IEEE 802.14 was very successful in standardizing the cable modem in particular and more specifically its architecture: the physical and the MAC layers and associated security aspects.

A cable modem is a technically complex component to standardize. Unlike any other appliances, it requires the talents of the analog, RF, protocol, traffic management, signaling, and shared medium control experts. Sharing access among multiple users creates security and privacy problems. One user connected to a cable network can possibly receive transmissions intended for another or maliciously make transmissions pretending to be another user. As such, all components of the cable modem are providing the hooks for the management and security of the system. The basic generic description of the cable modem below hides all that complexity, as it is intended to provide operational understanding of a cable modem.

4.3 Cable Modem Operation

The primary function of the cable modem is to transport high-speed data from the cable network to the user and from the user to the cable network. The similarity between the traditional PC modem and the cable modem ends there. Cable modems are much more complicated than the traditional telephone dial-up modem (MOdulates and DEModulates). At the channel level, the cable modem in the downstream direction must tune its receiver between 450 and 750 MHz to receive data digital signals. The digital data is modulated and placed into the 6-MHz channel (traditional TV signal). A cable modem, therefore, functions as a tuner. The QAM 64 modulation scheme was selected by IEEE 802.14 as the modulation technique for the downstream direction. In the upstream direction, the cable modem transmits the signal between 5 and 45 MHz. The data is modulated and placed in the 6-MHz channel using the QPSK modulation technique. At this frequency range, the environment is very noisy because of interference from CB and ham radios and impulse and ingress noise from home appliances. For that reason, QPSK was selected by IEEE 802.14 as the modulation scheme. QPSK is more robust in terms of its immunity to noise at the cost of delivering data at much lower speed than QAM modulation technique.

There are several ongoing cable modem field trials. Some are unidirectional, in which only the downstream link is provided. The upstream channel uses the POTS line to send data and control information to the connected server. In some cases, the one-way modem was a vendor-specific solution to lower cost, but in most instances the one-way limitation was due to the inability of the cable network to provide two-way communication. For the majority of the 11,000+ systems in the United States, it's been a one-way world. That, of course, does not mean that some of these systems are not HFC-based, but it takes more than HFC to make two-way communication.

4.3.1 Cable Modem Speed

Bandwidth of up to 36 Mbps can be delivered to the cable modem in the downstream direction with the QAM modulation technique. In the upstream direction, QPSK modulation can deliver a speed of up to 1.5 to 2 Mbps. In this upstream direction, however, it is expected that many users will be competing for access onto the same 6-MHz channel. The MAC in the modem will arbitrate access to this shared medium bus. The bandwidth allocated to a user in the upstream direction will invariably depend on the number of users sharing the bus. It will also depend, to a large extent, on the characteristics of the traffic being used by the others sharing the bus. A smart cable modem can retune its receiver and hop on to a different 6-MHz channel if congestion is encountered. This mechanism is referred to as *frequency agile* and will be described later in the chapter.

This sort of asymmetric function (i.e., high-bandwidth availability in the downstream direction and limited bandwidth for the upstream) is not necessarily limiting. The downstream high-bandwidth availability is more than adequate for downloading files, Internet traffic, WWW video images, or video clips. In the upstream direction, the bandwidth is also adequate to most Internet applications where users need to send commands such as mouse clicking or transmitting short control messages. The VOD application is also asymmetrical in terms of functionality. In the downstream direction, MPEG digital video of approximately 3 Mbps is a good-quality video picture, better than VCR quality. In the upstream direction, the VOD application is less demanding on the bandwidth. Users occasionally use the remote control to perform VCR-like functions.

Table 4.1 illustrates the time required to transmit a 500-kbyte image over the telephone network using a traditional modem, ISDN (64 Kb/s line), and 10 Mbps cable modem. Using a cable modem at higher speed should result in even better performance than the figures show in Table 4.1. It must be noted, however, that speeding up the access is not necessarily the only bottleneck in the network, especially in the Internet network. The backbone, and particularly the WWW servers, must also be upgraded with high-speed interfaces to handle high-speed traffic. If a server is still operating at 64 Kbps, then that rate will be the maximum speed that will be delivered to the cable modem.

4.3.2 Services over Cable Modem

The cable operators will undoubtedly deploy cable modem for high-speed Internet access as the first service offering. Telecommuting is another application that looks very appealing in terms of the market. Given the topology of the HFC in which it is limiting the cable modem to operate in an asymmetrical fashion, the cable operators will most likely focus on marketing applications that are asymmetrical in nature. The asymmetric applications that are in demand are:

- Internet access, including all Web access
- Multimedia information services

TABLE 4.1

Time Required to Transmit a 500-kbyte Image

Device	Speed	Transmission time
Traditional modem	28.8 kbit/s	6–8 minutes
ISDN	64 Kb/s	1–1.5 minutes
Cable modem	10 Mb/s	About 1 second

Source: CableLabs

- Directory services

- Home shopping

- Surveillance/home security

- News/entertainment

- Archives/Library

Limited upstream capacity does not in any way prohibit the cable modem from running symmetric applications. In fact, with good traffic engineering and an efficient MAC, a cable modem can just as easily operate in this more hostile environment. Symmetric applications and/or services that are or will be in demand are:

- Email

- Videoconferencing

- Video phone

- Voice

- Interactive gaming

4.3.3 Time to Market

Cable modem pilot tests have been under way for quite a while. Vendors and operators, under the leadership of CableLabs, have been experimenting with various modulation schemes, testing their behavior in the noisy cable environment. There are many proprietary cable modems in operation today with different modulation techniques. Wide-scale deployments should start some time in 1997 or 1998. The telco operators are deploying ADSL today in their networks. This is pressuring cable operators to accelerate the planning activities for deploying cable modems. The lack of cable modem standards may have delayed the deployment of this solution in 1996. The IEEE 802.14 working group is scheduled to release its standard in July 1997. Standards may not be the only obstacle facing cable operators. Modernization of the cable system is expensive, even for the large MSOs. Economic models have to be evaluated on a case-by-case basis. The fast pace of technology and the unpredictability of future applications are contributing to and confusing long-term planning.

4.4 Cable Modem and HFC Architecture

Chapter 2 described the details of HFC and the ongoing modernization effort in the cable network to upgrade it into a two-way communication. Figure 4.1 is a physical approach of representing the HFC and the cable modem. In the shared medium environment, the cable modem must be able to tune to any of the downstream 6-MHz channels in the range between 450 and 750 MHz.

In the upstream direction, the cable modem performs the transmitter function, transferring information to the headend using the 6-MHz band, between 5 and 45 MHz. In the upstream direction, the communication is many to one. The n number of cable modems attached to this subnetwork depends on the traffic. Five hundred to two thousand homes were specified in the IEEE 802.14 requirement document.

4.4.1 Spectrum Allocation

The cable spectrum, as shown in Fig. 4.2, is the permissible spread of frequency allocation for the cable system. IEEE 802.14 adopted this spectrum for standardization of the cable modem. The range of frequencies from 50 to 450 MHz is divided into 6-MHz channels. TV analog signals are modulated in each of the 6-MHz channels. Each 6-MHz channel represents the TV signal. The introduction of fiber optics to the cable network enables cable operators to put more and more channels in a cable network. A cable network can carry as many channels as the infrastructure will permit.

In the upstream direction, the range of frequencies from 5 to 45 MHz is dedicated to digital transmission. In this direction, the cable modem, as

Figure 4.1
Cable modem in HFC architecture.

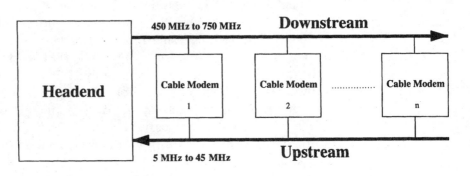

transmitter, uses this range to transmit digital information from the user to the headend. In the downstream direction, the frequency range from 450 MHz to 750 MHz is restricted for downstream digital transmission. Cable modems tune to these channels to receive digital data. The direction is from the headend to the user.

4.4.2 Frequency Agile

IEEE 802.14 in the requirement documents adopted the concept of frequency-agile capability for the cable modem. *Frequency-agile capability* means that the cable modem can tune in any one of the downstream frequencies the cable system uses and transmit on whatever frequency the cable system is equipped to handle. This capability is very instrumental in designing a flexible cable modem. It gives cable operators the tools to change the upstream and downstream bandwidth allocations in their system due to changing traffic demand without having to change the terminal equipment. Excessive noise due to ingress (temporarily or long-term) of an upstream channel can be dynamically isolated by simply retuning the cable modem to other downstream and associated upstream channels. A wider-range frequency-agile cable modem is implementation-dependent. IEEE 802.14 does not preclude it. The expense of providing more complex agility may not justify the development cost. It will, however, offer a very flexible and robust cable modem. Assume, for example, that in a cable architecture the cable modem could use a different part of the cable spectrum altogether for the upstream transmission. This accommodates possible future requirements of adding more upstream capacity to the cable system in the upper ranges of the spectrum. This is well and good, but this solution brings about a host of other complications. As more and more digital channels are added to the spectrum, it will be difficult for cable operators to coexist with traditional cable networks and will also be diffi-

Figure 4.2
Cable spectrum.

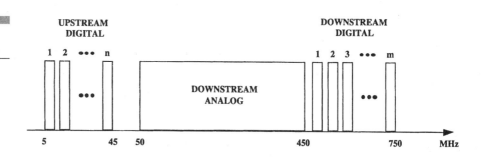

cult to reallocate downstream bandwidth to upstream. Existing upstream use might further constrain what upstream bandwidth is available for digital systems. Planning for more upstream channels assumes that future services need more flexibility. Cable operators cannot just assign unused channels, idling by without revenue. Hence, without assigning a range of frequencies for future upstream use, vendors are unlikely to develop these modems and stay competitive in the marketplace.

4.4.3 Digital Cable Network

All forms of communication today migrated or are migrating into digital format (e.g., CDs, cellular, voice, video MPEG). Future services and applications are likely to be in digital format. The cable companies are under a lot of competitive pressure to go digital all the way. The Digital Video Broadcasters are nibbling away their revenue base much faster than anyone expected. Digital transmission results in a noticeably good-quality picture, at least enough to be a differentiating factor for the consumer.

There is nothing inherent in the characteristics of cable or fiber pipes that prevents signals from being carried in digital format. Today's cable system can carry digital signals without modifications as long as the modulated signal fits within the bandwidth and power constraints that the cable system will carry. Digital communication can also coexist with analog TV signals as long as the digital signals are contained in their own 6-MHz channel. This is illustrated in Fig. 4.2, where the upstream channels from 5 to 45 MHz and downstream channels from 450 to 750 MHz are allocated for digital transmission using the 6-MHz channel. In the upstream direction, one can have up to 6 channels.

Using the cable network to transmit digital signals including broadcast video signals is of course possible. There is nothing in the cable network that specifically prevents such migration. Analog amplifiers in the system will be replaced with digital repeaters much like what the telco network uses today to recondition the T1 digital signals. The advantage of going digital, beyond improving signal quality due to noise, is the increased capacity of the cable system.

4.4.4 Potential Capacity on Cable

The capacity of the cable system will increase enormously if digital signals are transmitted instead of analog TV signals. The cable networks are built to support about 50 or more television channels. Cable channels should maintain 48 to 50 dB SNR in each 6-MHz channel. Modern modulation techniques such as QAM encoding can achieve 43 Mbps capacity in a 6-MHz channel. The compression scheme used in the MPEG2 standards for audio and video dramatically reduced the data rate required for transmission. A digitally compressed video signal of 2 to 3 Mbps can deliver a good-quality broadcast video. So, potentially, a digital capacity of the cable system can easily reach over 500 channels. MPEG2 over AAL5 is well understood and specified in the ATM Forum.

The question of upgrading the system to all digital is obviously a matter of economics: what revenue cable operators can get for the extra channel use versus capital investment. Near video on demand (NVOD) is one possible application where a set of late-release movies continuously plays every five minutes. We must keep in mind that digital repeaters, AD/DA converters, and MPEG hardware are still expensive components.

4.4.5 Central vs. Distributed Control

IEEE 802.14 specified the headend to be the common or central control in the HFC network. The headend controls all aspects including management of a session. For a shared medium network, such a decision is not obvious. Distributed control would be more efficient, directing the flow of information in a shared medium environment. It would be best suited if the traffic were confined to the local community. Some network technologies such as DQDB use fully distributed control, employing a powerful mechanism to control and distribute information very efficiently and fairly. Ethernet is designed without a single point of control, giving every device connected to the network equal access and accomplishing the sharing of network resources through a distributed mechanism. In a distributed environment, data from the upstream bus is echoed back to the downstream channel to all users. While the data is being echoed, the address label flags are compared with the intended destination, instructing it to receive data. The cable network, however, does not behave like the local area network. Users on a cable modem most likely want to communicate with the Internet or other external network, like voice, rather than communicate with the other users who are in the cable network. With that in mind, the headend as a central controller is the best technical alternative. Another advantage of having a central control is the ability of the headend to contain cache data that is frequently requested by users on the WWW. If the cache is to be placed somewhere in a distributed architecture, then the precious resources of the upstream transmission will be required to deliver the cache data back to the intended user. The headend, as a central controller, will have to act as resource manager and echo back messages in the downstream direction to resolve collision, identify stations upon initialization, and other control management information intended for local control.

4.4.6 Latency

Latency in a cable network presents several technical problems. The headend as a controller could be stationed as far away as 50 miles (or 80 km) from the access network where the cable modems, in the residential areas, are stationed. Assuming the signal propagates at half the speed of light, then the time delay could be as much as 100 μsec. This delay in and by itself is not a problem for real time or other services. The problem is that the system MAC fails to detect collisions within the time while a transmission is occurring. As a result, a transmitting node cannot immediately know if its transmission was successful. IEEE 802.14 selected small packets called minicells for transmitting information in the upstream. In this way, the transmission time of this mini ATM cell would be shorter than the propagation delay of the largest cable network.

4.5 Cable Model Layered Architecture

IEEE 802.14 defined a reference model for a cable modem and the headend controller. The reference architecture is the building block needed to define the layering structure. A reference model is the blueprint that is needed to build any device. It is as shown in Fig. 4.3 and contains:

■ Physical (PHY) layer
■ MAC layer
■ Upper layers

The PHY layer is broken down into two sublayers:

1. Transmission convergence (TC) sublayer
2. Physical media dependent (PMD) sublayers.

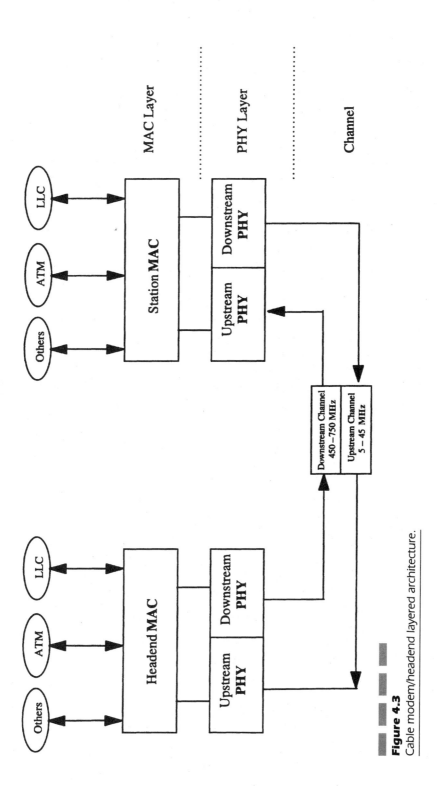

Figure 4.3

Cable modem/headend layered architecture.

4.5.1 Physical Layer

The physical interface for digital cable systems is the ordinary coax cable. This is unlikely to change, but older coax with severely damaged connectors may have to be replaced for users who subscribe to the multimedia service. The physical layer contains both the upstream and downstream channels. The characteristics of upstream channels (many to one) on cable networks make upstream transmission more difficult than downstream transmission due to the multitude of ingress noise that will find itself in the frequency spectrum below 50 MHz. This noise problem can be compensated for, to some extent, by using complex encoding technologies at the cost of a reduced data rate. Hence, a digital cable architecture may use only one downstream transmitter at the head and have several associated upstream transmitters, one at each terminal device. Transmitters are more expensive than receivers.

The two different complexities inherited by the downstream and upstream channels lead to the inevitable conclusion that different modulation schemes will be used. IEEE 802.14 came to this conclusion and defined the downstream PHY using a QAM modulation technique, while the upstream uses the QPSK technique. Chapter 5 will describe the physical layer and the various modulation techniques in detail.

4.5.2 MAC Layer

The MAC layer is one of the most technically complex layers that was developed by IEEE 802.14. The complexity came about because of the shared medium coupled with the requirement for maintaining quality of service for each application. In the downstream direction, the communication is one to many. In the upstream direction, the communication path is shared by many users, all transmitting data to the headend. A fundamental function of any MAC is to devise a mechanism that performs random access to the network, resolves contention, and arbitrates resources when more than one node wishes to transmit at the same time. The MAC is also burdened further due to the requirement to preserve the quality of service of any and all specific applications. If real-time video in being transmitted, then the jitter must be minimized and bandwidth of constant bit rate (CBR) must be allocated. If such a CBR cell is delayed, then it becomes useless.

Several MAC proposals were made in IEEE 802.14, promoting a mechanism that combines a MAC random access and immediate access into a single system. Slots can be allocated: some dynamically based on traffic demand and others subjected to random access transmissions with collision resolution. Chapter 6 describes the details of all MAC proposals made in IEEE 802.14.

4.6 Cable Modem Upper Layer

The MAC developed in IEEE 802.14 was designed to handle several protocols: IP interface, native ATM, and others.

4.6.1 IP Interface

The cable modem can connect directly to a PC handling IP traffic. The physical layer is most likely to stay Ethernet 10BaseT, which is the most predominant method. Although it probably would be cheaper to produce the cable modem as an internal card for the computer, this would require different modem cards for different computers and additionally would make the demarcation between cable network and the subscriber's computer more confusing.

4.6.2 ATM Native Interface

The cable modem in IEEE 802.14 was also standardized to handle native ATM services. That means ATM adaptation layers will be developed to handle ATM applications, including CBR, VBR, and ABR.

4.7 Future Trends in Telecommunication Networks

In this section, the telecommunication long-term architecture (LTA) is described. Cable modem future migration cannot be viewed in isolation. Its host, the cable network, is unique in its architecture, but it is not expected to stay that way despite possible opposition. The communication industry will converge because of business necessity. The 1996 Telecommunications Act will break down the barriers that preserved the traditional networks. It is believed that, within a short period of time (a decade or two), one will no longer be able to distinguish between telecommunication and cable network. The U.S. west continental cable relationship is a case in point. If that relationship survives, the emerging network will likely contain common services using a single core platform. PCS is expected to be a key opportunity for cable operators; the cable infrastructure is well suited for PCS. PCS needs IN-like services and its management platform. All these developments are hinting toward network convergence.

4.8 Long-Term Architecture

In this section, we will describe the vision of the telecommunication industry and focus specifically on the special study made by ITU regarding the long-term architecture (LTA). That study was inspired by the GII (global information infrastructure). LTA could be seen as a prognosticator to the future migration of the telecommunication industry. It also focuses on the social attitude of the people in the industrialized nations and draws conclusions on the types of services and applications that would be demanded. This should provide a hint about how cable modem fits into the future or what additional functionalities it must have to survive to see the future. The LTA is as summarized below.

4.8.1 Trends toward Future Telecommunications

There are indisputable rapid shifts in the telecommunication industry. The most notable are:

- Convergence of the separate communication industry, in particular the telco and broadcasting companies, due to integration of the common services and applications
- Increased emphasis on standards rather than proprietary solutions, thus facilitating interoperability and the open competitive market
- Shift from national to global approach concerning services and regulation
- Shift from monopoly supply to open competition in both services and equipment

End users are demanding operational simplicity in the applications. Falling costs of digital processing are leading to cheap information appliances such as:

- Computers
- Televisions
- Smart telephones
- Combinations thereof plugging into a multimedia network

Such factors have created an environment in which packaged applications obtainable from a single source become an economic necessity. As monopolies crumble and industries converge, telecommunication is gradually developing the wholesale/retail characteristic seen in other industries.

The costs for upgrading existing communications infrastructures to provide advanced value-added services are immense. For the private sector to make the required investments, the regulatory regime, global trade policies, and the commercial potential all have to be right and conducive to the market economy. It is expected that the legal requirements will be relaxed further as conformance to standards becomes a label of quality rather than a legal or national mandatory requirement.

4.8.2 Common Trends in Industrialized Countries

Customer demands are related to the social environment in which the individual resides. As each country has its unique lifestyle, one would expect a variety of trends within this global community. Fortunately, some trends seem to be shared by different societies belonging to the group of industrialized countries and can serve as a basis for predictions of customer demand in the telecommunications market. The common trends are:

Individualization. People all over the world are becoming more aware of their individual rights, demands, wishes, and so on. The consequence of this individualization is likely to be increasing demand for mobility. Telecommunication is expected to play a major role in providing this service. The Internet showed us that global communication was a catalyst for substituting personal mobility, minimizing differences between societies, and widening the scope of the individual to the global context.

Health. People are putting more emphasis on their health, so much so that they are willing to accept restrictions to their personal lifestyles in return for better fitness.

Personal security.   People are concerned about their personal security, including the protection of thoughts and ideas.

Environment.   People are aware that environmental developments can cause difficulties for the community but are often prepared to accept its impact on their individual lifestyle benefits.

Longevity.   Life expectancy has increased due to medical improvements, especially in the industrialized countries. Older people create a significant buying power.

Trustworthiness.   People are perplexed over the increased complexity of the products they purchase. People rely upon the seller of the products for its quality or harmlessness.

Work at home.   People are dependent on work or jobs. Global markets will change this attitude toward work dramatically. The cozy job in the office will no longer be; instead a virtual office will be created with fewer dependencies from the company and more emphasis on entrepreneurs even within the company. Work will be performed anywhere that is convenient.

The above provides us with people's perceptions and attitudes. These trends should help in deriving a list of the future demands in the telecommunication market.

4.8.3 Future Customer Demand in Telecommunications

Based on the above, the demands of the telecommunications customer of the future are likely to include:

- Individualized products at mass-market price levels
- Specialized products for particular age groups
- High-quality products (value for the money)
- Presentation of nonvoice information in accordance with human perception
- Multimedia
- Trustworthiness
- Secure use of the products (no harm to persons or the environment, no fraud or violation of privacy)
- Ease of use, self-explanatory product handling
- Support (or at least no restriction) of personal mobility

- Manageable access at different places by use of different products
- Support for the personal security of the user (e.g., emergency calls)
- Support of mobile teleworking

The rank of these services will no doubt vary with cultural difference in this global market. It is just as true to say that, even if the social driving forces are different, it can be expected that most other countries will be making similar calls for a wide variety of networks and services. Deployment costs are crucial in this development. Wireless services are attractive as a substitute for wireline technologies and services and will be applied on a broad basis, especially in the underdeveloped countries.

4.8.4 Network Trends

In the 1990s, telecommunications networks have become ever more complex. The network evolved from POTS lines to ISDN, and now broadband capabilities are also being introduced. Several different mobile networks have been developed to meet different market needs for mobility, GSM being one good example. We'll see rapid introduction of cable TV networks and network broadcast capabilities. Satellite systems will be used for personal communications services, and the distinction between private and public networks is becoming more indistinguishable with the introduction of virtual private networks.

In the multimedia services, one of the obvious trends is the rapid acceptance of the Internet. Its differences from traditional telecommunications networks are quite significant in many respects such as: tariffing, service quality, architecture, and technology.

Presently, there are many different types of service networks. Some operators are bringing about a convergence between fixed and mobile networks. IN (intelligent network)-based services are expected to play a major role in providing mobility in fixed networks. Customers will access their personal profiles via a terminal using a smart card; their personal profiles will define their identity and the range of services they are entitled to use as well as details of their billing records.

4.8.5 Technological Trends

The evolution of mobile and personal communication will be influenced by technological trends. New technologies and advances in existing ones will allow the realization of innovative system concepts. Several technological areas identified below will play a major role in the fixed mobile and personal communication systems (PCSs).

- Digital radio access
- Fixed-access network technology
- Data compression technology
- Antenna technology
- Satellite technology
- Communication technology
- Optoelectronics
- CPU technology
- Memory technology
- Disk and memory card technology
- Display technology
- Smart card technology
- Efficient energy management
- Software technology
- Portable object distributed software
- Interactive user interfaces

4.8.6 The Role of Standardization

Telecommunications standards will play an important role in this transitional period. Historically, the reason for standards was to enable the interconnection of national and international networks. Since then, other factors created the desire for harmonized networks and for cost reduction due to economies of scale. The standards, however, must be receptive to the market and attractive enough to be adopted by all (big and small organizations or forums) so that products, services, and applications can be used worldwide.

4.8.7 Requirement for LTA

The future telecommunications long-term architecture (LTA) should include the following requirement criteria:

Interoperability.   The architecture should allow architectural components in different administrations to interoperate in a consistent manner for seamless execution of services and management.

Reusability of architectural components.   The architecture should enable design reuse of architectural component specifications when new services or management capabilities are created. This includes design portability, i.e., the ability to reuse the design of architectural components on different computing environments.

Distributed execution.   The architecture should not dictate the location of architectural components. The architecture should allow transparent distributed computing. This means that architectural components may be distributed on nodes of the system.

Support of services.   The architecture should support but be free from the limitations of traditional call models.

Mobility services.   The architecture should support terminal and personal sessions and enable end-user services to be tailored to meet different customer requirements.

Support of management.   The architecture should enable coordinated management of architectural components and network resources.

Customer control.   The architecture should allow network providers to facilitate selection of QOS by the customers. It should also allow customers to choose security levels and durations for specific services such as:

- Multimedia libraries and interactive games in which the user can utilize the distributed processing capability of the telecommunication system.
- Automatic recovery from hardware and software faults
- Online upgrade of service or management architectural components without taking them off-line
- Degradation of the quality of service, if it occurs, shall be predictable and acceptable

Quality of service.   The architecture should support a range of service performance requirements.

Scalability.   The architecture should accommodate and allow the evolution of the scale of networks and service and management capability from very small to very large.

Security.   The architecture should support:

- Authentication and authorization of entities involved in an interaction

- Mutual identification and authorization
- Monitoring of malicious and unusual activities and adoption of countermeasures. This is required to maintain integrity, confidentiality, and availability within the system and compliance with the architecture
- Protection of information (messages or stored data); this includes maintaining the integrity of information and confidentiality within the system conformance to the architecture
- Realization of security controls across equipment provided by different vendors

Compatibility with existing telecommunication systems. The architecture should allow:

- Migration from existing architectures
- Interoperating with TMN, IN-based system, and B-ISDN
- System conformance to the architecture providing existing services to meet their QOS objectives

4.8.8 Services to Be Offered by Multimedia Services

A list of services is being considered as part of the G11 study. It is appropriate to refer to these services and consider their value and priority in relation to the enhancement of mobile and personal communications, namely:

- Interactive speech
- Real-time image transfer
- Electronic mail
- Multimedia document retrieval
- Video on demand
- Interactive video services
- Computer-supported services
- Broadcast TV/radio/data contribution
- Broadcast TV/radio/data distribution
- Real-time multipoint retrieval

The services mentioned are, at a lower level, eligible for dynamic provisioning to mobile users. For most of them, only a limited number of situations have to be considered when defining the service offering to mobile subscribers. For example, subscribers would be willing to access VOD services under their personal identifier and service profile when they are in a train or airplane but maybe not when driving their cars. However, one should note that there may be several ways of implementing such a service that do not necessarily require a full broadband air interface with the provision of handover, irrespective of the speed of the vehicle in motion.

NETWORK ASPECTS FOR MULTIMEDIA SERVICES. Multimedia services communications will be supported not by one system but by a set of systems and subsystems working together to provide multimedia services. Therefore, the approach is to define a framework within which different elements can coexist and work together. The network architectures for multimedia communication systems should meet the following objectives:

- Ability to cope with a diversity of services to be supported
- Enable service provision by a large diversity of network operators and service providers
- Allow the same application to be transparently accessible from different terminal equipment
- Offer services in different environments

MULTIMEDIA NETWORK ARCHITECTURE. The multimedia network architectures consist of four conceptual domains as illustrated in Fig. 4.4. They are:

1. Terminal equipment
2. Access networks
3. Core transport networks (including IN)
4. Application services

The access network domain contains several different access components that can be based on different existing and future technologies. To communicate with these access networks, several different types of terminals are introduced as the terminal equipment part of multimedia communication systems. These include mobile terminal equipment as well as fixed terminal equipment with and without a user identity module (UIM) to support personal-mobility-related services, operator-specific services, and subscriber- and home-network-specific logic and data. Multiple-mode

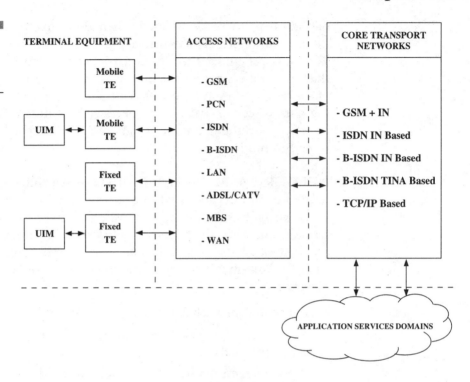

terminals are introduced to be able to access different access networks from the same terminal.

The core transport domain (which includes intelligence) can be based on existing and future network elements. Network functions and features are offered by the core transport network to support service processing within and between different core transport networks. In this report, the services and features are referred to as *network services* and *network features* to distinguish them from the application services of the application services domain. Examples of network services/features include:

- Support of mobility management
- Support of advanced routing of calls based on service logic in the core transport network
- Support of billing and accounting to the users' home accounts when roaming
- Support of home network services/features when moving into other networks

Various examples of access networks are mentioned in the above figure, some of which require no further explanation. ISDN would include basic access, primary rate access, and DSS-1 signaling or a variant thereof. Similarly, B-ISDN would include the B-ISDN UNI (user network interface) along with DSS-2 signaling. LAN could possibly be based on ATM technology. The multimedia access network in this context is the access network to be developed to support multimedia services, including high bit rate data and multimedia applications. A WAN is another example (e.g., as access to ISDN or B-ISDN). CATV might also be an access network (e.g., to access a TCP/IP-based data network such as the Internet). The last example, MBS, refers to the use of the new mobile broadband system as an access network in multimedia communication systems to cater to very-high-bit-rate wireless communications. This is being developed as part of the ACTS research program.

The first example of core transport networks is the use of the current GSM network and switching subsystem, including its evolution. *IN/ISDN* refers to the use of an ISDN-based core transport network, which would mean a digital network with SS7 ISUP signaling combined with IN capabilities. The other examples are a core transport network based on B-ISDN in combination with IN as a future solution and one based on TCP/IP (e.g., the Internet). The division of multimedia communication systems into four domains allows for the implementation of different types of multimedia services networks, as well as the introduction of different operator roles as outlined in Chap. 2. In the following sections, a more detailed description of the four domains is presented, together with a description of how current systems and subsystems can evolve toward multimedia communications systems.

TERMINAL EQUIPMENT DOMAIN. The terminal equipment domain of multimedia communication systems will consist of single- or multimode terminals able to access one or several different types of access networks. The UIM can be used to identify the user and support personal-mobility-related services (and possibly additional functions) in the different multimedia domains.

When accessing multimedia application services via different access and core networks, the mode of operation can be chosen depending on the application service requested. This can be arranged within the limit of the set of services the user has subscribed to and the access networks locally available.

The above approach implies that a user with a need for high-bit-rate multimedia application services can upgrade his or her terminal to support this mode of operation. This could be done possibly in combination with one or more of the existing access networks. Thereby, the user could use a high-bit-rate multimedia access network when available, as well as existing access networks in other areas.

ACCESS NETWORK DOMAIN.   In this domain, the basic concept of multimedia communication systems is that the network should offer several different components in parallel based on different or similar technology. The access network component will include current systems, the evolution of those systems, and new components offering capabilities beyond those of the present systems. It is expected that at least one new radio access network component will be needed in order to support high-bit-rate multimedia communication services with flexible allocation of communication capabilities during the call.

The access network is used to gain access to the application services domain via a core transport network. The use of several different access networks in parallel will allow for a gradual introduction of high-bit-rate multimedia application services in different areas where market demand and economics justify the provision of high-bit-rate core transport networks.

CORE TRANSPORT NETWORK DOMAIN.   The role of the core transport network is to connect different access networks to each other and to provide access to the application services domain. It provides network services and features such as mobility management and service logic for the intelligent routing of calls. An example is a core transport network based on the ISDN in combination with IN, consisting of elements such as LEs, TEs, SSPs, and SCPs. A further example could be a core transport network based on TCP/IP. The following are also foreseen:

- A core transport network based on ATM/B-ISDN in combination with SCPs for service provision
- A core transport network based on ATM/B-ISDN using other network elements for service provision

It is important to emphasize that many different combinations of the above network elements can be foreseen and that it would be up to the service provider to choose which core transport network capabilities should be made available locally. Various factors would figure in the decision such as market need, practicality, technical complexity, cost of implementation, and the history of the operator (e.g., a GSM network operator will try to evolve gradually from an existing GSM core transport network). This approach would also allow for an evolution from several different starting points (e.g., a GSM operator could start with GSM/IN, an ISDN operator with ISDN/IN, and so on). New operators could decide to go directly to new technologies such as ATM. The above approach would allow for competition between different operators where each could choose the preferred core transport network technologies according to his or her users' needs, the current investments in infrastructure (if any), and the regulatory regime in that area.

APPLICATION SERVICES DOMAIN. Application services in the context of this study are defined as applications that can be reached via user terminal equipment connected to a multimedia communication network. The application services are either transparent to the access and core transport networks or they make use of network-specific logic like location information. The only limitations would come from the capabilities of the networks and terminals used and subscription conditions.

Existing application services have typically been developed to be used in different fixed environments. It is expected that users would wish to access these services also when mobile, and therefore an important driver for future multimedia communication systems will be the development of the different access and core transport networks such that these application services can be run when mobile with as few limitations as possible and at a price acceptable to the user. An example of a transparent application service is the World Wide Web (WWW). This could be provided via multimedia communication systems, but the bit rate (for instance) will be dependent on the access and core transport networks used to connect to it. Such restrictions may limit the viability of such a service if it is offered. As an example, the possible bit rate for using the WWW via a GSM access network may be limited to somewhere between 9600 bit/s and 100 kbit/s (the highest rate would require an extension to the current GSM standard).

It is expected that most of the software used in the application services domain for multimedia communication systems will be written by third-party companies or companies not directly involved in the development of the multimedia access and core transport networks. To allow for this, it is important that application programming interfaces (APIs) are developed so that third-party applications services can run on the networks. JAVA technology supports this development.

OVERVIEW OF THE BUSINESS MODEL. The proposed business model is composed of six domains. Nothing in this model precludes any combination of these six domains in a single physical layer. Figure 4.5 illustrates the business model, showing the following domains:

Consumer. Enrolls for and consumes services. A consumer may be an individual, a household with multiple end users, or a small or large business with multiple end users. Major functions of the consumer are:

Figure 4.5
LTA business model.

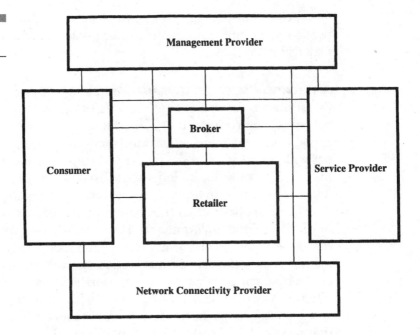

- Obtaining the location of retailers, service providers, and other consumers
- Creating service relationships that include service providers and other consumers
- Enrolling with retailers
- Indicating availability to retailers
- Accepting download from retailers to upgrade the interaction capability with the retailer

The consumer will initiate or accept control plane communication for the following purposes to the indicated business domains:

- For the purpose of locating retailers, communication to broker
- For the purpose of locating services or other consumers, communication to retailer (which in turn initiates control plane communication with the broker) or broker (directly)
- For the purpose of accessing services, communication to retailer
- For the purpose of accessing other consumers, communication to retailer
- For the purpose of receiving an invitation to join a session, communication from retailer

Broker. Provides location information that the business domains use to locate other entities for the purpose of establishing communications. Major functions of a broker include:

- In response to an identifier, provide a unique end point address

  In response to a name, provide the network address (e.g., E.164 number)

  In response to a name with a request for retailer, provide the network address (e.g., E.164 number) of the associated retailer

  In response to a network address (e.g., an IP address), provide an associated address from a different network (e.g., E.164 number)

  In response to a network address (e.g., 800 number) with a request for routing address, provide a routing address (e.g., E.164 number).

- In response to a service category, provide a list of identifiers associated with that service category

  In response to a service category (e.g., video on demand), provide the identifiers of service providers that provide that service

  In response to a service category (e.g., video on demand) with a request for retailer, provide the identifiers of retailers that provide that service

Retailer. Provides the contact point for the consumer to arrange for, and the contact point for the service provider to offer, services that employ communication services such as the delivery mechanism. Major functions include:

- Management of enrollment to obtain various services
- Management of enrollment to provide various services by service providers
- Authorization prior to service usage
- Maintenance of session-level user service profiles and treatment policies
- Session management: communication in establishing and maintaining the association list of parties and resources that partake in a session with the session owner and session policy information for the purpose of establishing access to the session
- Initiating download to consumers and service providers to upgrade the interaction capability with them
- Establishment of network connectivity associated with a session including: coding, rate, and QOS negotiations
- Collecting accounting information for the purpose of off-line billing in the general case for each invoked service (including network connectivity as well as for the services of the retailer)

Service Provider. Provides any of a variety of services that employ communication services such as the delivery mechanism. Note that a content provider is a subclass of a service provider that supplies information content to another service provider. Major functions include:

- Obtaining location of retailers
- Enrolling with retailers
- Indicating availability to retailers (logon function)
- Accepting download from retailers to upgrade the interaction capability with the retailer
- Collecting accounting information for the purpose of off-line billing for service usage; the party requesting the service is associated with this information

Network Connectivity Provider. Provides the communication services that transport information that may either be control-plane information or user-plane information. The service control information can be transported by either means depending on the business application. Major functions include:

- Setting up calls with or without connections
- Adding and modifying connections
- Collecting accounting information for the purpose of off-line billing for network connectivity; the billing point is associated with this information
- Providing advanced network features to the basic connectivity service
- The independent roles of the provision of infrastructure (both software and hardware) and its connectivity to support network provision

Management Provider. Manages one or more domains in whole or in part and provides functions needed for administrative and off-line billing capabilities. The management provider has the mission of providing necessary management functions for provisioning and maintaining system resources as well as for billing appropriate parties who use the system. A management provider manages one or more domains in whole or in part. It also may federate with other management providers to increase the management provider's coverage as viewed by the business domain it manages.

4.9 CATV/Cable Modem Evolution

This chapter intentionally mixed the cable network and the telco world so we can witness the differences between the two networks' potential migration strategy. As mentioned earlier, network convergence is the ultimate objective of operators so that the services and applications that are the most expensive to build will have a common platform. These differences on the surface seem immense. Cable operators understandably develop very-short-term strategies to meet their immediate business goals. The consequence in most instances is that cable operators will be provided with a proprietary product with little emphasis on interworking, reliability, redundancies, or network management. This has worked in the past because HFC behaved in a closed network and provided specific services.

4.9.1 Assumptions for Network Convergence

The differences between the telco and cable network would become achievable if certain assumptions and planning are adopted. The most crucial are:

- HFC to be considered as an access node
- Headend to assume central control
- Cable modem to assume UNI functionalities

HFC AS ACCESS NODE.   HFC as an access node will be the first obvious step to start network convergence. The Residential Broadband group of the ATM Forum places the HFC in the reference model architecture at the same level as all other access nodes, be it FTTC or FTTH or telco-based access. The standard interfaces from the HFC toward the network shown in Fig. 4.6 can then be developed to connect to:

- The voice network via specification TR-303 interface developed by Bellcore or a T1 trunk line
- Internet via routers
- Broadband core network via the ANI interface (Fig. 4.6 illustrates this connectivity)

These interfaces could, obviously, connect to local switches if the cable operators wish to become service providers. With this architectural approach, cable operators can then provide network services and evolve with the telecommunication network in steps and in accordance with their business model.

HEADEND CENTRAL CONTROL.   The HFC network connectivity alone, as shown in Fig. 4.6, does not guarantee service integration. If distributed control is assumed and cable modem users can communicate with each other, then service integration becomes much more difficult to realize. This headend centralization consolidates the controls and dictates band-

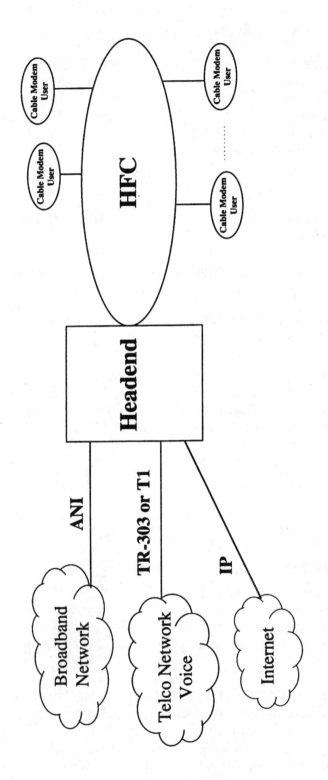

ANI = Access Network Interface

Figure 4.6
HFC network interfaces.

142

width allocation to all connected cable modem users. This also assures that each cable modem respects its service contract with the network. IEEE 802.14 recognized the importance of this, and the specifications were based on the headend as the sole controller in the system. The headend will also moderate the ABR traffic, signaling, network management, and so on.

CABLE MODEM. The cable modem should assume UNI-like functionality to the headend. This UNI functionality ensures that new applications developed for broadband will be applicable to users who subscribe to the cable network. IEEE 802.14 also recognized this important aspect, and the specifications, to the extent possible, were based on QOS as defined in the broadband network and MAC layer. To do otherwise would limit the cable modem functionality in the marketplace. Plug-and-play should not be a cable-modem-specific application.

4.9.2 Next-Generation Cable Modem

The IEEE 802.14 working group is planning to release the cable modem specification in July 1997. The specifications will not include all the features that will make it complete. Among the most crucial are:

- ABR
- ATM native service with UNI functionality
- Signaling
- Security updates
- Advance PHY development work
- In-home wiring
- POTS
- Layer management

Once these features are provided to the cable modem, convergence becomes more feasible and in line with LTA.

High-Speed Cable Modem Physical Layer

5.1 Overview

The cable network infrastructure is a one-way channel from the headend to the subscriber, referred to as the downstream direction. In such a system, a subscriber has no way of communicating back to anyone or even the network central control. It is like a firewall. When PPV (pay-per-view TV) was first introduced, the subscriber made a phone call to order a movie and an addressable command from the headend was sent to the set-top box disabling the filter that scrambled the selected TV program (movie). Before then, personnel were dispatched to the site to disable the scrambler manually. To establish a two-way communication, new physical layers, among others, need to be specified.

In this chapter, the focus will be on the physical layer and the sort of modulation scheme needed for the upstream and downstream layers. Efficient modulation techniques are obviously needed to work in the cable environment and deliver the speed and frequency agility necessary to obtain as much bandwidth as can be acquired from the cable physical environment.

In setting the stage for describing the physical layers, we will first briefly review the cable plant physical environment; then we will define the parameters and barriers that directly influence the physical layer upstream and downstream characteristics.

This chapter covers the following:

- Brief description of cable plant
- HFC reference model with PHY focus
- PHY proposal submitted to IEEE 802.14
- PHY adopted by IEEE 802.14
- U.S. vs. Europe (Annex A/Annex B)
- PHY selection criteria

5.2 Upstream/Downstream Cable Plant

The upstream cable plant shares the same coax cable that the downstream plant uses. Division of the frequency spectrum into two disjointed regions allows the upstream and downstream amplifiers in the system to work back to back, each tuned to its assigned frequencies. The upstream frequency is allocated below the 50-MHz band. In this frequency range, the signal encounters a variety of ingress noise and attenuation. This attenuation will reduce and play havoc on the upstream signal power. The signal present in the drop cable and in-home wiring is 10–30 dB stronger than that presented to the feeder cable. From the feeder cable, the upstream signal is delivered to the headend through a number of amplification stages, to the trunk cable, and finally to the headend.

A serious problem inherent in the upstream plant is ingress noise. Ingress noise is the pickup of electromagnetic energy. These noise signals enter the system anywhere in the network due to poorly shielded components. Cascading amplification in the upstream direction also causes noise funneling. As a result, the headend must contend with all of the ingress noise and must deal with it in a suitable manner to attain the maximum performance. Leakage in a cable network is also a source of egress noise. *Egress* is the radiation of electromagnetic energy from the cable plant. Since the cable plant acts as an antenna for the ingress, then it can also be an efficient radiating antenna for signals that appear on the cable. The main causes of the ingress and egress are poor shielding and/or worn-out connectors in the drop portion of the cable plant and inside the home. This upstream band in that frequency range is much more friendly to noise than the downstream. So it was obvious from the outset that different coding techniques are needed in the upstream and downstream physical layers to meet the characteristics and personality of each. The goal, specifically for the upstream channel, is to develop a modulation technique that deals with the ingress in an effective manner and also minimizes the amount of unwanted noise.

5.2.1 Conditioning a Cable Plant for Two-Way Communication

In order to provide a communication link to the upstream direction, three technical changes in the cable plant must be made:

1. The amplifiers present in the coax pipes must be upgraded to operate in both the downstream and upstream directions. They must be tuned to the specific downstream and upstream frequencies.

2. The upstream channels must be assigned in the cable spectrum. A commonly used frequency range for upstream is from 5 to 45 MHz (6 MHz each). At that range, these frequencies are susceptible to all kinds of electrical noise, e.g., ham radio, home electrical appliances, and many others. Some MSOs are considering adding more upstream channels to be allocated from the high end of the spectrum.

There is nothing to prevent them from doing so to meet the future need and beef up the upstream bandwidth. Such an undertaking, however, cannot be done dynamically. These frequencies today are hardwired to the amplifiers in the network. If reallocation is to be made, the amplifiers must be manually modified. Sending a command to amplifiers to add or take out a frequency band should not technically be a major undertaking and could be developed.

3. In the upstream direction, the communication is many to one as shown in Fig. 5.1. All stations will be competing on the shared medium bus to get the attention of the headend so permission can be granted to start sending the data.

A MAC will arbitrate the flow of the information from the stations to the headend. This MAC will be described in more detail in Chap. 6. MAC's main responsibility is to ensure that station A is granted permission to send data to the headend without colliding with station B or station C or other stations that want to do the same. A collision resolution mechanism is provided to facilitate the information flow.

5.2.2 Cable Spectrum Allocation

The upstream/downstream channels are allocated in different regions of the spectrum as shown in Fig. 5.2. The 400-MHz band, from 50 MHz to 450 MHz, is used to carry the downstream TV signals, including NTSC analog and FM audio. The 40-MHz band (between 5 and 45 MHz) is sectioned into multiple upstream radio frequency (RF) digital channels. Each channel can be 1 MHz to 6 MHz wide, and each is capable of carrying a

Figure 5.1
Sharing the upstream bandwidth.

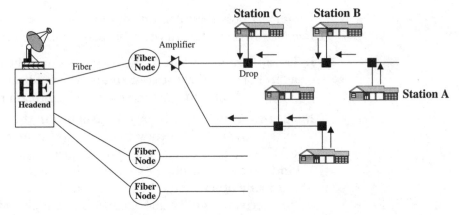

Figure 5.2

RF cable spectrum.

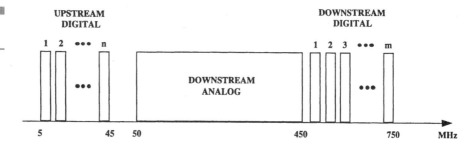

digital bandwidth in the range of 1.6 Mb/s to 10 Mb/s when the QPSK (quaternary phase shift keying) modulation scheme is used. The upstream RF channels are designed to carry control and data information to the headend. A larger number of downstream RF channels are located in the 300-MHz band between 450 and 750 MHz. These RF channels are used for control and data information broadcasted from the headend to all stations. These channels are referred to as *downstream*.

5.2.3 Spectrum Physical Mapping to Cable Modem

Communication between the headend and cable modem can best be described as shown in Fig. 5.3. A cable modem (also referred to as a *station*) receiver is connected to all the downstream 6-MHz channels (1 to m) of the spectrum. The cable modem must be able to tune in to any one of the 6-MHz bands to receive data from the headend. At the transmitting end, the cable modem must also be able to transmit at any of the upstream channels from 1 to n as shown. A dialogue between the cable modem and head-

Figure 5.3

Headend/station spectrum mapping.

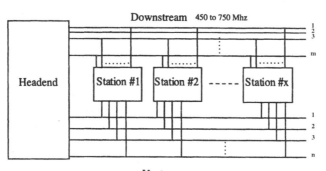

end is triggered when a modem gets attached. It automatically starts listening to the downstream channels, seeking entry to register its device. The network responds by assigning the appropriate upstream channels in which it can transmit its data. One downstream channel, depending on the headend configuration, is normally associated with several upstream channels. The dialogue between the headend and the stations is the responsibility of the MAC management layer and will be described in more detail in Chap. 8.

The ability of a station to tune and reconfigure its physical ports in the system is referred to as *frequency-agile cable modem.*

5.3 HFC Reference Model

IEEE 802.14 defined an architectural reference model as shown in Fig. 5.4. The layers are independent of each other such that a physical layer can be replaced without making changes to the MAC layer. This has a profound impact on developing both the physical layer and the MAC layer. The added complexities force the cable modem to be very robust and future-safe. IEEE 802.14 created an ad hoc group to look into advanced physical layers that will have superior performance to today's modulation technology. Moreover, physical layers developed especially for Europe or the Far East can easily adopt the standardized MAC layer. The MAC layer is one of the most complex modules to develop for the cable modem.

The cable modem layers defined by IEEE 802.14 are:

1. Channel

2. Physical (PHY) layer

3. MAC layer

The IEEE 802.14 committee developed three criteria documents. They were used to evaluate proposals that were submitted by the industry for standardizing the PHY layer. The three documents are:

1. Channel model for the upstream

2. Channel model for the downstream

3. Convergence specifications

These criteria documents are further described later in the chapter and in more detail in Chap. 7.

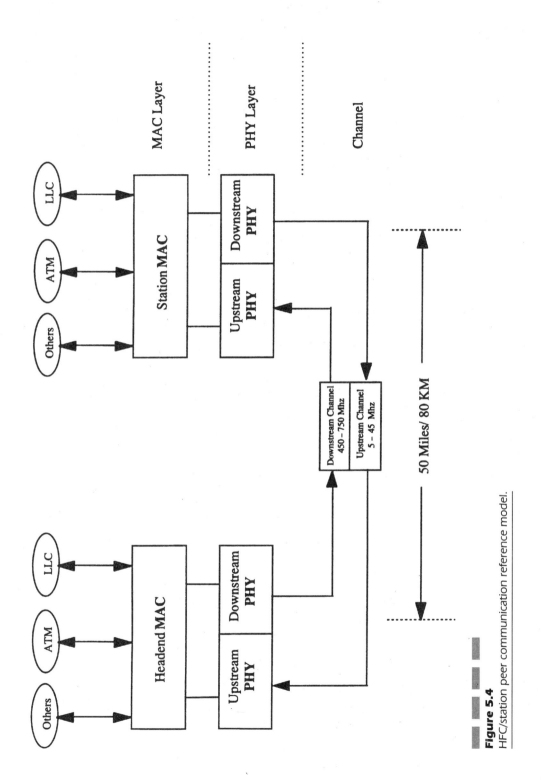

Figure 5.4
HFC/station peer communication reference model.

5.4 Physical (PHY) Layer

IEEE developed a PHY layer. As shown in Fig. 5.4, it contains both the downstream and upstream PHY. The physical layer is also broken down into two components:

1. Transmission convergence (TC) sublayer
2. Physical medium dependent (PMD) sublayer

5.4.1 Transmission Convergence Sublayer

The physical layer TC sublayer is mainly responsible for processing the data and adding any fields that are unique to the particular downstream PHY and passing the data to the lower layers of protocol to transfer on the physical media. Functions include:

- HEC (header error check) processing
- Performing encryption/decryption and scrambling
- Aligning the data into a structure used as the physical layer

5.4.2 Physical Medium Dependent Sublayer

The PMD sublayer is responsible for adapting to a particular transmission medium. The critical functions that must performed in this sublayer are:

- Synchronization
- Correct transmission and reception of bits
- Ensuring bit timing construction/reconstruction
- Modulation coding at the transmitter and receiver

5.5 PHY Proposals

Below are proposals that were officially submitted by the industry to IEEE 802.14. These proposals should provide the reader with an insight into the various technologies that are available in the development of the

PHY modulation techniques, and the clever mechanism used to overcome the noise and other barriers found in the cable environment. IEEE 802.14 adopted QAM-64 and QAM-256 for the downstream and QPSK and QAM-16 for the upstream.

After repeated attempts, a comparison chart between the various proposals did not seem appropriate; therefore, a chart was not created. Each proposal in and by itself was very suitable as a solution for the PHY modulation, and therefore each is described below on its own merit. IEEE 802.14 selected the best features of all the submitted PHY proposals that met the various assumptions adopted in both the technical and the business sense.

The official proposals submitted are:

1. Scientific Atlanta	A proposal to use QAM for HFC
2. Amati Communications	SDMT PHY layer
3. Alcatel Telecom	ATM-based PHY for both in-band and out-of-band HFC systems
4. CORTEC	INTRA PHY proposal
5. Stanford Telecom	Proposal for QPSK upstream modulation standard
6. Zenith	Proposal for a spread-spectrum upstream modem for CATV
7. General Instrument	QAM/QPSK
8. AT&T Lucent	Physical layer protocol for HFC networks in support of ATM and STM integration
9. Microunity	Proposal for multirate QPSK/QAM PHY protocols
10. IBM	Frequency-agile multimode for upstream transmission in HFC systems
11. Aware	Discrete wavelet multitone (DWMT) PHY layer

5.5.1 Scientific Atlanta: A Proposal to Use QAM for HFC

OVERVIEW. Scientific Atlanta (SA) proposed a quadrature amplitude modulation (QAM) and Reed-Solomon forward error correction (RS-FEC) based on physical layer design for HFC systems. SA developed the multirate transport (MRT) system for the delivery of digital video in hybrid fiber coax systems. This physical layer system specification was specified prior to the adoption of similar systems in recognized standards such as the Digital Video Broadcasting Project (DVB) and Digital Audio-Visual Council (DAVIC). Home communications terminals (HCTs) with silicon implementation of the SA-MRT are currently deployed in ongoing field trials. The SA-MRT defines its required data scrambling, interleaving, error correction, modulation format, and transmission.

Quadrature amplitude modulation (QAM) is the modulation format included in the SA-MRT physical layer. QAM is a nonproprietary and widely utilized modulation format. It is now efficiently implemented as an ASIC for the CATV market. The ubiquity and robustness of QAM has made it the logical choice for most equipment manufacturers. The SA-MRT includes 64 and 256 QAM with a 20 percent excess BW in 6-MHz channels. The resulting transmitted data rates are 30 Mbps for 64 QAM and 40 Mbps for 256 QAM. DVB includes 16, 32, and 64 QAM with a 15 percent excess BW in an 8-MHz channel. DAVIC proposes 16, 64, and 256 QAM with a 15 percent excess BW in both 6- and 8-MHz channels.

Reed-Solomon error correction coding is implemented in the SA-MRT and adopted by both DVB and DAVIC. In all three cases, the RS is (188,204) with $t = 8$. The payload of $N = 188$ is equivalent to 1 MPEG2 transport packet. The raw coding gain with RS $t = 8$ is approximately 4.8 dB at 1×10^{-9}. This amount of coding gain is adequate for most CATV subscriber drops. QAM and R-S FEC form the core of the physical layer for HTC systems.

SA-MRT PHYSICAL LAYER. To carry downstream content information, a combination of QAM and a multirate framing structure are specified. QAM is specified due to its performance characteristics with respect to spectral efficiency. Two levels of modulation, 64 QAM and 256 QAM, are defined to allow flexible implementation. The multirate transport protocol is optimized to efficiently carry MPEG transport stream packets in an error-free manner. A combination of Reed-Solomon coding bit interleaving is used in this protocol for error correction. A Reed-Solomon correction key is aligned with each transported MPEG packet.

QUADRATURE AMPLITUDE MODULATION (QAM). QAM is used as a means of encoding digital information over radio, wireline, or fiber transmission links. The method is a combination of amplitude and phase modulation techniques. QAM is an extension of multiphase phase-shift keying, which is a type of phase modulation. The primary difference between the two is the lack of a constant envelope in QAM versus the presence of a constant envelope in phase-shift keying techniques.

The technique is used as a result of its performance with respect to spectral efficiency. QAM can have any number of discrete digital levels. Common levels are:

- QAM-4
- QAM-16
- QAM-64
- QAM-256

QAM is based on suppressed carrier amplitude modulation of "quadrature" carriers, 90 degrees out of phase with each other. The modulation specifications for the SA-MRT are found in Tables 5.1 and 5.2 and Figs. 5.5 and 5.6.

MULTIRATE TRANSPORT FRAMING. The multirate transport (MRT) format is designed to work in a variety of transmission systems. It provides a means of matching error protection to application requirements. This format utilizes a combination of Reed-Solomon forward error correction (RS-FEC) and double-byte interleaving to accomplish both random and burst error protection. The MRT format allows efficient mapping of MPEG2 transport packets with an RS-FEC transmission envelope in preparation for QAM modulation.

TABLE 5.1

Modulation
Specification

Modulation	64 QAM or 256 QAM
Carrier frequency	151–749 MHz. 6 MHz spacing is applied.
Carrier frequency accuracy	20 parts per million (ppm).
Signaling rate Transmission rate	5.000000 MHz ± 50 Hz based. 30 Mbps (64 QAM), 40 Mbps (256 QAM).
Signal element coding	Differential quadrant coding and Gray coding within quadrant.
Transmitted spectrum	Square root raised cosine approximation. Alpha = 0.2.
Scrambler	See Section 5.5.1 on data field scrambling.
Modulation I/Q amplitude imbalance	Less than 0.2 dB.
Modulation I/Q timing misalignment	Less than 4 nanoseconds.
Modulation quadrature imbalance	Less than 1.0 degree.

TABLE 5.2

64 and 256 QAM
Element Coding

Inputs	Previous inputs	Phase change (degrees)	Outputs ($b5_n b4_n$)
A_n	A_{n-1}	0	00
A_n	B_{n-1}	270	10
A_n	C_{n-1}	180	11
A_n	D_{n-1}	90	01
B_n	A_{n-1}	90	01
B_n	B_{n-1}	0	00
B_n	C_{n-1}	270	10
B_n	D_{n-1}	180	11
C_n	A_{n-1}	180	11
C_n	B_{n-1}	90	01
C_n	C_{n-1}	0	00
C_n	D_{n-1}	270	10
D_n	A_{n-1}	270	10
D_n	B_{n-1}	180	11
D_n	C_{n-1}	90	01
D_n	D_{n-1}	0	00

MRT FORMAT. The MRT format is illustrated in Fig. 5.7. The definitions of the fields are described in the following sections.

SYNCHRONIZATION BYTE (SYNC). The SYNC byte is used to aid in obtaining synchronization for the MRT transport structure. SYNC = 11000011.

HIGH-RELIABILITY MARKER (HRM). The high-reliability marker (HRM) is used to accomplish two purposes:

1. To aid in obtaining synchronization for the MRT transport structure
2. To define the boundaries of the interleave function

The HRM consists of two fields:

1. An $(N+2t)/2$-byte linear feedback shift register (LFSR) generated pattern
2. An $(N+2t)/2$ network-specific field

Figure 5.5
64 QAM
constellation.

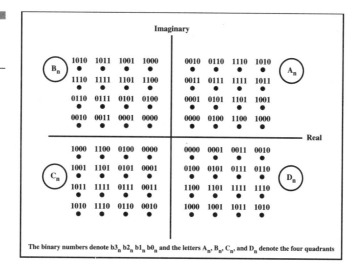

Figure 5.6
256 QAM
constellation.

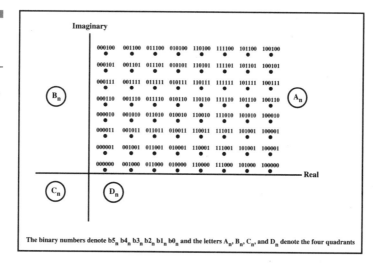

For DAVIC specification, the configuration for the network-specific field is a continuation of the LFSR-generated pattern. The HRM is sent in constant intervals defined by the high-reliability marker interval L. The HRM interval L defines the boundaries of the interleave function. The LFSR-generated pattern is defined by the polynomial $x^8x^6 + x^5 + x + 1$, with an initial seed value of 10,000,000. For the SA-MRT configuration, the HRM interval L is 205.

MRT DATA PACKET FORMAT. The MRT data packets are of the form shown in Fig. 5.8.

Figure 5.7
Multirate transport
structure.

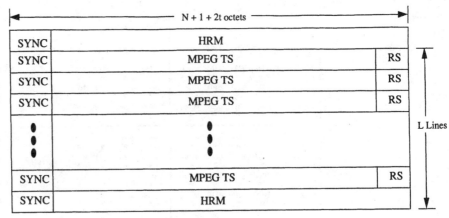

N = MRT data packet length (N=188 for MPEG TS)
t = error correcting power of Reed Solomon code
SYNC = 11000011
HRM = High Reliability Marker
L = High Reliablity Marker Interval

The data field is encoded using the Reed-Solomon algorithm to generate the Reed-Solomon (RS) parity field. The resultant MRT data packets are double-byte-interleaved to the defined interleave depth (I) and then placed into the MRT packet payload. This MRT packet payload is scrambled using an LFSR-generated pattern (0). The MRT packet size (N) is 188 bytes. Each MRT packet contains one MPEG2 transport packet.

DATA FIELD SCRAMBLING. The data field is scrambled using an LFSR pattern that is generated using the polynomial $x^{23} + x^{18} + 1$. The LFSR is reseeded every L data packets, beginning with the first packet that follows the HRM, with the value 10,001,000,000,000,000,000,000 or 1.0001×10^{22}.

REED-SOLOMON PARITY. There is need for error correction and detection techniques in any system where digital information is being transported across a medium where noise is present and the possibility of errors exists. The Reed-Solomon class of coding is part of a large class of

Figure 5.8
Multirate transport
data packet.

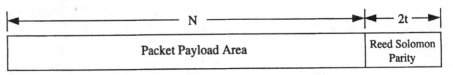

N = MRT data packet length (N=188 for MPEG TS)
t = error correcting power of Reed Solomon code

powerful random error-correcting cyclic codes known as the Bose, Chaudhuri, and Hocquenghem (BCH) codes. Reed-Solomon codes are a subclass of the nonbinary BCH codes.

The Reed-Solomon Parity is calculated over the preceding data field (N bytes). The Reed-Solomon code that is used for the parity calculation is RS(N, $N+2t$) code, which is implemented using a symbol size (M) of 8 bits and the primitive polynomial $p(x) = x^8 + x^7 + x^2 + x + 1$.

To generate a Galois field (GF) of 256, the generator polynomial for the code is:

$$g(x) = \prod_{i=120}^{119+2} (x - \alpha^i)$$

where t = error-correcting power of the Reed-Solomon code

The Reed-Solomon code error-correcting power (t) that is used in the MRT configuration is: $t = 8$, RS(188,204) code.

INTERLEAVE FUNCTION. Interleaving is utilized in conjunction with the Reed-Solomon code to provide protection against burst errors. To perform the interleave function, consider an array of packets $A[(N+2t) \times I]$, where $A[\,]$ has the form:

$$p_1(1),\ p_1(2),\ p_1(3),\ p_1(4),\ \dots\ p_1(N+2t-1),\ p_1(N+2t)$$

$$p_2(1),\ p_2(2),\ p_2(3),\ p_2(4),\ \dots\ p_2(N+2t-1),\ p_2(N+2t)$$

$$\vdots \qquad \vdots \qquad \vdots \qquad \vdots \qquad \dots \qquad \vdots \qquad \vdots$$

$$\vdots \qquad \vdots \qquad \vdots \qquad \vdots \qquad \dots \qquad \vdots \qquad \vdots$$

$$p_i(1),\ p_i(2),\ p_i(3),\ p_i(4),\ \dots\ p_i(N+2t-1),\ p_i(N+2t)$$

where $\quad p_1(1)$ = the first byte of packet $p_1(N+2t)$
$\quad p_i(N+2t)$ = the ith MRT packet with total length $N+2t$
$\quad I$ = the interleave depth

The result of performing double-byte interleaving on $A[\,]$ to an interleave depth I is the array $B[2I \times (N+2t)/2]$, where $B[\,]$ has the form:

$$p_1(1),\ p_1(2),\ p_2(1),\ p_2(2),\ \dots \quad p_i(1), \qquad p_i(2)$$

$$p_1(3),\ p_1(4),\ p_2(3),\ p_2(4),\ \dots \quad p_i(3), \qquad p_i(4)$$

$$\vdots \quad \vdots \quad \vdots \quad \vdots \quad \dots \quad \vdots \qquad \vdots$$

$$\vdots \quad \vdots \quad \vdots \quad \vdots \quad \dots \quad \vdots \qquad \vdots$$

$$p_1(N+2t-1),\ p_1(N+2t),\ \dots\ p_i(N+2t-1),\ p_i(N+2t)$$

The only constraint on the interleave depth is that it must be an integer multiple of the HRM interval $(L-1)$, i.e., $(L-1)/I =$ integer value.

For the DAVIC specification configuration, the maximum interleave depth (I) is 204 and double-byte interleaving is utilized. The SYNC byte is not interleaved.

For a summary of SA-MRT configuration, see Table 5.3.

SA-MRT-BASED HOME COMMUNICATIONS TERMINAL. The QAM and FEC path of an interactive home communications terminal (HCT) is a production unit utilized for interactive digital video trials. The HCT is capable of NTSC reception for analog video, QAM reception for digital video, QPSK reception for control information, and upstream QPSK transmission for interactivity. This discussion will focus on the architecture and performance of the QAM/FEC section. The HCT supports both 64 and 256 QAM constellations.

HCT ARCHITECTURE. See the HCT functional block diagram, illustrated in Fig. 5.9, of the QAM and FEC path of the HCT.

RF/IF Sections. The broadband RF input at the subscriber's premises is applied to the RF front-end assembly. The RF front end interconnects to the NTSC, QPSK, and QAM subassemblies of the HCT. The QAM broadband digital tier is connected to an up/down converter. The desired QAM signal might be anywhere in the 151—749-MHz RF band. In NTSC-based systems, the channel spacing will be 6 MHz. Dual conversion is utilized to down-convert the desired QAM RF signal to the typical 41—47-MHz TV IF

TABLE 5.3

SA-MRT Configu-rated Tested

Modulation Level	64, 256 QAM
Alpha	20%
Mapping	Differential Gray
FEC	RS $t = 8$ (204,188)
Interleave	204×204 block interleaver
Overhead	8.75%
Framing	HRM plus packet start code
Coding gain	4.8-dB raw coding gain at 1×10^{-9}
No noise burst Performance	445 msec, 64 QAM at 30 Mbps 326 msec, 256 QAM at 40 Mbps

Figure 5.9
HCT functional block diagram.

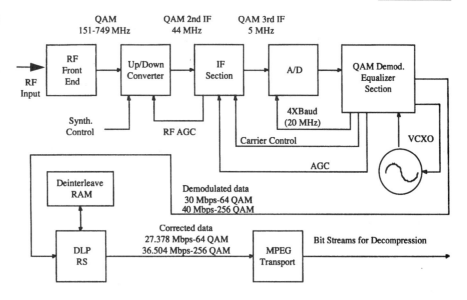

band. Frequency synthesized local oscillators are utilized in the up/down. The received QAM IF signal is centered on 44 MHz. Given the 20 percent alpha in this physical layer, the second IF signal will have a 3-dB bandwidth of approximately 5 MHz. Delayed RF AGC is applied to the up/down to maintain good tuner intermodulation performance with high-input signal levels. An IF surface acoustic wave (SAW) filter is utilized to provide out-of-band attenuation ahead of the third frequency conversion. Within the second IF section, AGC is applied. The gain-controlled 44-MHz IF is then converted to a 5.000-MHz baud rate IF. For 15 percent alpha applications, the third LO frequency will be changed in order to accommodate the required 5.217-MHz baud rate IF. As part of the carrier recovery process, the QAM demod controls the frequency of the third LO.

A/D and QAM Section. The 5-MHz baud rate IF signal is applied to a single analog-to-digital (A/D) converter. The occupied bandwidth of the baud rate IF signal is approximately 6 MHz. The A/D is sampled at four times the baud rate (20 MHz for 20 percent alpha and 20.868 MHz for 15 percent alpha). An 8-bit A/D provides near theoretical performance in 64 QAM mode. A 9-bit A/D is required for near-theoretical performance in 256 QAM mode. The QAM section for this HCT is set up as indicated in Table 5.4. The HCT tested included a three-chip set implementation for the QAM demodulator, equalizer, and synchronizer.

The digitized IF signal is applied to the QAM demodulator section. The functions of the QAM digital ASICs include automatic gain control,

TABLE 5.4

QAM Demodulator/
Equalizer
Configuration

Modulation format	64 or 256 QAM selected by HCT microcontroller based on cable channel mapping
Baud rate IF	5 MHz
Data rate	30 Mbps for 64 QAM 40 Mbps for 256 QAM
Digital matched filters	49 taps finite impulse response (FIR) linear phase low pass filter; square root raised cosine approximation
Alpha	20%
Adaptive equalizer	Feed forward equalizer (FFE) and decision feedback equalizer (DFE) structure; sixteen complex T(1/baud rate) spaced taps; eight FFE and eight DFE taps provide of up to 1.6 μsec of echo cancellation capability; blind equalization, no training sequence required; fast acquisition

clock recovery, carrier recovery, data filtering, and adaptive equalization. The QAM ASICs include DACs to drive the three analog control loops. The all-digital demodulator/equalizer solution requires minimal external analog support. A voltage-controlled crystal oscillator (VCXO) is utilized for clock recovery. Simple low-pass filters are used to interface the digital ASIC to the third LO, VCXO, and RF/IF gain control sections.

Forward Error Correction Section. The DLP/RS section, as illustrated in Fig. 5.10, receives the demodulated data from the QAM adaptive equalizer. The DLP/RS performs the quadrant decoding, constellation Gray decoding, data descrambling, data framing, block deinterleaving, and Reed-Solomon decoding in accordance with the SA-MRT format. The frame synchronization takes advantage of the HRM for additional framing robustness. The data output from the RS decoder is applied to the output control section. The clock and data are connected to an MPEG2 transport ASIC. The Reed-Solomon decoder and frame synchronizer are monitored by the HCT microcontroller. Frame synchronization status, RS corrections, and uncorrectable blocks are registered. The register values monitored provide useful diagnostics of both the transmission link and the HCT.

DAVIC PHYSICAL LAYER SUMMARY. The DAVIC 1.0 physical layer was agreed to in June of 1995. Table 5.5 summarizes the DAVIC 1.0 physical layer.

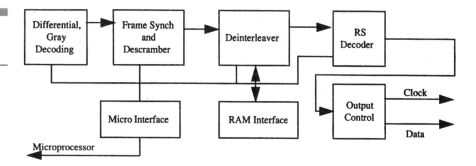

The DAVIC 1.0 physical layer includes 64 and 256 QAM with 15 percent excess bandwidth. Approximately, an additional 4.35 percent (1.2/1.15 – 1 in percent) of data is transmitted than with an alpha of 20 percent. The DAVIC interleaving is convolutional rather than block for SA-MRT. Convolutional deinterleaving requires approximately one-half the memory of a block deinterleaver for the same depth. The DAVIC framing does not require the use of the HRM.

TABLE 5.5

DAVIC 1.0 Physical
Layer Summary
for HFC

Channelization	6- and 8-MHz channels
Data rates	Based on 5.217 Mbaud rate for 6-MHz channels and 6.887 Mbaud for 8-MHz channels.
Data scrambling	Yes, polynomial is $x^{15} + x^{14} + 1$.
Modulation level	16, 64, 256 QAM.
Alpha	15%.
Mapping	Differential Gray. Differential encoding of two MSBs and Gray coding within each quadrant.
FEC	RS $t = 8$ (204,188).
Interleave	12 × 204 convolutional interleaver for 16 and 64 QAM. 204 × 204 convolutional interleaver for 256 QAM.
Overhead	8%.
Framing	MPEG packet start code. HRM carried as null packet with 0×1FFF PID.
Coding gain	4.8-dB raw coding gain at 1×10^{-9}.
No noise burst Performance	24.5 μsec, 64 QAM at 31.304 Mbps; 6-MHz channel. 313 μsec, 256 QAM at 41.739 Mbps; 6-MHz channel.

5.5.2 Amati Communications: SDMT PHY Layer

OVERVIEW. This proposal made by Amati was classified under the advance PHY development by IEEE 802.14. It uses a multicarrier system to transport data. The advance development group is studying various solutions for future PHY developments. A brief summary of the proposal appears below.

The DMT (discrete multitone) physical layer proposed involves 128 frequencies arranged in subgroups of 32. Assignment of bits to frequencies depends on the amount of noise in that frequency, ranging from 2 to 8. This is frequency agility at the micro level and one that deals with the noise environment very effectively, at least theoretically. IEEE 802.14 will be looking at this technology very seriously in the next phase of the PHY standard development.

DMT modulation uses an inverse discrete fourier transform (IDFT) to partition a transmission channel into a set of independent, equal-bandwidth subchannels such that the frequency response of each subchannel is approximately constant across its bandwidth. Consequently, the resulting subchannels are essentially memoryless, and the effect of the channel can be eliminated in the receiver by single complex multipliers per subchannel. The set of complex multipliers is known as the frequency domain equalizer. Changes in the channel magnitude and/or phase are accommodated by updating the frequency domain equalizer taps as the system operates. Hence, a traditional equalizer is not required in the receiver because there is no intersymbol interference in a DMT system. In the transmitter, a bitstream is encoded as a set of QAM subsymbols. Each QAM subsymbol represents a number of bits determined by the signal-to-noise ratio (SNR) of its associated subchannel.

Synchronization is very tight relative to the transmission time of one symbol. The symbol time is long because the bits are distributed over all the frequencies; that is, the symbol time is 128 times longer than it would be on a single carrier. Transmissions arrive at the headend as if they had come from a single source, even though several users may be transmitting at the same time.

Backward compatibility with QPSK systems can be achieved by leaving QPSK signals in the middle of the SDMT frequencies. Separate receivers would be necessary for the embedded QPSK signals. CBR can be accommodated by dedicated assignment of frequencies to particular users. The important parameters of the proposed SDMT system are:

Symbol rate	32 Kbaud
Subcarrier separation	34.5 kHz
Data samples per symbol	512
Guard period (cyclic prefix)	32 (approx. 2 ms)
Super-frame	204 symbols of data + 3 symbols of silence

There are two small overhead periods in the time domain: one guard period per symbol and one silent period per super-frame. The ratio of used to usable symbols in the time domain is the same as the ratio of used to usable frequency bandwidth. The total transmission overhead is 7.2 percent. The purpose of the guard period is to allow the transient response of the network to settle before collecting the 512 samples used for the receiver; the silent symbols are used for installation and training.

TRANSMISSION. The noise (including RFI, EMI, etc.) received at the headend receiver is the same for all remotes, and during training the transmit levels are adjusted to equalize the receive levels. Therefore, the SNRs of signals from all remotes are nearly equal, and there is almost no loss in throughput from making the upstream bit/subcarrier the same for all remotes. This simplifies both CBR and VBR (or ABR) transmission.

CBR. A single-carrier system such as QPSK can use only time division to multiplex many transmitters, and the method of providing a CBR service is controversial. An SDMT system, however, can use TDMA or FDMA, and it appears that a mixture is best for CBR. The symbol rate of the SDMT system is 32 Kbaud, so a remote granted exclusive use of a 2-bit subchannel would have a 64 kbit/s CBR service; this would be pure FDMA. On quiet, high-capacity systems, a remote could be granted every fourth symbol of an 8-bit carrier; this would be FDMA/TDMA. Regardless, however, of whether a subchannel assigned for CBR is dedicated or time-shared, it would not be used for VBR.

PACKETIZED VBR AND ABR. If the time (measured in number of symbols) to transmit a VBR or ABR packet is the same for all remotes, then a conventional QPSK-like TDMA system can be used. The only difference from a QPSK system is that the total upstream capacity may vary widely from system to system, and the number of subcarriers available for VBR may change as CBR services are added or dropped; therefore, the number of symbols required to transmit a packet may vary. The Amati MAC proposes the use of three subchannels, so there are 29 available for data. For a quiet system (very little RFI or EMI so that it supports an average of 8 bits per subchannel) with no CBR, the capacity per symbol is $29 \times 8 = 232$ bits. An ATM cell with a (57,53) RS-FEC contains 456 bits and could be transmitted in 2 symbols. On the other hand, a noisy system with 50 percent of the subchannels assigned for CBR might have a capacity as low as 30 bits and require 16 symbols for one ATM cell. For quiet, high-capacity systems, the symbol granularity is rather coarse, and there is some loss of efficiency (defined as aggregate throughput divided by capacity). This loss is much less when CBR removes subchannels from the variable bit-rate pool.

COMPATIBILITY WITH A QPSK SYSTEM. As an HFC system is upgraded, a transitional system will probably have a mix of QPSK and SDMT. This mix would be most convenient if the QPSK frequency spectra were contained within one SDMT subgroup. The headend would then have both SDMT and QPSK receivers operating in separate, noninterfering 1.1-MHz bands, and the frequency-agile remote transmitters would be assigned to appropriate RF bands depending on whether they were SDMT or QPSK. The total bandwidth available to each QPSK system would be 1.1 MHz. A symbol rate of 1.536 Mbaud would allow for 43.2 percent excess bandwidth—probably somewhat more than expected, but this would increase the amount of timing information in the signal and allow the faster acquisition.

5.5.3 Alcatel Telecom: ATM-based PHY for Both In-Band and Out-of-Band HFC Systems

OVERVIEW. Alcatel Telecom proposed the basic building blocks for modulation (e.g., 64 QAM) and framing. The proposal aligns with specifications of other standardization bodies (DVB, DAVIC 1.0, ETSI 300 429). The proposal assumes that delay-sensitive applications, POTS in particular, are in separate frequency channels. Support is provided for up to 41.7 Mbps in a channel with 64 QAM in a European 8-MHz channel or 6 MHz in North America. For the upstream channel, 384 kbps, 1.536 Mbps, and 3.072 Mbps are supported.

Since DAVIC calls for out-of-band signaling, this proposal can operate both with in-band and out-of-band signaling.

CHANNEL STRUCTURE.   Two main options exist for HFC physical layers:

1. In-band HFC systems
2. Out-of-band HFC systems

IN-BAND CHANNEL STRUCTURE.   In the in-band HFC system, as shown in Fig. 5.11, all downstream data flows (applications, TDMA control, and so on) to one particular subscriber are carried by only one downstream channel. These downstream channels are referred to as HFC-BB (HFC broadband).

OUT-OF-BAND CHANNEL STRUCTURE.   In the case of out-of-band systems, as shown in Figs. 5.12a and 5-12b, these data flows are carried on two different downstream channels: a low bit rate one (HFC-DS, typical 3 Mb/s in QPSK) and a high bit rate one (HFC-BB, typical 30–45 Mb/s in 64 QAM). The different data flows to one user have different VPI/VCI values such that a splitting to two different channels is possible. The allocation of a data flow type to one of these two channels is made on the basis of the content. An example of such an assignment is listed in Table 5.6.

Figure 5.11
Channel structure for in-band signaling HFC system.

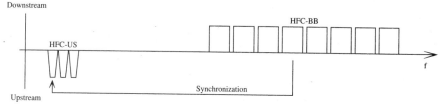

Figure 5.12
Channel structure for out-of-band HFC system: a, one separate downstream channel (HFC-DS) for each upstream channel (HFC-US); b, only one downstream channel (HFC-DS) for several (or all) upstream channels (HFC-US).

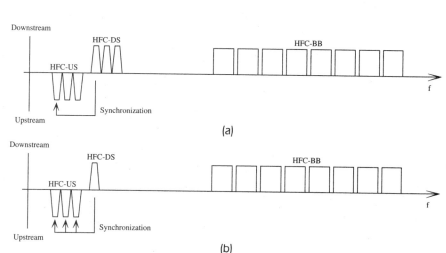

(a)

(b)

TABLE 5.6

Content of the
Channels for an
Out-of-Band and
In-Band System

Channel	Out-of-band HFC system	In-band HFC system
HFC-BB = HFC broadband downstream channel	• Data (ATM)	• Data (ATM) • ATM and application signaling • Upstream synchronization • TDMA control (grants)
HFC-DS = HFC downstream signaling channel	• ATM and application signaling • Upstream synchronization • TDMA control (grants)	Not applicable
HFC-US = HFC upstream signaling channel	• ATM and application signaling • Data (ATM)	• ATM and application signaling • Data (ATM)

HFC terminals (e.g., set-top boxes, cable modems) for out-of-band systems require two different receivers, while in-band systems need only one. In both cases, only one transmitter is needed.

In general, the number of HFC-BB channels will be higher than the number of HFC-US (and HFC-DS) channels, because the CATV return bandwidth is limited (e.g., up to 50 MHz). This means that some users tuned on different downstream HFC-BB channels will have to use the same upstream HFC-US channel. In the in-band case, this makes the synchronization of the upstream TDMA slots of different users more complex. However, in out-of-band systems, all TDMA slots in one upstream channel can synchronize on just one HFC-DS channel, which makes it much easier.

NARROWBAND SERVICES (POTS). Delay-sensitive services (e.g., POTS) are transported in separate RF channels. Upstream bursts are shorter than ATM cell length to fulfill delay performance requirements. For the same reason, these services are not transported in ATM cells in the HFC access network, except if sufficient echo cancellation is provided in the network.

DOWNSTREAM MODULATION. Both the European and North America systems are described below.

8-MHZ CHANNEL (EUROPE)

Symbol rate   6.000–6.952 Msymbols/s (6.952 recommended)

Bit rates for 64 QAM   36.000–41.712 Mbit/s

Filter	Square root raised cosine (compliant with DAVIC 1.0 spec)
Roll-off	0.15
Symbols	16 QAM / 64 QAM / 256 QAM symbols
Symbol mapping	2 MSBs of each symbol differentially encoded

6-MHZ CHANNEL (USA AND JAPAN)

Symbol rate	5.000–5.304 Msymbols/s
Bit rates for 64 QAM	30.000–31.824 Mbit/s
Filter	Square root raised cosine (compliant with DAVIC 1.0 spec)
Roll-off	0.13
Symbols	16 QAM / 64 QAM / 256 QAM symbols
Symbol mapping	2 MSBs of each symbol differentially encoded

FRAME OF THE HFC CHANNEL IN OUT-OF-BAND. The HFC-BB signal for out-of-band HFC system is DAVIC-compliant as follows:

- Mapping of 7 ATM cells in 2 DVB frames (aligned)
- Profile A: 16 QAM / 64 QAM
- Profile B: 16 QAM / 64 QAM / 256 QAM

UPSTREAM AND DOWNSTREAM MODULATIONS. The modulation format for the downstream signaling channels is DEQPSK (differentially encoded quaternary phase-shift keying). The modulation is differentially encoded. The demodulation can still use coherent or noncoherent detection based on the choice of the implementation. The downstream and upstream signals for the HFC-DS and HFC-US channels are filtered at the modulator, resulting in a square root of raised cosine filter with a roll-off factor of 0.5 or 0.3. The downstream bit rate of the HFC-DS channel is not linked with the downstream HFC-BB channel. The proposed bit rate for the HFC-DS channel is 3.072 Mbit/s. With a filter roll-off of $a = 0.3$, this maps exactly into a bandwidth of 2.000 MHz. This allows easy alignment with 6-MHz and 8-MHz slots. The value of 3.072 Mbit/s also equals 48×64 kbit/s. This allows a combination with different proposed upstream bit rates.

Figure 5.13
Modulator in the
access node with
framer and near-DVB
processing.

DOWNSTREAM SIGNALING MODULATOR

STAGE 1: SUPER-FRAME

STAGE 2: NEAR-DVB PROCESSING

47h — S
IC01;_ — IC
TEA1;_ — TCB
— ATM
00h — FEC dummy

A

Synch 1
Inversion
&
Random

Optional
Reed-
Solomon
Coder
(204,188)

Optional
Convol.
Interleaver
1=12

Byte to
symbol
(2 bit)
conversion

Differential
encoding

QPSK
Modulator
& IF PHY
Interface

The downstream modulator in the access node contains two stages as illustrated in Fig. 5.13. The first stage generates a superframe of 53 rows with 204 bytes per row. Each row is compatible with the input of a DVB encoder. The second stage executed near-DVB processing on a row-by-row basis (each row being treated as a 204-byte DVB-compatible frame). Figure 5.14 shows the downstream/upstream frames.

Figure 5.14
Downstream frame
for HFC signaling
channel with a FEC.

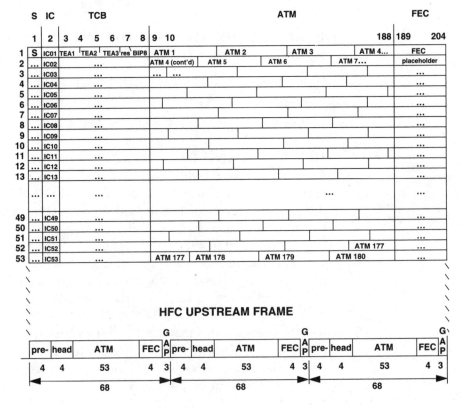

HFC DOWNSTREAM FRAME *WITH* FEC

HFC UPSTREAM FRAME

Code descriptions for the figures follow:

S   Synch byte (DVB) 47h

IC   Information channel and super-frame synchronization

TEA   Transmit enable address

TCB   TDMA control block

FEC   Dummy bytes for DVB forward error correction

IC   See assignment in Table 5.8

The super-frame, shown in Fig. 5.14, is defined at the interface A between the first and second stage (see Fig. 5.13). The data present on the line (after the second stage) includes extensive additional processing to the original data in the super-frame at interface A. The super-frame is generated in the first stage. It is divided into 5 blocks, with each block corresponding to a number of columns in the super-frame (see Table 5.7):

1. DVB-frame synch byte (47h)

2. Information channel and super-frame synchronization (IC)

3. TDMA Control Block (TCB)

4. Data (ATM cells)

5. Optional placeholder for forward error correction as specified in DVB (RS-FEC)

Interleaving is optional. For the more robust QPSK, it is probably not needed, and by preference it is not implemented, because it creates additional delays, especially at lower bit rates.

In the second stage (see Fig. 5.13), the rows of the superframe are further manipulated by a series of functions, near to those specified by DVB. The deviations from true DVB processing are:

TABLE 5.7

The Five Blocks in the Super-Frame

Columns	Content	Code
1	DVB-frame synchronization	S
2	Super-frame synchronization	IC
3 .. K (8)	TDMA control block	TCB
K+1 (9) .. L (188 or 204)	Data (ATM cells)	ATM
L+1 (189 or 204) .. 204	Optional placeholder for error correction	Optional FEC

■ The presence of the FEC is optional (in DVB it is mandatory). When the FEC is not present, the rows are filled with ATM data up to byte 204 and have no FEC. When the FEC is not present, the width of the ATM data block is expanded and the width of the FEC block is reduced to zero. The total width of the super-frame remains identical. The IC field contains an indication (the value L) of the width of the ATM data block.

■ The use of the interleaver is optional (in DVB it is mandatory).

■ The modulation format is DEQPSK (DVB specifies 16 QAM and 64 QAM).

■ The bit rate is lower (in the range of 2 to 3 Mb/s). DVB specifies around 30–40 Mb/s.

SUPER-FRAME. The super-frame contains 53 rows of 204 bytes. Each row of 204 bytes is compatible with the input of a DVB-compatible source encoder. The five elements described previously are each contained in a number of columns of the super-frame. The super-frame is shown in Fig. 5.14. The FEC is filled in by the near-DVB processing block. The advantage of defining a super-frame of 53 rows is that an integer number of ATM cells always fits exactly in the ATM block, independent of the size (i.e., number of columns) of this block, such that no dummy bytes are lost. In that case, the ATM cells are aligned to the super-frame, which makes cell alignment easy.

S. The value of the S field, as supplied to the DVB-compatible equipment, is defined as 47h (0100 0111).

TCB. This is the data required for control of TDMA of the upstream burst. The TCB blocks contain three transmit enable addresses (TEA) of 12 bits (total 36 bits), a reserved area of 4 bits and a BIP-8, even parity, error detection code over the combination of the IC and the TCB. The block over which the BIP-8 operates extends over seven bytes, as from column 2 (the IC) up to and including column 8 (the column with the BIP-8). The three transmit enable addresses (TEA1, TEA2, TEA3) sent in the current DVB frame (row of the super-frame) are applied to the three upstream bursts sent during the next downstream DVB frame. The TEAs, sent during row i of the frame, are used for the upstream bursts sent during row i+1 (this is approximately 200 bytes later). The content of TCB is:

TEA 1 (12 bits)	TEA for the first upstream burst during the next downstream DVB frame
TEA 2 (12 bits)	TEA for the second upstream burst during the next downstream DVB frame
TEA 3 (12 bits)	TEA for the third upstream burst dur- the next downstream DVB frame
Reserved (4 bits)	Reserved area (filled with 0000)
BIP-8 even (8 bits)	The bitwise parity over the 7 bytes in the IC and TCB is even

TEA RANGES (12-BIT).   A number of ranges are defined for TEAs. These limit the number of available addresses in this version and allow for later increases of the number of HFC terminals or additional functionality. An example is given below of how the TEAs could be assigned. This list is not final and constitutes only an example of implementation.

IC.   The value of the IC bytes (IC01 .. IC53) is defined as a function of their position in the super-frame. A complete list is given in Table 5.8. They are in groups of four. Each group of four contains the following fields:

- *IC01; IC05; IC09,...IC49.*   Synchronization of the super-frame (ranges from 240–252; from F0h–FCh).
- *IC02; IC06;....*   Position of the TCB and the ATM block. In an alternating fashion, the values of K and L are transmitted in this field.
- *IC53 (optional).*   Phase of the 8-kHz clock.
- *IC03; IC07;....*   Reserved for later use (initially set at 00h).
- *IC04; IC08;....*   Reserved for later use (initially set at 00h).

ATM.   The ATM cells in the ATM block are byte-aligned and aligned to the super-frame. The first byte of the first ATM cell of the super-frame starts in the first column and the first row of the ATM block (column K+1; row 1). The first row of the super-frame is indicated by the presence of the value 240 (F0h) in the IC column (column 2). The ATM cells are following each other continuously in the ATM block. No gaps are present between the ATM cells in the ATM block. On other rows, the first column of the ATM block does not necessarily coincide with the

TABLE 5.8

Assignment of Information Channel (IC) Bytes

Code	Value	Remark	Code	Value	Remark
IC 01	F0	Super-frame synch	IC 33	F8	Super-frame synch
IC 02	08	Value of K	IC 34	08	Value of K
IC 03	XX	Reserved	IC 35	XX	Reserved
IC 04	XX	Reserved	IC 36	XX	Reserved
IC 05	F1	Super-frame synch	IC 37	F9	Super-frame synch
IC 06	BC/CC	Value of L (188/204)	IC 38	BC/CC	Value of L (188/204)
IC 07	XX	Reserved	IC 39	XX	Reserved
IC 08	XX	Reserved	IC 40	XX	Reserved
IC 09	F2	Super-frame synch	IC 41	FA	Super-frame synch
IC 10	08	Value of K	IC 42	08	Value of K
IC 11	XX	Reserved	IC 43	XX	Reserved
IC 12	XX	Reserved	IC 44	XX	Reserved
IC 13	F3	Super-frame synch	IC 45	FB	Super-frame synch
IC 14	BC/CC	Value of L (188/204)	IC 46	BC/CC	Value of L (188/204)
IC 15	XX	Reserved	IC 47	XX	Reserved
IC 16	XX	Reserved	IC 48	XX	Reserved
IC 17	F4	Super-frame synch	IC 49	FC	Super-frame synch
IC 18	08	Value of K	IC 50	08	Value of K
IC 19	XX	Reserved	IC 51	XX	Reserved
IC 20	XX	Reserved	IC 52	XX	Reserved
IC 21	F5	Super-frame synch	IC 53	YY	Phase of 8-kHz clock
IC 22	BC/CC	Value of L (188/204)	—		
IC 23	XX	Reserved	—		
IC 24	XX	Reserved	—		
IC 25	F6	Super-frame synch	—		
IC 26	08	Value of K	—		
IC 27	XX	Reserved	—		
IC 28	XX	Reserved	—		
IC 29	F7	Super-frame synch	—		
IC 30	BC/CC	Value of L (188/204)	—		
IC 31	XX	Reserved	—		
IC 32	XX	Reserved	—		

first byte of an ATM cell. Because the number of rows in the super-frame is 53, the last byte of the ATM block (byte L on row 53 of the super-frame) contains the last byte of an ATM cell. The number of ATM cells contained in a super-frame equals L minus K. The two optional values are ($L = 188$; $K = 8$) 180 ATM cells in a super-frame and ($L = 204$; $K = 8$) 196 ATM cells in a super-frame. Because the ATM cells are aligned with the super-frame, no ATM cell synchronization is required. Synchronization of the super-frame is sufficient to recover the positions of the starting bytes of the ATM cells.

FEC. The FEC is calculated independently for each DVB frame (optionally after DVB-compatible interleaving) in the near-DVB processing block. At the interface A, there are only dummy bytes filled in positions 189 to 204 of the frame. The FEC bytes are calculated and filled in the near-DVB-compatible processing. The value of the FEC is not related to the position in the super-frame, but it is local to each individual row.

NEAR-DVB FRAME PROCESSING. The DVB frame (ETSI spec 300 429) is supplied to a near-DVB-compatible modulator/line/demodulator chain and contains:

- 1 byte   A synch byte value 47h (= DVB frame synchronization)
- 187 bytes   The payload of the DVB frame
- 16 bytes   Dummy bytes, placeholders for a Reed-Solomon FEC

This totals 204 bytes. In the second processing block, additional modifications to individual rows of the super-frame are introduced (scrambling, periodic inversion of the synch byte, FEC calculation and writing in bytes 189 to 204, optional interleaving, differential encoding, QPSK modulation). These functions are entirely specified in the DVB ETSI spec and are transparent to the format of the super-frame, except for an increased latency.

The format of the super-frame described here is the format of the data as supplied at interface A to the near-DVB-compatible equipment. The format is supplied parallel, byte-wide. If also the interleaving of DVB is implemented, this will create a latency of approximately 15 frames (i.e., approximately 3000 bytes).

MODULATION. The modulation format for the upstream signaling channels is DEQPSK (differentially encoded quaternary phase-shift keying). The modulation is differentially encoded. The demodulation can still use coherent or noncoherent detection based on the choice of the implementation. The downstream and upstream signals for the HFC-DS and HFC-US channels are filtered at the modulator, resulting in a square root of raised cosine filter with a roll-off factor of 0.5 or 0.3. The bit rate of the upstream HFC-US channel is linked with the downstream HFC-DS channel. The proposed bit rate for the HFC-DS channel is 3.072 Mbit/s. The value of 3.072 Mbit/s equals 48×64 kbit/s. This allows a combination with different proposed upstream bit rates. The proposed bit rates for the upstream are 3.072 Mbit/s, 1.536 Mbit/s, and 384 kbit/s (1/1, 1/2, and 1/8 of the downstream bit rate). With a filter roll-off $a = 0.3$, this maps to a bandwidth of 2 MHz, 1 MHz, and 250 kHz, respectively. The low bit-rate case is of interest in CATV systems with high noise levels (the ingress noise will be lower in a narrow channel, and the channel can be more optimally positioned in between interfering carriers). The high bit rate will allow services with T1/E1 useful bit rates.

FRAME OF THE UPSTREAM HFC-US CHANNEL. The upstream frame is shown at the bottom of Fig. 5.14. The upstream frame is compatible with the DAVIC and contains:

- Preamble of 4 bytes
- Header of 4 bytes
- ATM cell of 53 bytes
- FEC of 4 bytes (upstream interleaving is not used in HFC)

PREAMBLE. The following preamble is proposed (*not* differentially encoded):

```
11 11 11 11 00 00 11 11 00 11 00 11 00 11 00 11
```

This is FF0F3333h.

HEADER. A possible use of the header is (dependent on the selected MAC protocol):

Queue filling level	4 bits
Source ID	12 bits
Reserved	16 bits for other system-specific information

ATM CELL. The ATM cell is filled in here (53 bytes) without any modification.

FEC. The FEC is per DAVIC 1.0. It is calculated over bytes 4 to 60, using a (61,57) Reed-Solomon code. Up to two bytes per upstream frame may be corrected. The Reed-Solomon code generator is the generator polynomial $x^8 + x^4 + x^3 + x^2 + 1$.

GAP. The gap has a nominal length of three bytes. The minimum required gap between arriving burst on the upstream burst mode receiver on a plant with HFC terminals positioned at different distances should never be smaller than one byte (= 8 bits = 4 symbols in QPSK). This dictates that the differential distance between the HFC terminal at smallest distance and the HFC terminal at farthest distance should be maximally equal to two bytes (two-way delay). When a larger differential distance must be handled, a ranging protocol must be used that adds delay for the closest HFC terminal to an accuracy of better than ± 1 byte.

SCRAMBLING OF THE UPSTREAM BURST. After R/S calculation, the upstream data is scrambled as shown in Fig. 5.15. The scrambler starts its operation from the first bit following the four-byte preamble. In the in-band system, the downstream broadband channel carries ATM cells and synchronization and MAC control information for the upstream channels. The modulation is identical to the modulation used for the HFC-BB channel in the out-of-band system. The downstream frame is very similar to the frame in the downstream HFC-DS channel in the out-of-band system.

MODULATION OF THE HFC-BB CHANNEL. The modulation of the downstream HFC-BB channel in the in-band system is identical to the modulation of the HFC-BB channel in the out-of-band system.

Figure 5.15
Upstream data descrambler.

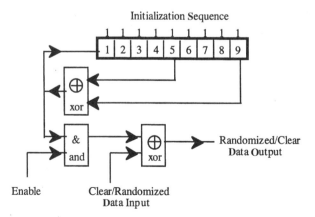

FRAME OF THE HFC-BB CHANNEL. The frame generated for the HFC-BB channel in the in-band system as shown in Fig. 5.16 is very similar to the frame of the HFC-DS channel in the out-of-band system. The only difference is that, due to the higher downstream bit rate, fewer columns of the downstream frame must be used to transport an identical number of transmit enables. The TCB block will thus become only two bytes wide. It contains a single transmit enable address and four reserved bits.

IC. The five blocks in the super-frame of the in-band system is as shown Table 5.9.

TCB. This is the data required for control of the TDMA of the upstream burst. The TCB blocks contains one transmit enable address (TEA) of 12 bits and a reserved area of 4 bits. An additional error correction is not required, since the downstream 64 QAM channel is already protected by an extensive Reed-Solomon and interleaver (for the DEQPSK HFC-DS channel in the out-of-band system and the R/S and interleaver, where it is optional—however, for the HFC-BB channel, it is mandatory).

TEA 1 (12 BITS). This is the TEA for the first upstream burst during the next 3 or 4 downstream DVB frames (three downstream frames with a 6-MHz downstream channel and 8 MHz with an 8-MHz downstream QAM).

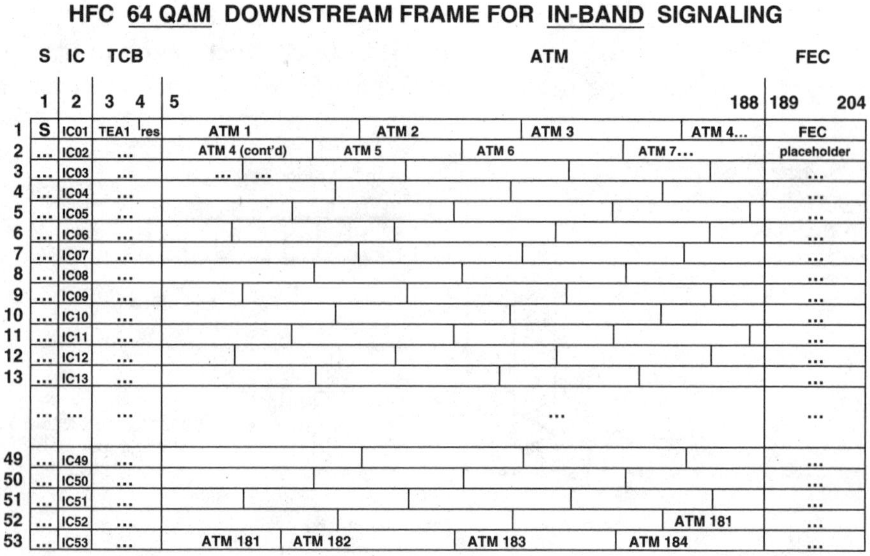

Figure 5.16
Downstream frame for the HFC-BB channel in the in-band system.

HFC **64 QAM** DOWNSTREAM FRAME FOR **IN-BAND** SIGNALING

TABLE 5.9

The Five Blocks in the Super-Frame for the In-Band System

Columns	Content	Code
1	DVB frame synchronization	S
2	Super-frame synchronization	IC
3 .. K (5)	TDMA control block	TCB
$K+1$ (6) .. L (188 or 204)	Data (ATM cells)	ATM
$L+1$ (189 or 204) .. 204	Placeholder for error correction	FEC

ATM. The ATM block is larger in the case of the in-band signaling (compared to the ATM block in the HFC-DS channel). It now contains 184 ATM cells per super-frame.

MODULATION. The modulation format for the upstream signaling channels is identical to the format used for the upstream signaling channel in the out-of-band system (DEQPSK).

THE UPSTREAM SIGNALING CHANNEL (HFC-US). The upstream bit rate of the HFC-US channel must be aligned with the downstream bit rate of the HFC-BB channel. This is a difference with the out-of-band system. In an in-band system, the bit rate of the upstream signaling channel (HFC-US) is linked to the bit rate of the downstream broad band channel (HFC-BB).

FRAME OF THE UPSTREAM HFC-US CHANNEL. The frame of the upstream HFC-US channel in the in-band system is identical to the frame of the HFC-US channel in the out-of-band system. The example for the in-band system that is proposed here is based on the practical case of 64 QAM modulation over an 8-MHz channel and an upstream bit rate of 3.072 Mbit/s.

DOWNSTREAM HFC-BB CHANNEL IN 8-MHZ CHANNEL. Bit rate: 27×1.536 Mbit/s = 41.472 Mbit/s (very close to the maximum value of 41.712). Bandwidth: $(41.472/6) \times 1.15 = 7.948\ 800$ MHz. Grant position rate: 1 grant position per 204 bytes.

UPSTREAM HFC-US CHANNEL. Bit rate: 1.536 Mbit/s. Bandwidth: $(1.536/2) \times 1.3 \approx 1$ MHz (exact 0.9984 MHz). Upstream packet rate: 1 upstream packet per 9 downstream DVB frames.

GRANT SYNCHRONIZATION. Synchronization of the upstream is as shown in Fig. 5.17. Three upstream packets = 204 upstream bytes = same time as 27 downstream DVB frames:

- Per 27 downstream DVB frames transmitted, 3 upstream frames are transmitted.
- Per 9 downstream DVB frames, one active TEA (grant) is given.

5.5.4 CORTEC: INTRA PHY Proposal

OVERVIEW. CORTEC proposed INTRA. INTRA stands for *in*formation *tra*nsform. INTRA considers data transmitted by the modem to be vectors. Modulation is a rotation; demodulation is the reverse rotation. FM modulation should be used because of its interference rejection. The gain in signal/noise occurs because of modulation gain, preemphasis gain, and gain from a companding process that is applied after modulation.

THEORY OF LINEAR INTRA. The concept of baseband INTRA can be visualized as an apparent coordinate rotation. The rotations are performed by baseband multirate filter banks. In modems, the information being conveyed exists in two distinct but equivalent forms, namely, as digital data and as a band-limited analog signal. A group of N samples of the signal at the D/A converter comprises the coordinates of an N-dimensional vector. The data must be another representation of the same vector. That is, the data must be the same vector projected onto another coordinate system.

The modulation operator $[M]$ rotates the representation of the information vector from data coordinates into signal coordinates. The demodulation operator $[D]$ rotates the A/D converter vector back to data. As with any modem, $[D][M] = [I]$, where $[I]$ is identity. By insisting that the operators are rotations, they must also commute so that $[D][M] = [M][D] = [I]$. At large N, a band-limited Gaussian signal G can be demodulated and

Figure 5.17
Synchronization of upstream burst with downstream frames for the in-band system (TEA = transmit enable address).

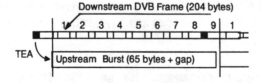

Downstream DVB Frame (204 bytes)

1 2 3 4 5 6 7 8 9 1

TEA

Upstream Burst (65 bytes + gap)

remodulated, since $[M][D]G = G$. This proves that the modulator is capable of sending a signal with the maximum possible entropy, which is the prerequisite for an optimum modem receiving at Shannon's limit.

Because modulation and demodulation are commuting operators in an INTRA modem, it is possible to send any band-limited analog signal through a demodulator to rotate it to data, then digitally encrypt that data, and finally use a modulator to rotate back into an analog signal. This provides the long-sought means to encrypt without expanding the bandwidth, or, stated another way, to encrypt digitally without compression. The practical significance of this, aside from security, is that voice or NTSC video, for example, can be digitally networked, error-protected, and so forth without compression. Compressionless transmission can be competitive in actual bandwidth to compression algorithms because the latter react poorly to bit errors unless there is considerable error protection.

Quadrature mirror filter banks (QMFs) have commuting polyphase matrices so they can perform the rotations. Note that the two-dimensional QMF performs the discrete wavelet transform by multiresolution analysis. The perfect reconstruction (PR) property of QMF pairs makes an exact identity operator $[I]$, although near-perfect (NPR) QMF banks, particularly those with linear phase, are a better choice for modems. Because the rotations are performed by subband filter banks, it is easy to determine which Data coordinates will rotate into in-band frequencies for the signal representation. The filter overlap and rolloff is a design parameter in QMFs (unlike an FFT, which has a fixed 13-dB stop-band attenuation), so the commuting property can exist over any in-band region to make an optimal modem.

AM MODEMS.   An AM modem is a linear mapping from baseband to passband. For CATV, a megahertz linear baseband modulator can be followed by a headend translator in a manner similar to channel translation of baseband NTSC to any downstream CATV channel. With all forward channels phase-locked, any video carrier can be used by an INTRA AM modem receiver as a frequency reference for coherent demodulation back to baseband. The design of translator modules is straightforward, so only the linear baseband modem is examined here. The AM modem would carry digital video and data, possibly in an ATM format, plus network management. It could also provide uncompressed NTSC through the same modem, if desired.

The full 6-MHz baseband channel can be defined by a 12-MHz sample rate. An eight-dimensional rotation has six coordinates that are usable without a DC response. That is, the inner six subbands will provide an overall response profile that fits within the translator's SAW (surface acoustical wave) filter. Since the symbol rate is 12/8 of the sample rate, the bit rate for B bits per channel is $6 \times B \times 1.5$ Mbps. Using $B = 6$, there are 36 bits per symbol at 54 Mbps. The 3-dB points of the outermost subbands will be at 1/8 and 7/8 or 750 kHz and 5.25 MHz. The AM profile is then given by the filter response curves. (Using 12 dimensions would allow 51 Mbps with $B = 5$ and 6 dB less S/N.) An optional synch bit in the upper subband provides synch at 5.25 MHz. The lower subband can transmit a synch at 625 kHz and/or at 750 kHz. The sample rate would be phase-locked to the passband translator, so the receiver can also use any video carrier as a pilot. The nominal delay is equal to 36 bits at 54 Mbps for framing, plus 72 samples in the 12 Msps rotation filter or $D = 36/54 + 72/12 = 6.67$ microseconds.

THEORY OF NONLINEAR INTRA.   The FM modem is an example of a nonlinear mapping of baseband into passband that has good immunity to impulse noise and narrow-band interference. An INTRA FM modem can have gain in its noncoherent receiver. This gain permits an FM receiver to use more signaling levels than an AM receiver for the same carrier-to-noise ratio and BER.

Modulation gain comes from collapsing the inherent band-spreading of FM. For the FM-SSB modem, the spreading is negligible. The other two gains employ the principle that a baseband nonlinear amplification (log-amp) at the transmitter will have no net effect on the signal if there is a matching (inverse-log) attenuation at the receiver. Noise at the receiver will be attenuated, thereby giving an improvement in the signal-to-noise ratio (i.e., gain). Noise reduction uses a log-amp in the signal coordinate representation to suppress link noise, whereas deemphasis uses logarithmically related amplifiers in the data coordinate representation to suppress parabolic noise out of the FM discriminator.

SIGNAL DOMAIN NONLINEARITY.   For noise reduction, a nonlinear signal look-up table, the mu-law function, is proposed with mu = 1023 where the mu-law output is related to the input by

```
OUTPUT=SIGN(INPUT)*Vc*LOG2{1+mu*ABS{INPUT/Vp})/1LOG2{1+mu}
```

The peak magnitude of the INPUT is V_p and the peak magnitude of the OUTPUT is V_c so that fewer bits of resolution are needed for the analog components than for the rotation. The mu law provides noise reduction; since it amplifies small signals more than large signals, it improves the peak-to-average ratio of the baseband signal that coincidentally increases the modulation gain.

FM MODEMS. The FM-DSB efficiency for $N = 4$ at $E = 25$ dB is 3.564. So 5.61 MHz is needed for 20 Mbps. The profile for FM-DSB is specified by its Bessel functions. Less than 1 percent per side of its power is outside the Carson BW.

FM-DSB CHANNEL PROFILE. The efficiency of an FM-SSB modem is $2B/[N(m+1)]$. For $N = 8$, the channel profile is as shown in Table 5.10.

Switched Carrier. Some multiple access protocols require switched carrier operation. Upstream TDMA is an example. FM modems are particularly well suited because of:

- Noncoherence (no carrier acquisition time)
- FM capture effect (turn-on and turn-off can overlap)

SUBSPLIT SYSTEMS. Subsplit systems have difficult narrow-band noise in the upstream direction. FM-DSB is proposed for such systems because it should be more robust than any other modulation. Impulse noise or ingress from shortwave radio should have little effect on constant-envelope FM.

HIGH-SPLIT SYSTEMS. For high-split systems, the upstream channel can use the less robust FM-SSB and still have the benefits of FM for switched-carrier operation. High-split upstream channels are presumed to have a noise floor of 45 dB. This permits 51 Mbps in a 6-MHz upstream channel.

RANGE AND BER MEASUREMENT. To obtain range timing, one end must turn around the signal in hardware to provide a fixed delay. Proposed MAC layer TDMA protocols place ranging measurement in the headend so the turnaround is done by the NIU. FM links can use that method, but another possibility is to have the headend translate the upstream traffic to a downstream 6-MHz channel. In that case, all modems can "see" their own encrypted upstream transmissions, and each modem can measure not only its slot timing but also the effects of ingress noise and request-

TABLE 5.10

FM-DSB Channel Profile

10 Log (E) (dB)	FM-SSB (efficiency)	Benchmark
0	2.81	QPSK (4-QAM)
7.0	3.85	16-QAM
13.2	4.93	64-QAM
35	9.36	Approx. 16K-QAM

slot collisions. Local measurements by the NIU of the upstream noise funneled to the headend (and translated back to the NIU) should simplify closed-loop adaptive noise cancellation in the NIU. This arrangement also permits each modem to monitor the upstream bit-error rate and report it to the MAC layer.

CHANNEL SELECTION. Each modem would be capable of tuning to any 6-MHz channel in the designated direction. The lowest upstream channel would be T-7 (5.75–11.75 MHz). Installation of the PHY layer would begin by a search of downstream channels for a recognized protocol and, once recognized, the downstream protocol would identify the upstream channel number.

MANAGEMENT. All modems would be capable of digital and RF loopback, power adjustment, and BER measurement in response to network management commands on the forward channel recognized at power-on initialization or in response to the MAC layer. The use of robust FM modems in 6-MHz upstream channels simplifies the cable plant and avoids the need for extensive dynamic channel assignment due to ingress in subsplit systems.

BLOCK DIAGRAMS. A rotation is equivalent to a vector filter. The shift registers contain vectors. The weights are matrices. This is shown in Fig. 5.18 for the four-dimensional FM modem. In that proposed FM-DSB design, there are eight matrices, and each is 4×4. Because of symmetries, there are actually only 32 numbers needed to construct these eight matrices and another eight matrices that are used for the counterrotation filter. The eight-dimensional modems can use shift registers with nine matrices of 8×8. Both vector filters are linear-phase in their overall response, even though the polyphase (down-sampled) coefficients are not symmetric.

For the AM modem, the input data is grouped in eight coordinates as six words of six bits each, plus two one-bit synch words. For the FM-DSB modem, the input data at the transmitter is grouped into 18 bits divided

Figure 5.18
Vector filter.

VECTOR

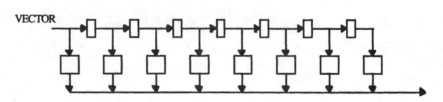

ROTATED VECTOR

into three words of seven, six, and five bits accordingly and a toggling synch bit for the fourth coordinate. The word for each coordinate is then converted to PAM form; namely, $+-1, +-3, +-5, +-7$, and so on. The conversion from serial to parallel, subdivision into words, and conversion to PAM is called *partitioning*. It is as shown in Fig. 5.19.

Preemphasis is obtained by multiplying the coordinates of the data after partitioning by the amplification factor $H(I)$. The resulting vector is then rotated by a *vector filter* into its *signal* representation. The signal coordinates can be converted from parallel to serial to form a sequence prior to being nonlinearly amplified by a mu-law lookup, and the result is input to the FM modulator as a digital sequence. This is illustrated in Fig. 5.20.

The building blocks are combined to make a FM modem transmitter as shown in Fig. 5.21. The FM modulator is an all-digital voltage-controlled oscillator (VCO) followed by a channel translator. The VCO could be implemented digitally with a phase accumulator. A Hilbert transformer and exponential lookup table can be appended to the VCO to generate SSB digitally.

5.5.5 Stanford Telecom: Proposal for QPSK Upstream Modulation Standard

OVERVIEW. Stanford Telecom proposed QPSK modulation on the CATV upstream channel. The signaling format includes differential QPSK

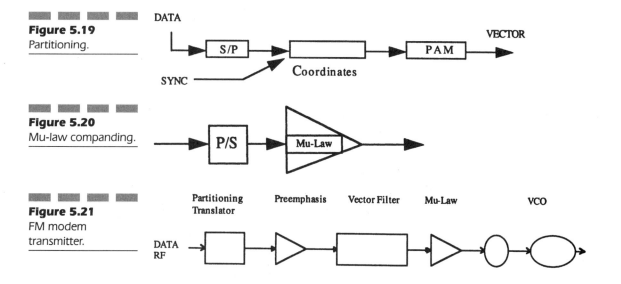

Figure 5.19
Partitioning.

Figure 5.20
Mu-law companding.

Figure 5.21
FM modem transmitter.

TABLE 5.11

QPSK Upstream
Transmission
Parameters

Parameter	Value	Notes
Burst bit rate R	2.56 Mbps	
Burst symbol rate $(R/2)$	1.28 Msps	
Modulation format	QPSK	
Pulse shaping	Root raised cosine, rolloff $\alpha = 0.4$	
Pulse duration	Truncated to 8 symbol times (4 symbols before and after peak)	
Channel spacing	1.8 MHz	
Transmission efficiency	1.42 bps/Hz	
Detection method	Coherent	
Phase ambiguity coding	Differential (see Table 5.14)	
Frame structure	See Figure 5-22	
Burst length	1—80 bytes for clock frequency accuracy of 100 ppm 1—256 bytes for clock frequency accuracy of 30 ppm	Receiver is informed of burst length before arrival of burst Clock accuracy is maximum offset permitted between subscriber and headend clock frequencies
Guard time between bursts	10 symbols nominal; 3 symbols minimum	
Preamble length	28 bits (14 symbols)	
Preamble bits	11 11 11 00 00 11 00 00 00 00 00 00 00 00	Not differentially encoded or otherwise modified
FEC coding	Application specific; compatible with Reed-Solomon	
Power control	Receiver measures power of bursts; controller adjusts subscriber power via downstream link	
Burst timing control	Receiver measures time of arrival of bursts; controller adjusts subscriber timing via downstream link	
Carrier frequency offset	±10 kHz	Maximum offset permitted between subscriber and headend carrier frequencies
Transmit frequency tuning resolution	25 kHz	
Input signal level range for acquisition	±10 dB from nominal	
Input signal level range for data transmission	±5 dB from nominal	
Output level with transmitter off	60 dB below nominal carrier level from 5—42 MHz	

modulation and bursts with defined preamble structure. QPSK is the existing de facto standard in the industry for upstream modulation, is proven in multiple field trials, provides a good balance between bandwidth efficiency and robustness to channel impairments, and offers low-cost implementation with open architecture.

QPSK TRANSMISSION PARAMETERS.   Table 5.11 summarizes the parameters for QPSK upstream transmission.

DIFFERENTIAL CODING.   Differential coding provides the following benefits:

- Resolves phase ambiguity at receiver (this may alternatively be done by observing the phase of the preamble).
- A phase slip in the receiver causes only a single symbol error. Without differential coding, a phase slip causes the remainder of the burst to be in error.

Differential coding has the following drawbacks:

- A loss in error rate performance occurs since a single symbol error is converted into two consecutive symbol errors. This is equivalent to a loss of about 0.25 dB at BER = $1E^{-6}$ uncoded. The loss may be increased with certain types of coding.
- The differential encoder in the subscriber transmitter converts information data bits into phase changes of 0, +90, 180, or –90 degrees between successive transmitted symbols according to Table 5.12. The differential decoder in the headend receiver does the inverse process, converting phase changes of: 0, +90, 180, or –90 degrees between successive received symbols into information data bits.

FRAME FORMAT.   Figure 5.22 is a diagram of the frame format. Each burst consists of a preamble and payload. There is a gap between bursts. See Table 5.11 for the definition.

TABLE 5.12 Differential Coding Matrix	Phase change between successive channel symbols (degrees)	I data	Q data
	0	0	0
	+90	0	1
	180	1	1
	–90	1	0

5.5.6 Zenith: Proposal for a Spread Spectrum Upstream Modem for CATV Networks

OVERVIEW. Zenith proposed a physical layer that utilizes vestigial side-band (VSB) modulation for the downstream modem and direct-sequence spread spectrum modulation for the upstream modem. Direct-sequence spread spectrum (DS-SS) modulation is efficient for providing code division multiple access (CDMA) in a shared network.

APPLICATION OF CDMA TO CABLE MODEMS. The goal of any good communication system design is to maximize the signal-to-noise ratio at the output of the receiver. A designer must attempt to minimize the impact of the noise sources while maximizing the signal power at the output of the receiver, subject to some constraint imposed on the system (e.g., maximum allowable transmitted power). The spread-spectrum system maximizes the received signal power and minimizes the impact of noise. Recall that ingress is a substantial problem associated with the upstream plant.

COMPARISON OF SPREAD SPECTRUM.
TDMA. The TDMA system spreads its energy over the frequency axis, but it is confined in the time domain. Each user occupies the full transmission channel for a slotted time period. Use of the channel is shared by time-multiplexing the channel resources among the users. The energy per bit is confined to a small time window, and therefore the power spectral density is large.

FDMA. The FDMA system spreads the signal energy out in time and confines it to a narrow frequency band. Hence, the available frequency allocation is divided up into narrow channels. In this case, the energy per bit is spread out in time but restricted in the frequency allocation.

Figure 5.22
Frame structure.

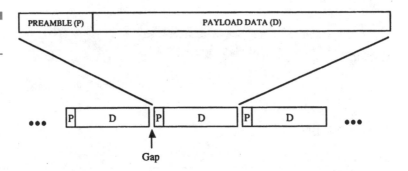

CDMA.   The CDMA system spread its energy out in both frequency and in time. That is, the available channel resources are shared simultaneously (in both time and frequency) among many users of the system. This is accomplished by the design of efficient signal waveforms that do not interfere with each other. With such a system, the energy per bit is spread out in both frequency and time, and, consequently, the limit on power spectral density yields a higher transmitted energy per bit.

Consequently, CDMA has an advantage over either TDMA or FDMA since it permits a higher energy per transmitted bit.

COMMUNICATING WITH ORTHOGONAL CODE VECTORS.   In the CDMA approach, the energy is spread out in both frequency and time. Each user's transmission is multiplied by a unique code vector from a set of orthogonal vectors. Data bits in pairs are mapped to the four-point QPSK constellation; the output is then multiplied by the code vector; hence, equalization is performed.

This approach eliminates interference from ingress peaks and avoids transmission in those bands. Therefore, if the ingress is a local ham operator, for example, there would be no interference to his transmissions caused by the CDMA system. Likewise, no interference would be produced on the international short-wave bands. Even with -10 dB signal/noise, it is possible to operate, although half of the possible code vectors must be dropped due to interference. The number of code vectors in use at any time is 64 minus the number that are dropped because they correlate with the interference sources. All of these are decoded simultaneously by a set of 64 filters.

SYSTEM ARCHITECTURE.   Communication within the upstream network requires access to a downstream modem so that the headend controller can manage the network access. This proposal uses a high-speed 16-VSB data modem that has been specified as a standard by both the ATSC (Doc. A/53, September 1995) and the ITU (COM 9-R 6-E, March 1995). The downstream modem provides a symbol clock frequency reference and frame timing to the upstream modem. It also allows a control channel to be implemented between the headend receiver and the upstream modems. Figure 5.23 illustrates the interaction between the downstream VSB demodulator and the upstream modulator.

SCDMA TRANSMITTER.   All clocks used by the transmitter are frequency-locked to the symbol clock reference supplied by the downstream modem. The carrier frequency used by the quadrature modulator is also frequency-locked to the symbol clock. The control data that is supplied to

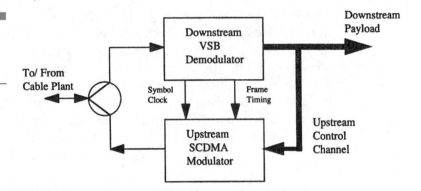

Figure 5.23
Interaction between
downstream and
upstream modems.

the upstream modem via the downstream link contains the set of code vectors used by the network, equalizer coefficients for the transmitter, and an output gain control for the transmitter.

The modem accepts a data input of 168 kbits/sec. These data bits are scrambled using a pseudo-noise (PN) sequence. This PN sequence is synchronized to the framing information supplied in the downstream and is uniquely defined for each of the code vectors. The purpose of this scrambling is to whiten the data pattern and reduce the cross-correlation between code vector channels. The scrambled data bits are then grouped in pairs and mapped to one of four points in a complex constellation.

The complex data symbol is then interpolated by L (where is L is the length of the code vectors) and passed through the signaling vector filter. This accomplishes the multiplication of the data symbol by the signaling vector. The output of the signaling vector filter is a data vector having L samples. The data vector is interpolated by two and then enters a T/2 preequalizer. The coefficients of the preequalizer are computed by the headend receiver and transmitted to the upstream modem through the control channel.

The preequalized data vector signal is then passed through a rootraised cosine filter with a bandwidth of 5.4 MHz. This filtered signal is then modulated with a quadrature modulator. The power output of the transmitter should be accurately controlled, and the output gain control is remotely adjustable through the control channel.

SCDMA RECEIVER. In the SCDMA receiver, illustrated in Fig. 5.24, received signal is applied to the receiver at an intermediate frequency. This signal is sampled by a single A/D converter and is digitally demodulated. The output of the demodulation process is supplied to the bank of correlation receivers. Each correlation receiver despreads the signal associated with a single code vector. The decoded data rate is 168 kbits/sec.

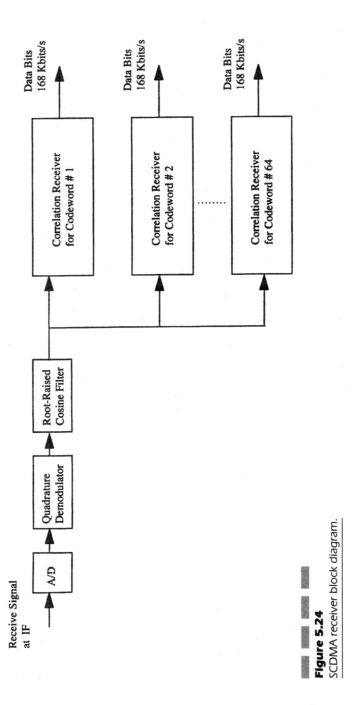

Figure 5.24
SCDMA receiver block diagram.

191

The correlated receiver block accepts the complex signal from the demodulation and performs a correlation of the signal against one code vector. This correlation is performed by:

- Signaling vector matched filter
- Derotated to correct the carrier phase error
- Decimated to sample the matched filter output at the correct sampling instant
- Sliced to the original four-point complex constellation—this provides two bits per symbol

The data bits are then multiplexed back to the original 168 kbits/sec and descrambled by the PN sequence generator.

The headend receiver must also monitor and correct the operation of the upstream transmitters. It executes an equalization algorithm to compute tap update values for the equalizer. Additionally, it measures the received signal power and supplies corrections to the transmitter.

UPSTREAM DATA PACKET FORMAT. Data transmitted in the upstream network is formatted into data packets. The data packet format is shown in Fig. 5.25. It has a two-byte preamble followed by the header and data payload. The data packet has eight bytes of Reed-Solomon parity.

FORWARD ERROR CORRECTION. The proposed system utilizes eight bytes of Reed-Solomon parity, yielding a $T = 4$ code. Other schemes are possible, including the addition of trellis-coded modulation.

5.5.7 General Instrument Corporation: QAM/QPSK

OVERVIEW. General Instrument proposed QAM for downstream and QPSK for upstream with a variety of flavors including: forward error control (FEC), a data stream framing structure, modulation, and the theoretical performance for proposed upstream signaling for HFC (hybrid fiber coax) systems.

Figure 5.25
Data packet format.

Preamble	Header	Data Payload	RS- Parity

QPSK/QAM FOR UPSTREAM. Table 5.13 describes the eight modes that are available in the proposed approach: four QPSK modes and four 16-QAM modes.

In general, it is expected that the majority of the frequencies in the 5- to 40-MHz band in the HFC systems will have sufficiently low noise plus interference and sufficiently high passband fidelity to allow C/N for successful 16-QAM operation. QPSK modulation is provided for those bands where the passband distortion and/or available C/N is not capable of supporting 16-QAM. QPSK has less throughput capacity than 16-QAM but is more robust and allows excellent service to be widely available at low risk. Different portions of the frequency band can be operated in different modes.

For both modulation types, three data rates are available (see Table 5.13):

1. A "slow" rate (133.33 ksym/sec)

2. Twice this rate (266.67 ksym/sec)

3. A "high" rate four times faster (533.33 ksym/sec)

The slower rates have less bandwidth and thus offer the potential of "fitting" in between severe ingress interferes situated such that the wideband mode will not operate. The slower rates also offer less latency without sacrificing throughput when many low-rate users vie for service. In the faster rates, ATM cells are bundled together into codewords and TDMA bursts containing multiple ATM cells. This offers advantages in increasing coding gain and lessening TDMA overhead per ATM cell (thus

TABLE 5.13

Characteristics of the Eight HFC System Upstream Transmission Modes

Mode	Modulation	Symbol rate (ksym/sec)	TDMA burst duration (microsec)	ATM cells per burst	Data rate (of the ATM cell payload)
1A	QPSK	133.33	2000	1	192 kbps
1B	QPSK	266.67	2000	2	384 kbps
1C	QPSK	533.33	2000	4	768 kbps
1D	QPSK	533.33	666	1	576 kbps
2A	16 QAM	133.33	2000	2	384 kbps
2B	16 QAM	266.67	2000	4	768 kbps
2C	16 QAM	533.33	2000	9	1.728 Mbps
2D	16 QAM	533.33	333	1	1.152 Mbps

providing higher throughput). Finally, for optimum (minimum) latency performance and for optimum cell delay variation (CDV) performance at both the slowest and fastest symbol rates, modes with a single ATM cell per TDMA burst are available.

Table 5.14 shows the performance of the various modes of the upstream signaling in the 5- to 40-MHz bandwidth of the HFC system. If all of the spectrum is available for 16-QAM operation, then allocating all-users mode 2C provides 86.4 Mbps upstream throughput. If low latency is a priority, then a user can be guaranteed latency no greater than the duration of each of its sourced ATM cells. This assumes the user is assigned burst times in the TDMA scheme evenly spread and matching or exceeding the user's rate. Operation in mode 2D will provide low latency performance, for example, while also providing 1.152 Mbps upstream for a user and yielding a composite system capability of 57.6 Mbps upstream for this mode. Note that additional latency is incurred for propagation delay and processing delay (not included in the table), a portion of which will depend on receiver implementation. If a constant bit rate (CBR) source desires a low CDV through the physical layer, modes 1A, 1D, or 2D will provide this low CDV, while sacrificing little throughput.

TABLE 5.14

HFC System Performance for Upstream Transmission Modes

Mode	Channel bandwidth (kHz)[*]	ATM cells per sec	Max cell delay variation	Max upstream PHY latency[**]	Data rate using all of the 5 to 40 MHz spectrum
1A	175	500	T_{gen}	T_{gen}	38.4 Mbps
1B	350	1000	$2T_{gen}$	$2T_{gen}$	38.4 Mbps
1C	700	2000	$4T_{gen}$	$4T_{gen}$	38.4 Mbps
1D	700	1500	T_{gen}	T_{gen}	28.8 Mbps
2A	175	1000	$2T_{gen}$	$2T_{gen}$	76.8 Mbps
2B	350	2000	$4T_{gen}$	$4T_{gen}$	76.8 Mbps
2C	700	4500	$9T_{gen}$	$9T_{gen}$	86.4 Mbps
2D	700	3000	T_{gen}	T_{gen}	57.6 Mbps

T_{gen} = avg. time between new ATM cells from a source = 1/(source ATM cell rate).

[*] Channel center frequencies are agile: 5 MHz to 40 MHz in 25-kHz steps.

[**] Propagation delay and processing delay in the receiver will add to these table values.

TRANSMISSION PROCESSING. Figure 5.26 shows the layered block diagram of the upstream transmission processing. The data entering the transmitter is formatted as individual packets intended for communication upstream through a HFC plant in a burst TDMA/FDMA system. In order to support the transmission of this bursty data through the HFC plant, appropriate data conditioning, modulation, acquisition, demodulation, and restoration must be performed. Depending on which of the eight modulation modes are selected, the ATM cell stream is grouped into information Blocks of 1, 2, or 9 53-byte ATM cells. To support automatic repeat request (ARQ) at a higher layer, an identification byte is added to each information block. The resulting information block is randomized by modulo-2 addition with a data pattern to increase the occurrence of near-equally likely symbols and symbol transitions in the data stream. At this point, the FEC encoding is performed wherein the randomized ATM cells, with prepended ID byte, are encoded with a Reed-Solomon code over GF(256) (i.e., 8 bits per Reed-Solomon symbol). Each mode uses its own particular Reed-Solomon code. The TDMA bursts contain one or two codewords, depending on the selected mode.

Figure 5.26

The upstream transmission processing block diagram.

	ATM Cell Stream Input to Transmitter
	⇓
Block the data	Separate ATM cells into information blocks: 1 cell, 2 cells, or 9 cells per block
	⇓
Append ID byte	Append ID byte to each information block
	⇓
Randomize	Randomize each information block
	⇓
FEC encode	FEC encode (Reed-Solomon) each information block (yielding a codeword per information block except in Mode 2C, where the two codewords are used)
	⇓
Interleave	Interleave the symbols of the two codewords in the time slot (for modes with two codewords per time slot: 1C, 2B, 2C)
	⇓
Symbol map prepend unique word	Symbol map the map stream into modulator symbols
	⇓
	Prepend unique word symbols
	⇓
Filter	Filter symbol stream for spectral shaping
	⇓
Modulate	Modulate at precise times (QPSK for Modes 1A, 1B, 1C, 1D; 16-QAM for Modes 2A, 2B, 2C, 2D)
	⇓
	Output RF waveform bursts

The resulting data stream for each TDMA burst is augmented with a prepended unique word sequence that is crucial to the acquisition and synchronization processes at the receiver. This data stream is prepared (symbol-mapped) for either QPSK or 16-QAM modulation depending on mode (and there is also the option for differentially encoded modulation). This sequence of symbols is filtered (pulse-shaped) for 31.25 percent square-root-raised-cosine spectral shaping and QPSK or 16-QAM modulated and transmitted at the prescribed time and carrier frequency allocated in the TDMA/FDMA system. The carrier frequency assignments are agile, with center frequencies detectable from 5 MHz to 40 MHz in 25-kHz steps; this allows the system to assign channels in a fashion to step around particularly bad interference entering the HFC system.

Figure 5.27 shows the structure of the transmitted burst for each mode, while Tables 5.15 and 5.16 provide the particular timing of each segment of the burst structure for each mode.

The receiver is tuned to the appropriate frequency and continually scans the received signal at this frequency for the occurrence of bursts of data. A demodulator in the receiver will process all detected bursted data appearing in its tuned channel (i.e., all the time slots in a given channel are processed by a demodulator assigned to that frequency; the bursted data in these time slots will originate at different transmitters in general). After detecting the presence of a burst of data via special filtering tuned to the unique word pattern, the demodulator will further process the received waveform to determine the end of the unique word sequence and the beginning of the data field, perform precise carrier acquisition, and begin carrier tracking to facilitate demodulation and detection of the data. The demodulator will perform estimation of the time of arrival of the burst relative to a reference timing signal if it is provided, and it will estimate the power of the received burst and an estimate of signal-to-noise ratio (with good resolution in the range of SNR usable for signaling). In the absence of a detected burst, the demodulator estimates the power in the tuned channel (so that a noise power estimate for that channel is available). Further processing at the receiver consists of inverting the differential encoding induced upon the data portion of the cell (if the differential encoding option is selected). Then the data is deinterleaved and passed to the Reed-Solomon decoder, and finally the randomization is wiped off the data.

PREPEND INFORMATION BLOCK ID BYTE. ATM cells are prepared in the client unit ahead of the transmitter. These cells may be for network messaging, or they may contain data. For data transmission, the

Figure 5.27
Structure of bursts: a,
for each QPSK mode;
b, for each 16-QAM
mode.

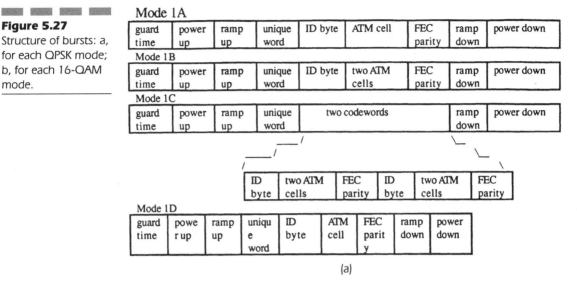

Mode 1A

| guard time | power up | ramp up | unique word | ID byte | ATM cell | FEC parity | ramp down | power down |

Mode 1B

| guard time | power up | ramp up | unique word | ID byte | two ATM cells | FEC parity | ramp down | power down |

Mode 1C

| guard time | power up | ramp up | unique word | two codewords | | | ramp down | power down |

| ID byte | two ATM cells | FEC parity | ID byte | two ATM cells | FEC parity |

Mode 1D

| guard time | power up | ramp up | unique word | ID byte | ATM cell | FEC parity | ramp down | power down |

(a)

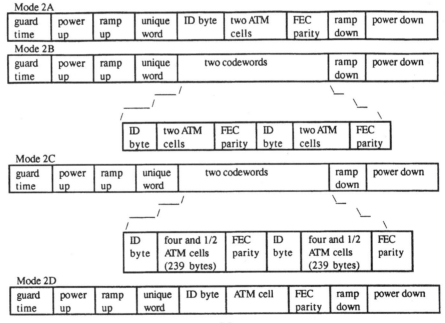

Mode 2A

| guard time | power up | ramp up | unique word | ID byte | two ATM cells | FEC parity | ramp down | power down |

Mode 2B

| guard time | power up | ramp up | unique word | two codewords | | | ramp down | power down |

| ID byte | two ATM cells | FEC parity | ID byte | two ATM cells | FEC parity |

Mode 2C

| guard time | power up | ramp up | unique word | two codewords | | | ramp down | power down |

| ID byte | four and 1/2 ATM cells (239 bytes) | FEC parity | ID byte | four and 1/2 ATM cells (239 bytes) | FEC parity |

Mode 2D

| guard time | power up | ramp up | unique word | ID byte | ATM cell | FEC parity | ramp down | power down |

(b)

TABLE 5.15

Structure of Burst for Each QPSK Mode

Mode	1A		1B		1C		1D	
Symbol rate (ksym/sec)	133.33		266.67		533.33		533.33	
Symbol duration (μsec)	7.5		3.75		1.875		1.875	
Bits/sym	2		2		2		2	
Frame segment		**time (μsec)**		**time (μsec)**		**time (μsec)**		**time (μsec)**
Preamble	16 sym	120	24 sym	90	28 sym	52.5	28 sym	52.5
Data	432 bits	1620	856 bits	1605	1712 bits	1605	432 bits	405
Parity	48 bits	180	128 bits	240	320 bits	300	128 bits	120
Ramp up/down	3.5 sym	52.5	5 sym	37.5	5 sym	18.75	5 sym	18.75
Power up/down	12 μsec	12	12 μsec	12	12 μsec	12	12 μsec	12
Guard time	12 μsec	12	12 μsec	12	10 μsec	10	10 μsec	10
Total (μsec)		1996.5		1996.5		1998.2		618.25
Unallocated (μsec)		3.5		3.5		1.75		48.4

cells are arranged in information blocks of 1, 2, or 9 ATM cells depending on mode, and this data is prepended with an identification byte. An information block will comprise the information bits of a codeword (or two codewords, in the case of mode 2C).

RANDOMIZATION. The randomization consists of modulo-2 adding the information bits of each codeword with a known bit pattern. This is to increase the likelihood of using each of the possible modulation symbols a nearly equal number of times and of approaching equality for the transition probabilities between the symbols (as opposed to a dominance of zeros, for example, wherein the modulated waveform would contain a very disproportionate use of some symbols and particular transitions between symbols, thus allowing the possibility of atypical performance not apparent when random patterns are transmitted).

TABLE 5.16

Structure of Burst
for Each 16-QAM
Mode

Mode	2A		2B		2C		2D	
Symbol rate (ksym/sec)	133.33		266.67		533.33		533.33	
Symbol duration (μsec)	7.5		3.75		1.875		1.875	
Bits/sym	4		4		4		4	
Frame segment		**time (μsec)**		**time (μsec)**		**time (μsec)**		**time (μsec)**
Preamble	18 sym	135	24 sym	90	24 sym	45	24 sym	45
Data	856 bits	1605	1712 bits	1605	3824 bits	1792.5	432 bits	202.5
Parity	96 bits	180	256 bits	240	256 bits	120	96 bits	45
Ramp up/down	3.5 sym	52.5	5 sym	37.5	5 sym	18.75	5 sym	18.75
Power up/down	12 μsec	12	12 μsec	12	12 μsec	12	12 μsec	12
Guard time	12 μsec	12	12 μsec	12	10 μsec	10	10 μsec	10
Total (μsec)		1996.5		1996.5		1998.2		333.25
Unallocated (μsec)		3.5		3.5		1.75		0.083

The same randomization pattern is used for each codeword, but rather than store the pattern, it is generated with a feedback shift register.

FORWARD ERROR CONTROL (FEC) REED-SOLOMON CODING. The forward error control consists of four different Reed-Solomon codes, each over GF(256) (i.e., eight bits per Reed-Solomon symbol). Two of the codes are shortened by differing amounts for use in different modes, to give a total of seven different coding schemes among the eight modes.

Table 5.17 shows the code used for each of the eight modes.

Table 5.18 summarizes the theoretical performance of the modulation and FEC in AWGN for the eight modes (nondifferential modulation option is assumed).

Table 5.19 summarizes the theoretical performance of the modulation and FEC in the presence of impulse noise for the eight modes (again, the nondifferential modulation option is assumed).

TABLE 5.17

Reed-Solomon FEC
for Each Mode

Mode	RS code [GF(256)]	Error correction	Codewords per burst	ATM cells per codeword
1A	(60,54)	3 bytes	1	1
1B	(123,107)	8 bytes	1	2
1C	(127,107)	10 bytes	2	2
1D	(70,54)	8 bytes	1	1
2A	(119,107)	6 bytes	1	2
2B	(123,107)	8 bytes	2	2
2C	(255,239)	8 bytes	2	4 1/2
2D	(66,54)	6 bytes	1	1

Table 5.20 shows the coding gain and bandwidth efficiency of the modulation and FEC for each mode. In the most bandwidth-efficient mode, mode 2C, with 2.5 bits/sec/Hz, there is a 7.2-dB coding gain. In mode 2B, with 2.2 bits/sec/Hz, there is a 7.8-dB coding gain. The bandwidth efficiency is calculated using only the 48 payload bytes of each ATM cell of the mode and as such includes the penalty of the 5 ATM overhead bytes in each cell as well as the TDMA burst overhead, the frequency guard band, and the parity bits of the FEC. In conventional communication schemes where bit error rate (BER) is the appropriate performance metric of the link, coding gain is calculated as the difference in signal-to-noise

TABLE 5.18

Theoretical Performance of Modulation and FEC in AWGN for Each Mode

Mode	Modulation	RS code [GF(256)]	C/N required for $<10^{-6}$ codeword error rate	C/N required for $<10^{-10}$ codeword error rate
1A	QPSK	(60,54)	11.2 dB	12.4 dB
1B	QPSK	(123,107)	9.8 dB	10.7 dB
1C	QPSK	(127,107)	9.4 dB	10.2 dB
1D	QPSK	(70,54)	9.3 dB	10.2 dB
2A	16 QAM	(119,107)	15.9 dB	17.1 dB
2B	16 QAM	(123,107)	15.2 dB	16.3 dB
2C	16 QAM	(255,239)	15.9 dB	16.9 dB
2D	16 QAM	(66,54)	15.2 dB	16.6 dB

TABLE 5.19

Theoretical Performance of Modulation and FEC in Impulse Noise for Each Mode

Mode	Modulation	RS code [GF(256)]	Impulse rate for $<10^{-6}$ codeword error rate	Impulse rate for $<10^{-10}$ codeword error rate
1A	QPSK	(60,54)	<37 imp/sec	<4 imp/sec
1B	QPSK	(123,107)	<528 imp/sec	<160 imp/sec
1C	QPSK	(127,107)	<1669 imp/sec	<640 imp/sec
1D	QPSK	(70,54)	<1867 imp/sec	<640 imp/sec
2A	16 QAM	(119,107)	<266 imp/sec	<66 imp/sec
2B	16 QAM	(123,107)	<989 imp/sec	<333 imp/sec
2C	16 QAM	(255,239)	<1066 imp/sec	<320 imp/sec
2D	16 QAM	(66,54)	<1978 imp/sec	<488 imp/sec

ratio (Eb/No or C/N, etc.) required to achieve the goal BER. In our application, however, where ATM cells are the basic delivery vehicle, the appropriate performance metric is the ATM cell error rate. As such, Table 5.19 shows the difference in C/N required to achieve 10^{-10} ATM cell error rate with the selected modulation and FEC as compared to the required C/N to achieve the same ATM cell error rate with the modulation (QPSK for modes 1× and 16-QAM for modes 2×), but without the FEC. Schemes that show the coding gain based on a BER equivalence are overestimating the coding gain by a large amount (>2 dB) in this ATM cell delivery application.

TABLE 5.20

Coding Gain of Modulation and FEC in AWGN for Each Mode at 10^{-10} ATM Cell Error Rate

Mode	Modulation	RS code [GF(256)]	Bandwidth efficiency (bits/sec/Hz)	Coding gain for $<10^{-10}$ ATM cell error rate
1A	QPSK	(60,54)	1.1	1.8 dB
1B	QPSK	(123,107)	1.1	3.5 dB
1C	QPSK	(127,107)	1.1	4.0 dB
1D	QPSK	(70,54)	0.8	4.0 dB
2A	16 QAM	(119,107)	2.2	7.0 dB
2B	16 QAM	(123,107)	2.2	7.8 dB
2C	16 QAM	(255,239)	2.5	7.2 dB
2D	16 QAM	(66,54)	1.6	7.2 dB

INTERLEAVING. Interleaving is performed in three of the eight modes, as shown in Table 5.21. The interleaving is a simple byte-wide $2 \times N$ block interleaver. The interleaving is performed in each mode where two codewords are contained in each TDMA burst (as opposed to the five modes with a single codeword per burst). The interleaving provides additional mitigation against burst interference.

5.5.8 Lucent Technologies: Physical Layer Protocol for HFC Networks in Support of ATM and STM Integration

OVERVIEW. Lucent proposed a physical layer for HFC networks in support of ATM and STM integration: 7.68 Mbps downstream; QPSK in 6 MHz or 16 QAM in 3 MHz. Framing includes a 125-microsecond frame, 2-msec multiframe, and 8-msec masterframe. On the upstream side, the arrangement is QPSK to provide 2.56 MHz in a 1.8-MHz band. Upstream utilizes a 2-millisecond frame size to minimize delay for telephony with an 8-msec masterframe. The boundary between STM and ATM in the frame is variable. Downstream, the frame size is 960 bits over 125 microseconds. Upstream, the frame is 5120 bits over 2 milliseconds. The STM burst size upstream is 16 bytes for a given user. In the burst, the guard time is 10 bits, the preamble is 28 bits, followed by the payload, with another 10 bits of guard time at the end.

TABLE 5.21

Byte-Wide Interleaving for Each Mode

Mode	Interleaving	Size
1A	No	—
1B	No	—
1C	Yes	2×127
1D	No	—
2A	No	—
2B	Yes	2×123
2C	Yes	2×255
2D	No	—

PROPOSED HFC ARCHITECTURE. The HFC network consists of a NIU (network interface unit), a FN (fiber node), an ODS (optical distribution shelf), and headend equipment. The NIU terminates the coax interface from the network and provides access to various services to end users. The FN and ODS perform simple medium conversion between lightwave and RF signals and do not terminate any digital signals. The headend equipment consists of video, data, and/or telephony equipment interfacing to the HFC network on the end-user side and to the respective backbone networks on the network side. The physical layer protocol proposed specifies how the digital signals generated/terminated by the NIU and headend equipment should be modulated and formatted.

HFC EXPECTED SERVICES. It is expected that HFC networks will provide end users with access to many kinds of services using three transport modes: ATM, STM, and MPEG2 transport.

ATM. Expected end-user services include:

- ATM data service
- Video services
- Video-service-related data only
- Multimedia service

STM. Expected end-user services include:

- VF service, ISDN service
- Leased-line service
- HO-based video telephony service
- Multimedia service

MPEG2.

- Video service (video content only)
- Multimedia service

In order to meet various service needs, it is anticipated that a high-speed channel (30 Mbps or above) and a separate medium-speed channel (less than 10 Mbps) will be used in the downstream direction. The former may be used by video NIUs to carry video content in MPEG2 transport streams, and the latter may be used by video, data, and telephony NIUs to carry various ATM and STM traffic in an integrated fashion. In the upstream direction, it is anticipated that one medium-speed channel (less than 10 Mbps) will be used by video, data, and telephony NIUs to carry various ATM and STM traffic in an integrated fashion. Thus, typical HFC video NIUs will need two RF demodulators and one RF modulator, whereas typical HFC data and telephony NIUs will need one RF demodulator and one RF modulator.

Figure 5.28 shows expected near-term HFC NIU configurations. Note that if the RF channels do not carry ATM and STM traffic in an integrated fashion, NIUs providing multimedia applications requiring both ATM and STM traffic at the same time will be very costly since they will need two RF demodulators and two RF modulators to handle ATM and STM traffic separately. Furthermore, integration of ATM and STM traffic on the same RF channel can provide more efficient use of the RF spectrum.

Figure 5.28
Expected near-term HFC NIU configurations.

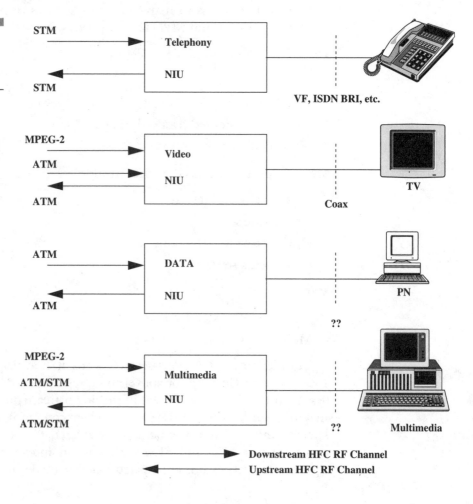

PROPOSED DOWNSTREAM/ UPSTREAM PHY.   It is proposed that TDM (time division multiplexing) and FDM (frequency division multiplexing) be used in the downstream direction and TDMA (time division multiple access) and FDM be used in the upstream direction. A medium-speed downstream channel running at 7.68 Mbps and a medium-speed upstream channel running at 2.56 Mbps are needed to carry both ATM and STM traffic in an integrated fashion. The 7.68-Mbps downstream channel is hierarchically organized as 125-µsec frames, 2-msec multiframes, and 8-msec masterframes; whereas the 2.56-Mbps upstream channel is organized as 2-msec frames and 8-msec masterframes. The 7.68-Mbps downstream channel is modulated either via QPSK on a 6-MHz band or via 16 QAM on a 3-MHz band. The 2.56-Mbps upstream channel is modulated via QPSK on a 1.8-MHz band. Both channels have been cost effectively implemented with today's technology.

PHYSICAL MEDIUM DEPENDENT SUBLAYER.   In this section, QPSK and 16 QAM modulation techniques for the 7.68-Mbps downstream channel and QPSK modulation technique for the 2.56-Mbps upstream channel are discussed. Tables 5.22 through 5.26 provide a summary of the modulation techniques. They are self-explanatory.

TABLE 5.22

Specifications for Downstream QPSK Modulation

Modulation	Differentially encoded QPSK
Signal element coding	Gray coding
Symbol rate	3.84 M symbols per second
Data rate	7.68 Mbps
Channel shaping	Square root raised cosine with 50% rolloff
Carrier frequency spacing	6 MHz
Carrier frequency accuracy	20 ppm

TABLE 5.23

Specification for Upstream QPSK Modulation

Carrier frequency range	50 MHz to 1 GHz range

TABLE 5.24

Specification for Downstream 16 QAM Modulation

Modulation	16 QAM
Signal element coding	Gray coding
Symbol rate	1.92 M symbols per second
Data rate	7.68 Mbps
Channel shaping	Square root raised cosine with 50% rolloff
Carrier frequency spacing	3 MHz
Carrier frequency accuracy	20 ppm
Carrier frequency range	50 MHz to 1 GHz

TABLE 5.25

Specification for Upstream QPSK Modulation

Carrier frequency range	5 to 40 MHz
Burst preamble	14 symbols: 1 1 11 1 1 00 00 1 1 00 00 0000 00 00 00 00
Guard time between adjacent bursts	10 symbols

TABLE 5.26

Specification for Upstream QPSK Modulation

Modulation	Differentially encoded burstmode QPSK
Signal element coding	Gray coding
Symbol rate	1.28 M symbols per second
Data rate	2.56 Mbps
Channel shapping	Square root raised cosine with 40% rolloff
Carrier frequency spacing	1.8 MHz
Carrier frequency accuracy	20 ppm when downstream symbol clock is of stratum 3 (4.6 ppm) quality

TRANSMISSION CONVERGENCE PROTOCOL.   In this section, the transmission format is specified for:

- 7.68-Mbps downstream and 2.56-Mbps upstream channels
- Downstream/upstream channel synchronization
- Upstream burst boundary alignment
- Upstream transmit power control methods

Transmission Format for 7.8 Mbps.   The 7.68-Mbps downstream channel on each carrier frequency is hierarchically organized as frame, multiframe, and masterframe, as shown in Fig. 5.29. Each downstream frame shall consist of 960 bits over 125 μsec. Each downstream multiframe shall consist of 16 downstream frames. Each downstream masterframe shall consist of 4 downstream multiframes.

STM traffic will be mapped at the 125-μsec-long frame level (e.g., 64 kbps VF service occupies 8 bits per frame). ATM traffic will be mapped at the 2-msec-long multiframe. Each HFC telephony access system should minimize round-trip signal delay so that echo effects are not objectionable. Bellcore's TR-303 integrated digital loop carrier system requirements specify a 1.25-msec round-trip signal delay budget, and any HFC telephony access system implementation should approach this budget as close as possible. The choice of a 125-μsec-long downstream frame structure is to minimize STM telephony time slot delay at the headend in the downstream direction.

Framing bit field in downstream frame.   The framing bit field in downstream frames contains eight bits to delineate downstream frame/multiframe/masterframe as shown in Fig. 5.29.

Figure 5.29
Format of 7.68-Mbps downstream channel.

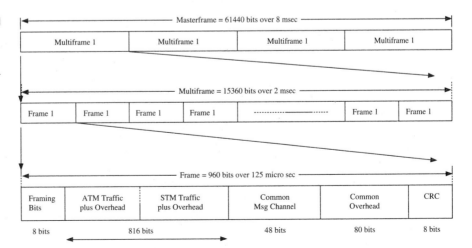

ATM/STM traffic field in downstream frame. A total of 816 = 8 × 102 bits shall be allocated to transport 6.528 Mbps worth of ATM/STM user traffic and associated overhead in the downstream frame. This is shown in Fig. 5.29. Associated overhead for ATM traffic includes medium-access control feedback information, and associated overhead for STM traffic includes bit-oriented signaling.

This field can be partitioned between ATM and STM traffic by a movable boundary at the MAC layer. All ATM and STM configurations will be allowed.

Common message channel field in downstream frame. Forty-eight bits shall be allocated to provide a 384-kbps message channel in the downstream frame as shown in Fig. 5.29. All NIUs served shall receive the same common message channel, regardless of which 7.68-Mbps downstream channel they tune to. The 384-kbps message channel is used to carry OAM&P messages and STM call connection control messages.

Common overhead field in downstream frame. Eighty bits shall be allocated to carry overhead for common message channel contention control, upstream transmit power control, upstream frequency plan, initial ranging offset, NIU low power operation control, privacy control, and some reserved bits for future use. All NIUs served by the same FN shall receive the same common overhead, regardless of which 7.68-Mbps downstream channel they tune to.

Cyclic redundancy check field in downstream frame. Eight bits shall be allocated to carry a cyclic redundancy check in the downstream frame as shown in Fig. 5.29. The cyclic redundancy check is used by the NIU to monitor the downstream transmission performance.

Transmission Format for 2.56-Mbps Upstream. The 2.56-Mbps upstream channel on each carrier frequency is as illustrated in Fig. 5.30.

Upstream transmission format. The 2.56-Mbps upstream channel shall be hierarchically organized as frame and masterframe as shown in Fig. 5.30. Each upstream frame shall consist of 5120 bits over 2 msec. Each upstream masterframe shall consist of 4 upstream frames. All burst slots in the upstream frame may be occupied at one point in time. Even when all burst slots are occupied, there will be a 20-bit (10-symbol) guard time placed between bursts where no upstream bit transmission is allowed as described below. Therefore, the 5120 bits above should be interpreted as virtual bits including guard-time bits.

The HFC telephony access system should minimize round-trip signal delay so that echo effects are not objectionable. The TR-303 integrated digital loop carrier system requirements specify a 1.25-msec round-trip signal

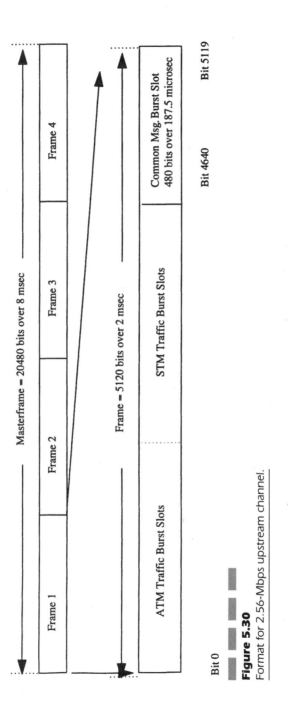

Figure 5.30
Format for 2.56-Mbps upstream channel.

209

delay budget, and any HFC telephony access system implementation should approach this budget as close as possible. The choice of a 2-msec-long upstream frame structure is to minimize STM telephony signal delay at the NIU in the upstream direction while providing a large enough upstream frame size to work efficiently in a TDMA environment.

Upstream frame organization. The upstream frame shall consist of one common message burst slot and various ATM/STM traffic burst slots as shown in Fig. 5.30. The common message burst slot shall always start at the 4640th bit. ATM/STM traffic bursts will be transmitted between the zero bit and the 4639th bit. The common message burst slot is used to send OAM&P messages, STM call connection control messages, and ranging burst, and it is shared by the NIUs on a contention basis. ATM/STM burst slots can be partitioned between ATM and STM traffic by a movable boundary at the MAC layer. All ATM and STM configurations will be allowed.

Generic burst format. All bursts shall consist of 10-bit-long guard time placed on both ends of each burst, a 28-bit-long preamble, and variable-length burst contents as shown in Fig. 5.31.

During the guard time, the NIU will not transmit any bits. However, some low-level residual RF energy resulting from RF transmitter power-up/power-down may exist during the guard time. The total guard time between back-to-back bursts is 20 bits long. The guard time accounts for transmitter turn-on time, transmitter turn-off time, ranging accuracy, temperature variation, and so on.

Downstream/Upstream Channel Synchronization. The headend will be timed from a Stratum 3 (4.6 ppm) or better clock available within the building. The following requirement specifies how the downstream channels generated by the headend and upstream channels generated by the NIUs are synchronized.

All downstream channels served by the same FN shall be frame/multi-frame/masterframe-synchronous with each other. All upstream channels served by the same FN shall be loop-timed from the downstream symbol clock and shall be frame/masterframe-synchronous with each other. This will allow NIUs tuning to different 7.68-Mbps downstream channels to share the same 2.56-Mbps upstream channel.

Figure 5.31
Generic burst format.

Guard Time	Preamble	Burst Contents	Guard Time
10 bits	28 bits	Variable Length	10 bits

Upstream Burst Boundary Alignment. Each NIU needs to control its upstream transmit timing so that the adjacent bursts sent by different NIUs cannot collide with each other as they arrive at the headend. Ranging is used to align upstream burst boundaries upon NIU power-up or after experiencing problems (e.g., loss of signal).

Upstream Transmit Power Control. Each NIU needs to control its upstream transmit power on a per-RF channel basis so that the headend can receive bursts from all NIUs at the same nominal power level for best data recovery.

5.5.9 Microunity: Proposal for Multirate QPSK/QAM PHY Protocols

INTRODUCTION. Microunity proposed multirate QPSK/QAM PHY protocols and described the rationale for selecting the associated parameters and characteristics as follows:

- Multiple rates and modulation are ideal applications for cable modems.
- Cable modem products will become highly integrated (e.g., transmit, receive, and possibly video services will be provided on same chip).
- Integrated implementations can be simplified by using a single clock source and rates related by simple integer ratios.
- Upstream rates will be set at 1/2, 1/4, 1/8 of downstream rate.
- 81/6 = 13.5 (the basic clock rate of CCIR-601 digital TV std)
- Less aggressive pulse shaping (relative to DAVIC) decreases implementation cost in several ways.

DOWNSTREAM CHANNEL. Table 5.27 summarizes the parameters of the downstream PHY and framing formats:

DOWNSTREAM PHY FRAMING AND FEC. Figure 5.32 illustrates the downstream PHY framing and FEC.

The header has a separate and short FEC; this makes it easy to discover that a packet is for someone else and should be ignored. The payload has a larger 240, 224 Reed-Solomon FEC, which takes longer to apply. Data rates on the upstream are 1/8, 1/4, or 1/2 of the downstream rate. This can make implementation easier. The less aggressive pulse-shaping relative to DAVIC decreases implementation cost in a number of ways. A summary of the field identification follows:

TABLE 5.27

Downstream
64/256 QAM
Characteristics

Parameter	Value	Notes
Modulation	64-QAM or 256-QAM	
Symbol rate	5, 062, 500 symbols/s	
Symbol mapping	Same as DAVIC for HFC	Differential H.O. 2 bits rotated Gray code w/in quad
Excess bandwidth	18%	
Framing	16-bit field/header; symbol-frame alignment	
Scrambling	none	Requires higher layer to scramble data
FEC	see Fig. 5.32	
Equalization	No training sequence; must cover echoes for which $T/(-A-1.6) < 75$	T = echo delay in ns A = echo power in dBc
Channel spacing	6 MHz	
Carrier frequency	54–750 MHz	
Carrier offset	<100 Khz	
Transmit level	–5 dB relative to NTSC carriers	
OOB spurious	–60 dBc	

Frame synch word. 16 bits, value is arbitrary. Frame synch and payload are always symbol-aligned. For 64-QAM, this is done by stuffing 0, 2, or 4 bits after the header FEC field, as header length mod 3 is 0, 2, or 1.

Header FEC. RS(12, 8) code over GF(256); $t = 2$; does not cover frame synch word.

Data FEC. RS(240, 224) code over GF(256); $t = 8$. Last block of payload shortened to RS(n+16, n), where n = length of payload mod 224.

Figure 5.32
Downstream PHY
framing and FEC.

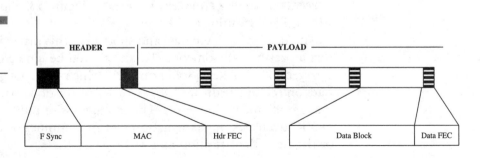

UPSTREAM CHANNEL. Table 5.28 summarizes the parameters of the upstream PHY and framing formats.

UPSTREAM PHY: FRAMING AND FEC, DATA BLOCKS. See Fig. 5.33.

Single FEC. RS(240, 224) code over GF(256). Last block of payload shortened to RS(n+16, n), where n = length of payload mod 224.

Upstream PHY Framing and FEC, Short Blocks. It has been proposed to use an FEC scheme run at a lower rate so that the task can be done in software. The short FEC on the header is decoded on all packets, but full FEC is performed only on data addressed to that remote. See Fig. 5.34.

Here is a definition of the fields:

- Preamble/synch pattern and length same as for data blocks
- Single FEC: RS(n+4, n) code over GF(256)
- Same definitions for GF(256) and generator polynomial

5.5.10 IBM: Frequency-Agile Multimode for Upstream Transmission in HFC Systems

OVERVIEW. IBM proposed the use of frequency-agile multimode (FAMM) modems for upstream transmission in HFC systems. Channel characteristics for upstream transmission can differ significantly among

TABLE 5.28

Upstream QPSK or 16 QAM Characteristics

Parameter	Value	Notes
Modulation	QPSK R 16-QAM	
Symbol rate	632, 812.5, 1,265,625, or 2,531,250 symbols/s	1/8, 1/4, or 1/2 of downstream rate
Symbol mapping	Same as downstream	
Excess bandwidth	35%	
Scrambling	none	Requires higher layer to scramble data
Carrier frequency	5–40 MHz	
Carrier offset	<100 kHz	
Transmit level	34–55 dBmV in steps of 2 dB or less	From modem
OOB spurious	< –50 dBc	

Figure 5.33
Upstream framing
structure.

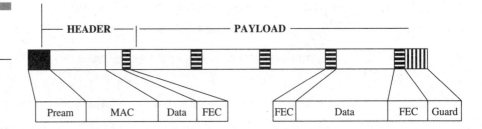

Figure 5.34
Framing and FEC,
short blocks.

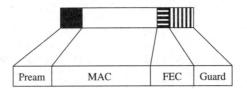

HFC plants and for a given plant across the available upstream signal bandwidth. A certain amount of flexibility in the choice of modulation parameters for upstream transmission and provisions in the MAC protocols for adapting these parameters to a priori unknown channel characteristics are needed to achieve the best possible utilization of the upstream channel resources.

FAMM modems are proposed to support data rates between 0.5 Mbit/s and 16 Mbit/s and four choices of spectral modulation efficiency: 2, 4, 6, and 8 bit/s/baud. The carrier frequency can be selected in integer multiples of one-half the bandwidth required for the smallest modulation rate. To uncoded QPSK signaling mode at the higher modulation rates and/or higher spectral modulation efficiencies, simple forms of trellis-coded modulation (TCM) are proposed, providing a 3-dB coding gain compared to their uncoded counterparts, with minimal decoding complexity.

FAMM MODEM CHARACTERISTICS. The FAMM modem operates at a modulation rate of baud. The pulse-shaping at the transmitter can be realized with a 20 percent excess bandwidth square root raised cosine filter. With 20 percent excess bandwidth, each channel occupies a bandwidth of Hz. The carrier frequency can be selected with an appropriate resolution (e.g., in integer multiples of one-half the bandwidth required for the smallest modulation rate). In addition to the uncoded QPSK signaling mode, to achieve more robustness against noise, simple forms of trellis-coded modulation (TCM) are proposed:

1. 8-PSK

2. 32-QAM

An upstream frame begins with a preamble and ends with a tail sequence, as shown in Fig. 5.35. Between these modem-specific sequences, which should be as short as possible, the frame body, comprising fields for protocol and user data, is transmitted.

PREAMBLE. A preamble is inserted at the beginning, and it ends with a tail sequence. The frame body field is used for management messages and user data. There are two preamble sequences:

1. *A long preamble used for channel probing and equalization.* It begins with a tone. After a period T, the phase of the tone is reversed. In the receiver, the phase reversal can be detected even in the presence of strong distortion that may be encountered without equalization. The phase reversal serves as a time marker, indicating the beginning of a subsequently transmitted sequence of equalizer training symbols.

2. *A short preamble used for fast resynchronization.* It is a short sequence of modulation symbols chosen such that a receiver can establish timing phase and carrier phase with suitable processing methods.

FRAME BODY. In the frame body, protocol and user data are transmitted. The data are scrambled.

CRC. For error detection, a cyclic redundancy check (CRC) field is provided.

TAIL. When TCM is used, a tail sequence must be appended to the frame body to reach a known code state before the end of signal transmission.

MODULATION OPTIONS OF FAMM MODEM. Table 5.29 shows the proposed set of modulation options (some are optional). Data rates between 0.5 Mbit/s–16 Mbit/s are supported. Higher data rates are also possible. Four different spectral efficiencies are provided:

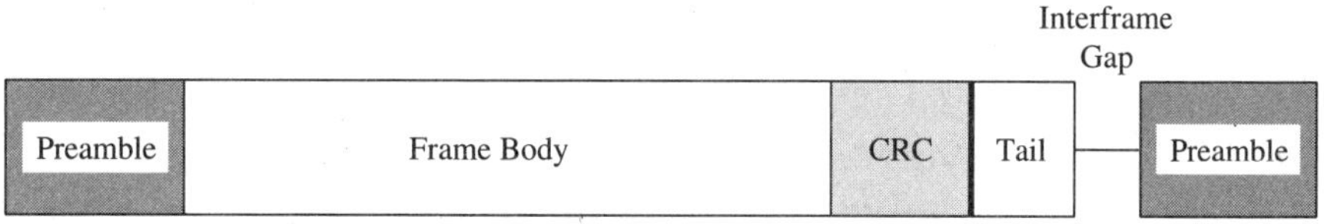

Figure 5.35
Upstream frame format.

TABLE 5.29

Modulation and
Data Rate (Mbaud)

	0.25	0.5	1	2	4
QPSK 8, PSK TCM (2 bit/s/baud)	0.5 Mbit/s	1 Mbit/s	2 Mbit/s	4 Mbit/s	8 Mbit/s
32-QAM TCM (4 bit/s/baud)	1 Mbit/s	2 Mbit/s	4 Mbit/s	8 Mbit/s	16 Mbit/s
128-QAM TCM (6 bit/s/baud)	1.5 Mbit/s	3 Mbit/s	6 Mbit/s	12 Mbit/s	
512-QAM TCM (8 bit/s/baud)	2 Mbit/s	4 Mbit/s	8 Mbit/s	16 Mbit/s	

Shaded areas indicate optional modulation modes.

- 2 bit/s/baud
- 4 bit/s/baud
- 6 bit/s/baud
- 8 bit/s/baud

The data rates, signaling schemes, and modulation rates contained in the shaded portions of Table 5.29 indicate the set of optional modulation modes. The rest of the table indicates the minimum required set of modulation modes.

FAMM TRANSMITTER. The block diagram of the transmitter is illustrated in Fig. 5.36. The MAC at the end-user node provides the data and the control signal, which determines the mode of the transmitter. The control signal contains the following information: type of preamble to be generated, modulation rate, number of bits per modulation interval, carrier frequency, and coefficients of transmit equalizer if necessary. Given this information, the transmit control block sets up the necessary modes for the transmitter.

The digital pulse-shaping filter is sampled at twice the modulation rate. This filter has square root raised cosine characteristics with 20 percent excess bandwidth. After the pulse-shaping filter and transmit equalizer, an interpolation filter is required so that the input to the D/A converter is at a constant rate for different modulation options.

FAMM RECEIVER—HIGH LEVEL VIEW. The receiver elements may be arranged as shown in Fig. 5.37.

Upstream signals arriving at the headend at various carrier frequencies are converted by RF receiver frontend units into digital complex base-

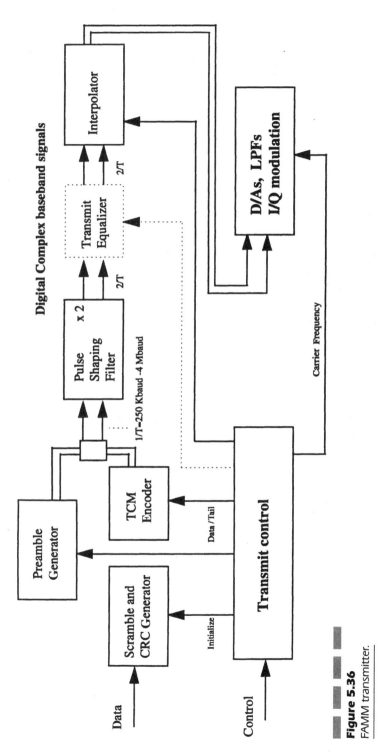

Figure 5.36
FAMM transmitter.

217

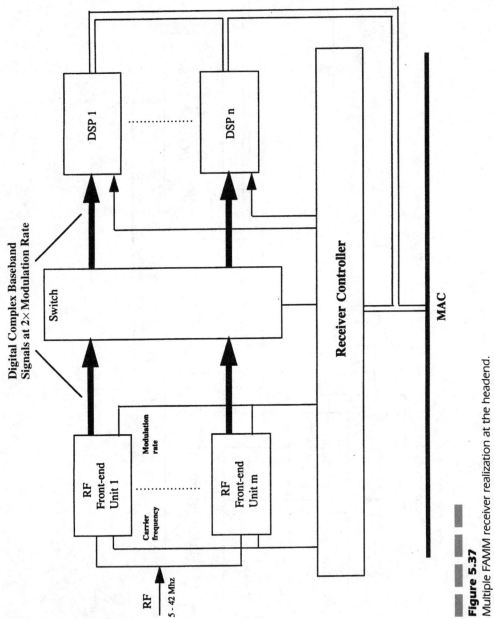

Figure 5.37
Multiple FAMM receiver realization at the headend.

218

band (inphase and quadrature) signals, which are sampled at twice the modulation rate. A receive controller provides the frontend units with the carrier frequencies and the modulation rates of the signals received at these frequencies. The receive controller, through its interaction with the MAC layer, is also aware of the times at which upstream frames are expected to arrive and knows the modulation parameters and formats used for these frames. It dispatches the digitized baseband signals corresponding to one frame to a digital signal processor (DSP) along with control information about modulation type and frame format. The DSPs may be any suitable programmable and/or hardwired processing elements. They perform the required receiver signal-processing functions and deliver the decoded information conveyed in the body of a frame to the MAC layer, together with side information on signal quality and correctness of a cyclic redundancy check. The DSPs do not necessarily have to process signals in real time. For example, more time can be spent on the processing of preambles for fast synchronization or channel probing and equalization than for subsequent receiver operations. However, bounds on decoding delay will have to be observed.

5.5.11 Aware: Discrete Wavelet Multitone (DWMT) PHY Layer

INTRODUCTION. Aware proposed DWMT. The proposal includes a multitone carrier; therefore, it was classified under the advance physical layer by IEEE 802.14. This and SDMT are work items to be discussed in the advanced PHY ad hoc group. A summary of the proposal is shown below.

Deployment of new services and applications such as telephony, Internet access, and other high-speed data services over HFC plants requires two-way transmission. Typically, the 5—40-MHz band will be allocated for the upstream return channel. This spectral region is subject to high noise levels due to narrowband interference (ingress) and funneling effects from the multipoint-to-point architecture. Multicarrier systems offer an attractive method for digital transmission over channels subject to distortion, multiple reflections, and narrowband interference

DWMT SYSTEM DESCRIPTION. The features of DWMT are as follows:

- Modulation is accomplished using a discrete wavelet transform.
- Multicarrier provides flexible dynamic bandwidth allocation using time frequency.

- It is spectrally efficient:
  Subchannels overlap in frequency
  Symbols overlap in time

- Transmission capacity is matched to channel quality.

- Robust performance in the presence of ingress and impulse noise is expected.

Referring to Fig. 5.38, the input data at the transmitter consists of a serial data stream that is divided into frames. The data bits are encoded into multilevel PAM symbols, and the PAM symbols are then orthogonally mapped to individual frequency subchannels via the inverse wavelet transform modulator.

The subchannels are grouped into pairs and modulated with the equivalent of a QAM constellation. The wavelet transform consists of a bank of N digital filters that have N_g coefficients, where $g = 6$ is the overlap factor. The properties of the transform are such that the subchannels overlap spectrally and the pulses transmitted in a subchannel overlap in time, while orthogonality among all symbols is maintained. The frequency overlap of adjacent subchannels results in a spectrally efficient transmission, and the time-overlapped pulses provide spectral shaping of the individual subchannel filters. The transform modulator produces a time domain sequence that is output to the D/A converter. In the downstream direction, the signal is broadcast to all modems, and, in the upstream direction, the analog signals from the subscriber modems are power-combined and then transmitted to the headend. The experimental setup consists of one headend transceiver and three subscriber transceivers as shown in Fig. 5.39.

Figure 5.38
Discrete wavelet multitone (DWMT) system for transmission over HFC.

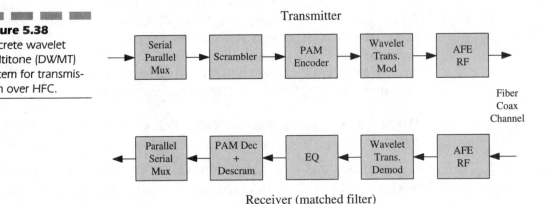

Transmitter

Fiber
Coax
Channel

Receiver (matched filter)

Figure 5.39
Setup with one head-
end modem and
three subscriber
modems.

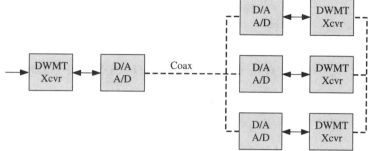

Figure 5.39
Setup with one head-
end modem and
three subscriber
modems.

Baseband Implementation

At the receiver, the baseband analog signal is filtered and digitized. The digital time domain signal is transformed back into PAM symbols via the wavelet transform. The demodulation transform consists of a set of matched filters with respect to the modulation transform. An equalizer is used to correct for channel distortion and timing errors. After equalization, the PAM symbols are decoded into bits, and the original data stream is reproduced by the parallel-to-serial multiplexer. The receiver can be constructed to demodulate all of the subchannels to recover the complete data stream or a specific subset of the subchannels for information addressed to individual users. Dynamic allocation of subchannels and transmission capacity can be implemented based on user demand.

A multicarrier wavelet system (DWMT) has been demonstrated that is applicable for telephony and data transmission over hybrid fiber-coax networks. Measured results confirm that the system achieves excellent transmission capacity and is robust against narrowband interference, clipping induced impulse noise, and timing errors.

5.6 PHY Modulation Techniques Adopted in IEEE 802.14

The modulation schemes selected for standardization by IEEE 802.14 were derived from the proposals described above. Some proposals address the downstream only, some the upstream only, and some both. Each of the proposals has strengths and weaknesses. The weaknesses have more to do with the assumptions and the criteria documents developed by IEEE 802.14 rather than technical flaws in the proposals. The criteria documents are described in Sec. 5.7. The modulation schemes adopted by IEEE 802.14 are described below.

5.6.1 QAM Downstream Modulation

In January 1996, IEEE 802.14 voted on the downstream PHY after officially presenting all the proposals. The five competing proposals were:

1. QAM
   - 16
   - 64/256
   - 16/64/256
2. INTRA-AM
3. VSB-16
4. QAM + QPSK
5. QPSK

QAM was selected as the preferred modulation technique. QAM 256 was thought to be too ambitious at the time, given the state of the U.S. cable network readiness to handle such a rate. IEEE 802.14, therefore, adopted QAM 64 as mandatory for the downstream modulation technique but did not preclude QAM-256. Release 1 of the specification will include QAM-256.

5.6.2 QAM-16 and QPSK for Upstream Modulation

IEEE 802.14, at the March 1996 meeting, selected QPSK as the modulation technique for the downstream PHY. The six competing proposals were:

1. QPSK
2. QPSK/QAM/FAMM
3. DWMT
4. SDMT
5. INTR
6. SCDMA

Single-carrier modulation was market-ready, and the technology was well understood and developed. Multicarrier modulation, although theoretically superior in performance, was considered under study for future application. Hence, IEEE 802.14 officially formed a study group to address multicarrier advanced PHY for future application for the cable modem.

As a result, IEEE 802.14 voted for QPSK and QAM-16 as the modulation technique for upstream channel transmitter.

5.6.3 IEEE 802.14 PHY Specification Timeline

The timeline for releasing version 1.0 of the IEEE 802.14 PHY specifications is as shown below. The minimum four steps and timeline are:

1. A working group letter ballot (LB) is scheduled for release in July 1997. The LB will be sent to all IEEE 802.14 working group voting members for comments.

2. At the September 1997 meeting, all negative comments will be addressed. A confirmation LB will be sent again to the working group voting members for a vote.

3. If approved (no negative votes), the specification will no longer be under IEEE 802.14 control. It will be sent out to IEEE member experts (computer society) for approval. This cycle is expected to last until March of 1998. The IEEE 802.14 working group will again be delegated to resolve the comments received from the industry.

4. If all goes well, the specification will be forwarded to the IEEE standards board for official approval in June of 1998.

5.6.4 IEEE 802.14 PHY Specification Preview

The specification contains all the necessary detail information that is needed to build a standard cable modem. A summary of its content is shown below.

DOWNSTREAM PHY SPECIFICATION (QAM-64 AND QAM-256).
At the minimum, the specification will include details on the following topics:

■ Downstream detailed specifications of both type A and type B PHYs (see Sec. 5.7)

■ Downstream transmission characteristics

■ Downstream spectrum allocation

- HFC topology (cabling, transmission characteristics/components)
- Transmission link noise characteristics and propagation delay
- Downstream framing structure
- Downstream coding, scrambling, and interleaving
- Downstream constellation for both type A and B
- Baseband filter characteristics
- QAM-64 and QAM-256 trellis coding modulation (constellation, modulation bit rate, baseband filter characteristics, spectral roll-off, carrier frequencies, impairment, timing distribution, etc.)

UPSTREAM PHYSICAL LAYER SPECIFICATION (QAM-16/QPSK). At the minimum, the specification will include details on the following topics:

- Upstream spectral allocation
- Upstream transmission link, components characteristics, and cabling
- Upstream propagation delay and insertion loss
- Upstream noise characteristics
- Upstream channel spacing
- Upstream carrier frequency and scrambling
- Upstream QAM-16 and QPSK preamble
- Upstream CRC polynomial used
- Upstream timing and synchronization
- Upstream modulation including bit rate, coding, and FEC
- Upstream QAM-16 and QPSK constellation, modulation bit rate, and Gray coding
- Upstream modulation impairment

PHY SPECIFICATION AVAILABILITY. The specification will not be officially released to the IEEE main office until late 1998. IEEE 8-2.14 is open to all. Anyone can have access to the draft when attending any of the IEEE 802.14 meetings. Once released, the document can be obtained from the standard documents for the IEEE office:

445 Hoes Lane
P.O. Box 1331
Piscataway, NJ 08855-1331
908-562-3380
908-562-1571 (fax)

5.7 Criteria Documents for PHY Standardization

A convergence specification was developed by IEEE 802.14. It was used as a guide to select the best features of all proposals described in Sec. 5.5. It is unlikely that the output PHY standards will deviate much from the parameters shown in the convergence specifications. The convergence specifications shown below were also based on available specifications and the two-channel models created in IEEE 802.14. This section will briefly describe:

1. The CATV channel model for the downstream
2. The CATV channel model for the upstream
3. Convergence specifications for the upstream

5.7.1 CATV Channel Model for the Downstream

The purpose of the CATV channel model for the downstream is to define parameters that must be determined before selecting a specific modulation scheme. The model shown below was also used to assess the performance of a modem by averaging the performance of the modem over a

Figure 5.40

Impairments in the downstream CATV channel.

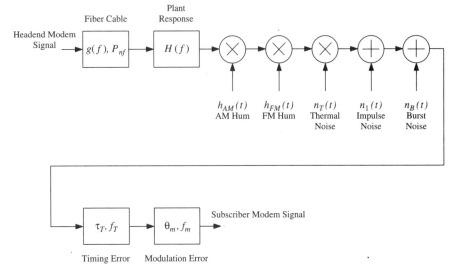

large number of channels picked randomly from the ensemble. Figure 5.40 illustrates is the downstream CATV channel model.

This channel model evaluated in detail each of the functions shown and illustrated the type of noise and other factors defining the characteristics of the downstream CATV channel. Details are described in Chap. 7. A brief summary is as shown below:

- The fiber affects the digital signal in two ways: group delay due to the high modulation frequency of the signal in the fiber and white Gaussian noise added of power.

- Plant response, which is created between the headend and a subscriber modem. Impulse response is defined using *tilt* and *ripple. Tilt* is a linear change in amplitude. *Ripple* is a sum of a number of sinusoidal varying amplitude changes riding on top of the tilt.

- AM hum modulation is amplitude modulation caused by coupling of 120-Hz AC power through power supply equipment onto the envelope of the signal.

- FM hum modulation is frequency modulation caused by coupling of 120-Hz AC power through power supply equipment.

- Thermal noise is modeled as white Gaussian noise with power defined relative to the power at the output of the plant response. Intermod is caused by nonlinearities in the system, generating harmonics of other channels.

- Burst noise is due to laser clipping, which occurs when the sum total of all the downstream channels exceeds the signal capacity of the laser.

- Phase noise is created in the headend and subscriber modems. Its characteristics will depend on the RF components in the modems.

- Channel surfing causes microreflections to appear and disappear. Because the significant sources of channel surfing are close to the receiver, a large but slowly changing ripple in the frequency domain will appear or disappear. This means each 6-MHz channel observes a fairly constant change in gain.

5.7.2 CATV Channel Model for the Upstream

Similarly, the purpose of the CATV channel model for the upstream CATV channel is to evaluate physical layer specifications. It was used as a guide by IEEE 802.14 as a basis for evaluation of physical layer proposals and development of the convergence specifications.

The models shown below were used to assess the performance of a modem by averaging performance over a large number of channels picked randomly from the ensemble. Figure 5.41 best illustrates the type of noise and its characteristics of the upstream channel. These impairments are described in the subsections shown below.

This channel model evaluated in detail each of the functions shown and illustrated the type of noise and other factors defining the characteristics of the upstream CATV channel. Details are described in Chap. 7. A brief summary is as shown below:

- Hum modulation is amplitude modulation due to coupling of 60-Hz AC power through power supply equipment onto the envelope of the signal.

- Microreflections occur at discontinuities in the transmission medium, which cause part of the signal energy to be reflected.

- Ingress noise is the unwanted narrowband noise component that is the result of external, narrowband RF signals entering or leaking into the cable distribution system. The weak point of entry is usually the drops and faulty connectors. Ingress noise creates apparently random

Figure 5.41
Impairments for upstream CATV channel.

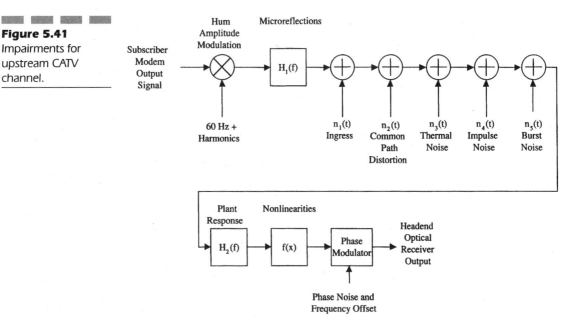

transmission interruptions usually on a single channel or closely associated channels. It also appears on a spectral display as an unwanted carrier, typically of a short time duration, and is a narrowband signal, less than 100 kHz in bandwidth.

■ Transmission impediments. There are several types of RF transmission impediments (noise) found in the return path of a two-way cable distribution network. Each type provides a "unique" contribution to the problem. However, transmission impediments are not easily identified and often not properly differentiated. Thus they are often bundled together under the generic category of ingress noise.

■ Common-path distortion. Common mode rejection is due to nonlinearities in the passive devices in the cable plant. The nonlinearity is created as a result of corroded connectors. It is common in distribution legs when oxides form between two metal surface, creating a point-of-contact diode. This diode may appear in the ground portion of the connector, creating common mode distortion and allowing penetration of ingress noise to the system. The effect of this diode creates distortion in both the upstream and downstream paths.

■ Thermal noise. White noise is generated by random thermal noise (electron motion in the cable and other network devices) of the 75-ohm impedance, terminating resistance, operating at 68°F, with a 4-MHz channel bandwidth.

■ Impulse noise is one major problem in the two-way cable system and the most dominant peak source of noise (a short burst duration less than 3 seconds). Impulse noise is mainly caused by 60-Hz high-voltage lines and any electrical and large static discharges such as lightning strikes or AC motors starting up, car ignition system, televisions, radios, home appliances like washers, and so on. Loose connectors also contribute to impulse noise. Impulse noise creates random transmission interruptions on multiple return channels on both the upstream and downstream.

■ Corona noise is generated by the ionization of the air surrounding a high-voltage line. Temperature and humidity play a major role in the contribution of this event. Corona is best described as a high-voltage line of at least 300 kV of energy discharging in the air and often located on the same poles or conduits as the CATV cable.

■ Gap noise is generated when the insulation breaks down or via corroded connector contacts. Such failures pave the way to the entry of lines discharge of 100-kV lines.

- Burst noise is similar to impulsive noise but with longer duration of each event. It typically results from several sources: corona discharges from power lines; discharges across corroded connector contacts; automobile ignition; and household appliances such as electric motors.

- Phase noise arises in frequency-stacking multiplexers that occur in some return path systems. It may also occur in laser cavities but is less pronounced. Phase noise is also produced in the subscriber and head-end modem oscillators.

- The cable plant contains linear filtering elements, which are dominated by the diplex filters that separate upstream frequencies from downstream frequencies.

- Nonlinearities include limiting effects in amplifiers, the laser transmitter in the fiber node, and the laser receiver in the headend.

5.7.3 PHY Convergence Specification

The upstream PHY convergence specification was developed by IEEE 802.14, from the various PHY proposals, the channel models described above, and standards developed in other organizations. The eventual IEEE 802.14 PHY specification is not likely to deviate much from it. Tables 5.30 to 5.34 and Figs. 5.42 to 5.45 summarize the PHY parameters.

5.8 J.83 Annex A or B

IEEE 802.14 developed the MAC, so it accommodated various PHY interfaces. The release 1 specifications will include two types of interfaces: type A and type B. IEEE 802.14 agreed to develop the type-B interface to accommodate U.S. MSOs and CableLabs. Type-A and type-B interfaces relate to ITU J.83 recommendation annex A and annex B, respectively. Annex B was the favored among the U.S. cable operators. A summary of annex A and annex B is as shown below:

Annex A. Derived from DVB (Digital Video Broadcast standard) originated in Europe. It incorporates Reed-Solomon block coding of a codeword aligned with a MPEG2 Transport Stream (TS) Packet for FEC. DAVIC selected annex A in its specifications.

TABLE 5.30

Upstream PHY Parameter

Parameter	Value	Comments/Source
Modulation format	Headend demodulator: QPSK (BPSK and 16 QAM optional) Subscriber transmitter: BPSK, QPSK and 16 QAM; see Table 5.31	80214 decision; plus BPSK added as robust mode
Constellation	See Fig. 5.44 and Fig. 5.45	
Data rates	See Table 5.31	DAVIC
Channel spacing	See Table 5.33	DAVIC
Burst length modes	1. Fixed, long (data slot, 1–2048 bytes) 2. Fixed, short (minislot) 3. Variable (1–2048 bytes, demod detects end of burst) 4. Inactive (detection gated off)	Supports ATM packets, IP packets, and minislots
Minislot format	Modulation: BPSK Extra guard time provided for ranging and power setting errors on net entry; contains CRC for error detection	BPSK less sensitive to echoes and noise with no FEC
Guard time between bursts, minimum	Fixed data slot: 4 symbols (guard time defined in Fig. 5.43) Minislot: 6 symbols Variable burst length: 10 symbols	Existing hardware (fixed length); analysis (variable length)
Preamble/unique word for BPSK and QPSK modes	Pattern: programmable Modulation: BPSK, except last symbol may be QPSK Length: 16 symbols total including preamble and unique word Default preamble: 12-symbol BPSK clock preamble followed by 4-symbol unique word; first 3 symbols of unique word are BPSK, last symbol QPSK	Default preamble is DAVIC; supports random preamble for adaptive equalizer training if needed
Preamble for 16 QAM mode	Pattern: programmable Modulation: BPSK Length: 16 symbols (TBD)	Need to support adaptive equalizer training
FEC coding	Mode 1: no coding; use for minislots and certain types of data traffic Mode 2: RS(x,y), $t = 3$. x, y are TBD depending on MAC definition. Example is RS(59,53). Use for ATM packets. For concatenated ATM packets, repeat code blocks Mode 3: RS(255,239), shortened if required; $T = 8$. Use for IP packets	Mode 1: maximum economy Mode 2: medium economy, DAVIC compatible Mode 3: maximum protection for long packets

Parameter	Value	Notes
Reed-Solomon code parameters	GF(256) (8-bit bytes). Code generator polynomial: $g(x) = \prod_{i=0}^{N-K-1}(x - a^i)$; Primitive generator polynomial: $p(x) = x^8 + x^4 + x^3 + x^2 + 1$	DAVIC
Cyclic redundancy code (CRC)	Not addressed in PHY spec	MAC layer function
Scrambling	Mod 2 addition to payload exclusive of preamble/unique word; Seed: all ones, restarted on each packet	O151 standard and Fireberd 6000; Same technique as DAVIC but longer; provides better energy dispersal
Byte/serial conversion order into scrambler	MSB first	DAVIC
Phase ambiguity coding	See Table 5.35	DAVIC; protects against cycle slip
Multiuser signal structure	FDMA/TDMA	
Pulse shaping	Root raised cosine approximation, rolloff $\alpha = 0.3$	DAVIC; industry standard
Transmit spectral mask	See Table 5.34 and Fig. 5.42	DAVIC; commensurate with 6- and 8-MHz channels
Carrier center frequency range	5—40 MHz	
Carrier frequency step size	50 kHz	DAVIC; <1/3 of narrowest BW
Carrier frequency accuracy	QPSK: ±0.5% of upstream symbol rate, relative to headend reference; 16 QAM: ±0.2% of upstream symbol rate, relative to headend reference; see Table 5.32	Demod does not need to estimate frequency on each burst (assumes carrier loop BW = 10% of symbol rate)

TABLE 5.30 CONTINUED

Parameter	Value	Comments/Source
Upstream PHY Parameter		
Additional carrier frequency inaccuracy, e.g, unlocked frequency converter in channel	Frequency stacking upconverter must be phase locked to headend pilot	Selected in tradeoff against (1) headend demod searching on each burst and (2) measuring each subscriber during setup
Transmit clock frequency inaccuracy	±50 ppm absolute or may be locked to headend reference via locking to downstream	DAVIC
Transmitter output signal level	Range: 30 to 55 dBmV into 75Ω Step size: 2 dB	CableLabs spec
Demod input signal level	±15 dBmV into 75Ω	Existing equipment
Demod input signal level, burst-to-burst variation	±5 dB	Existing equipment
Return loss	Demod input: >15 dB from 5 to 40 MHz Transmitter output: >15 dB from 5 to 40 MHz	5 dB better than DAVIC to support 16 QAM
Carrier suppression when transmitter active	35 dB below modulated signal	
Output when transmitter idle	5—40 MHz: 60 dB below "on" power in any 2-MHz band 54—1000 MHz: −40 dBmV in any 6-MHz band	
I/Q amplitude imbalance	<0.25 dB	Simulation, 16 QAM
Quadrature phase error	<1.5 degrees	Simulation, 16 QAM
Setup time to change programmable parameters	10 ms	

TABLE 5.31

Data Rate

Burst symbol rate (Msps)	Allocated bandwidth = carrier spacing (MHz)	Burst BPSK bit rate (Mbps) (Note 1)	Burst QPSK bit rate (Mbps)	Burst 16 QAM bit rate (Mbps)	Symbol rate divisor from 16 × 1.544 = 24.704 MHz	Symbol rate multiple of 8 kHz
1.544	2	1.544	3.088	6.176	16	193
0.772	1	—	1.544	3.088	32	96.5
0.128	0.17	—	0.256	(Note 2)	193	16

1. BPSK (form of QPSK with $I = Q$ data) permits robust operation in initial installations before plant upgrade. Choice of 1.544 Mbps makes BPSK appear transparent to MAC layer other than requiring wider carrier spacing.

2. 16 QAM at lowest symbol rate (128 kbps) is not included as it would require approximately 2.5× greater frequency accuracy.

TABLE 5.32

Carrier Frequency Accuracy Table

Burst symbol rate (Msps)	Carrier frequency accuracy required (kHz) QPSK	Carrier frequency accuracy (ppm @ 40 MHz) QPSK	Carrier frequency accuracy required (kHz) 16 QAM	Carrier frequency accuracy (ppm @ 40 MHz) 16 QAM
1.544	7.72	193	—	—
0.772	3.86	96.5	1.544	38.6
0.128	0.64	16	—	—

TABLE 5.33

Default Preamble for BPSK and QPSK Modes

Parameter	Value
Application	Recommended for standard BPSK/QPSK without adaptive equalization
Origin	DAVIC
Description	12 symbols BPSK 1010...+ 4-symbol unique word (BPSK except last symbol is QPSK)
Preamble + unique word contents, QPSK bits, binary	1100 1100 1100 1100 1100 1100 0000 1101
Preamble + unique word contents, QPSK bits, hex	CC CC CC 0D

NOTE: Bit pattern shown is on channel after differential coding.

TABLE 5.34

Transmit Spectral
Mask Table

Frequency	Frequency (symbolic)	Lower power spectral bound (dB)	Theoretical power spectrum (dB)	Upper power spectral bound (dB)	Lower group delay bound	Upper group delay bound
0	0	0.25	0	−0.25	$-0.07/f_N$	$0.07/f_N$
$0.2\,f_N$		0.25	0	−0.25	$-0.07/f_N$	$0.07/f_N$
$0.5\,f_N$		0.25	0	−0.4	$-0.07/f_N$	$0.07/f_N$
$0.7\,f_N$	$1-\alpha$	0.25	0	−0.5	$-0.07/f_N$	$0.07/f_N$
$1\,f_N$	1	−2.5	−3	−3.75	$-0.07/f_N$	$0.07/f_N$
$1\,f_N$	1	−2.5	−3	−50	$-0.07/f_N$	$0.07/f_N$
$1.3\,f_N$	$1+\alpha$	−20	−∞			
$1.9\,f_N$ to 2 MHz	$1+3\alpha$ to 2 MHz	−40	−∞			
>2 MHz		−45	−∞			

1. Up to $1.9\,f_N$ mask frequencies are normalized to Nyquist frequency (mask scales with symbol rate). Above 2 MHz, mask frequencies are absolute. This is to permit additional filtering outside largest data rate bandwidth.

2. f_N = symbol rate / 2.

3. Source of group delay bound is DVB spec.

Figure 5.42

Template for signal spectrum mask at modulator output represented in baseband frequency domain (0 = carrier center frequency): x axis = frequency normalized to Nyquist (half symbol rate) f/f_N; y axis = relative spectral density (dB).

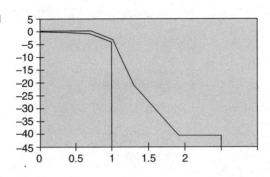

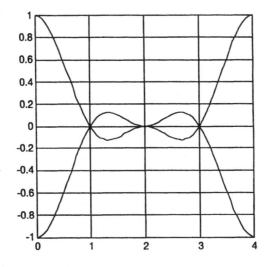

Figure 5.43
Interbust gap symbols are shown here as the envelope of the ringing signal between the last symbol ($t = 0$) of the previous burst and the first symbol ($t = 4$) of the next burst.

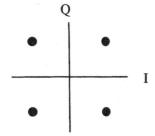

Figure 5.44
QPSK constellation.

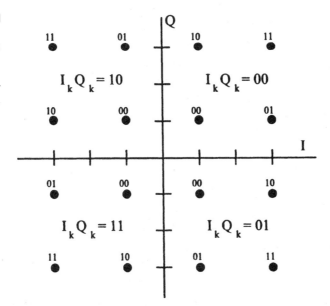

Figure 5.45
16 QAM constellation (taken from DAVIC spec).

Annex B.   A transmission format widely accepted in North America and Canada. Concatenated Reed-Solomon block coding with trellis coding modulation is used. The baseband MPEG-2 TS data is decoupled from the FEC and hence no alignment is needed for framing.

CableLabs promoted a modified version of annex B in IEEE 802.14. The measurement they performed showed that approximately 1.5-dB improvement in the carrier-to-noise ratio in J.83 annex B over A. CableLabs and the MSOs wanted to see a modified version of annex B with variable adjustable interleaving to allow QOS and address latency issue. CableLabs' rationale for promoting a version of annex B over and above its performance also has to do with future alignment with digital broadcast standards that was adopted for North America.

At the November 1996 meeting, IEEE 802.14 adopted both annex A and modified annex B for standardizing the cable PHY layer. Opponents of standardizing both annex A and annex B for cable modem had two important concerns:

1. *Interoperability.*   The consumer has to purchase a cable modem with either an annex A or annex B interface so it can be used with the cable operator's headend. If the user moves to another region, the cable modem may not operate if the equipment interfaces are mismatched.

2. *International standard.*   The economies of scale, globally, will bring the cost down.

It was the aim of IEEE 802.14 members to specify a cable modem that can be adopted internationally. It is unlikely that Europe will adopt J.83 annex B because of the DVB infrastructure based on annex A. It appears now that Europe and North America will have two different physical-layer standards. Some vendors feel that a programmable chip set can be developed to handle both annex A and annex B. In any case, as described above, IEEE 802.14 developed two PHY layers for the cable modem: an interface compatible with ITU J.83 annex A for Europe and a cable modem with J.83 annex B (modified) for North America.

Chapter

6

Cable Modem
MAC Layer

6.1 Introduction

In the previous chapter, the architecture and physical layers of the cable modem were described. In this chapter, we will focus on describing the MAC layer. The MAC is the most challenging to ascertain and has been and will continue to be the topic of discussion among the highly educated for what seems to be forever. The MAC for the cable modem offers even more challenging opportunities because it must operate under many more hostile environments than any previous MAC developed so far. Unlike other MACs, the cable network MAC must operate in a public environment where the quality of service and user expectations are of paramount importance. The MAC for SMDS was developed for the public domain but has to deal only with data service or very limited quality of service offerings. The cable modem MAC must deal with interactive and multimedia services with bandwidth-hungry applications requiring a multitude of service requirements.

There are several cable modems on the market today, and most if not all deal with service-specific applications. Market pressure also yielded MAC specifications dealing exclusively with that segment of the market, particularly for the Internet applications. This is fine and good, but just speeding the access interface, say for a cable modem, does not solve the multimedia service requirement. The concept of quality of service must be embedded in the development of a MAC protocol so it can operate to the satisfaction of the end users. The cable MAC must also be future-safe and reliable. The market dynamics are very unpredictable and unfriendly, confusing short-term planners. ATM technologies deployed in the network and sophisticated multimedia components being developed in PCs are paving the way toward the information society. This is no longer a choice society has; the information age has penetrated our society at all levels: education, business, government, marketing, entertainment, and global competition. All these factors are clear to the long-term planners, both for the telco and cable operators. The business reality, however, may take precedence over long-term strategies, and, as always, the free market will determine the outcome.

The IEEE 802.14 committee undertook the challenge of developing a standard for the cable modem that accommodates the present and more demanding future services and applications. The MAC developed in IEEE 802.14 is a formidable task that took more time to develop and provided the hooks for developing all future services.

The organization of this chapter is as follows:

- Overview of the cable network focusing on the MAC requirement
- Summary of available and working MACs
- Criteria for selecting a MAC for cable modem
- Proposed MAC submitted by the industry and educational institutions
- Detail performance evaluation and overview of the IEEE 802.14—selected MAC

6.2 Cable Network and MAC

The cable network has traditionally been a one-way communication. The direction was from the headend to the subscriber. This CATV model served the consumer well. The market recognized the potential of the cable network, and cable operators evolved and became content providers. The cable medium, in many ways, suits the high-speed multimedia market very well. The available bandwidth in the cable network is enviable. The task of providing a two-way interactive communication in the cable network is menacing but attainable. At the network end, three major network reconstructions must take place:

1. The amplifiers in the coax cable must be upgraded to operate reliably in both the downstream and upstream directions.

2. Allocation of the upstream frequencies must be flexible and clean. The range allocated now from 5–45 MHz is susceptible to noise generated by all sorts of electromagnetic interference. The noise is injected in the cable network through all sorts of old, low-cost coax, worn-out connectors, and inferior maintenance practices. Cable operators must clean this noise environment before a credible service offering can begin. Developing a robust physical layer as described in the previous chapter helps, but it is not enough by itself. In this shared medium environment, the network is susceptible to all sorts of misuse. Among the most critical ones are:
   - Security and eavesdropping
   - Service theft
   - Network reliability, both in terms of alternate routing due to amplifier failures or AC power outage
   - Network maintenance; a misbehaving user may be able to cause a network outage

The MAC for the cable modem will be responsible for solving some of these technical problems.

3. In the upstream shared medium environment, the communication is many to one. All subscribers will be competing on the shared medium bus to get the attention of the headend so it can be granted permission to start sending the data. MAC controls the behavior of users who want to access the network. Figure 6.1 illustrates one of the main functions of the MAC (e.g., station A wants permission from the headend to acquire the bus and send data). The MAC arbitrates the communicating path and ensures and/or resolves any collision that occurs. Station A and station B must go through the headend controller to be granted permission to access the bus. The headend acts as a central office in that it controls and mediates all communications between, and/or from, all connected stations. The MAC must also honor the service contract promised by the application and network.

It is clear that a powerful MAC must be developed for the cable modem to meet the stringent requirements.

6.3 Legacy MACs

There are several MACs developed and specified in the public and private networks. The two popular and simple mechanisms are time division multiplexing and collision resolution like CDMA.

Figure 6.1
Sharing the upstream bandwidth.

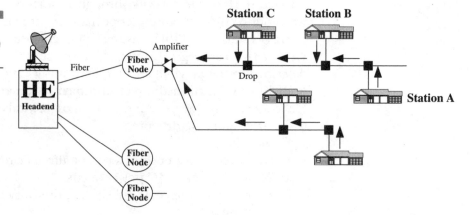

6.3.1 Time Division Multiplexing (TDMA)

In the TDMA approach, each connected station is allocated a time slot in a specified time frame. The frame contains a fixed number of time slots, and each will be dedicated exclusively to a station: station A, station B, station C, and so forth. This is illustrated in Fig. 6.1. When station A has data to send, it will simply use its dedicated time slot to send the information at its leisure. The obvious advantages of this mechanism are:

- No collision is experienced in the shared medium.
- It is the ideal solution for constant bit rate traffic like voice or video telephony. Station A can use its time slot all the time in a synchronous manner without interfering with any other station on the shared bus.
- TDMA gives fair access to all connected subscribers.

The disadvantages of this TDMA mechanism are also obvious:

- An idle user's time slot is unduly wasting network resources.
- In most multimedia applications, the traffic is bursty and unpredictable, whether the subscriber is on the Internet, sending email, on the WWW clicking the icons, requesting a Video On Demand (VOD) movie via his or her remote control, or reading the utilities meters. In this case, the allocated time slot is used occasionally. Providing full-time access to stations in such premium and limited upstream channels is a waste and may become prohibitively expensive.

6.3.2 Contention and Collision Resolution Mechanism

In this contention and collision approach, the mechanism assumes that stations, when they need to, must vie for accessing the bus so it can send data. MAC responsibility is to arbitrate the access, resolve contention, and control the traffic flow so the shared bandwidth is used only when needed. Hence, efficient use of the upstream resources is achieved.

Collision is certain to occur, especially in a folded architecture. In a folded architecture, the headend acts as the common control node, and messages are fed back to all stations in the downstream direction, informing all of the statutes. A simplified description of the methodology of acquiring the bus is as shown in Fig. 6.2. The figure assumes that time slots are passing in the upstream channel in the direction of the headend.

Station 1 writes its request in that time slot on the upstream with its identification (ID). The headend echoes it back in a portion of the downstream channel in a broadcast fashion to all users. An acknowledgment is hence obtained by the station. If collision occurs when station 1 and station 2 both write in the same time slot, the headend echoes these time slots in the downstream channel for all stations to monitor. The colliding stations will back off for a random period of time and retry the request. Collision is aggravated further due to the folded topology of the cable network and the distance between stations and between stations and headends (which could be as much as 50 miles or 80 km). Depending on the distance, it may take station n a few more microseconds to detect that the bus is not in use than it will take station 1. If station 1 starts sending, station n will also send during that propagation delay period. This results in a collision. Stations recognize the collision when listening on the downstream channel to confirm the acknowledgment.

One disadvantage of this topology is that the station nearest to the headend has the advantage of detecting the information sooner than all other stations. The longer apart they are, the worse the collision and unfairness become. Another problem associated with this mechanism is that the quality-of-service concept is not addressed. Each station must conform to the quality-of-service contract negotiated by the station during the call set-up state. Stations must accommodate the traffic parameters they promised the network and must perform accordingly. This problem is further compounded when dealing with ABR, CBR, VBR, and UBR traffic in the same station.

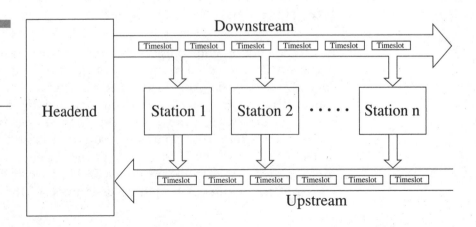

Figure 6.2
Station monitoring the downstream and transmitting upstream.

The ideal solution is to develop a MAC that is a hybrid of both the TDMA and reservation mechanism. Most IEEE 802.14 MAC proposals submitted by the industry alluded to such an approach.

6.3.3 MAC Objectives

The ATM Forum and particularly IEEE 802.14 developed the objects for development of a MAC for the cable industry that will conform to the multimedia service requirements (described in Chap. 2). IEEE 802.14 also created a MAC criteria document that was used as guide to evaluate proposals that were submitted for standardization. The ATM Forum also conveyed their MAC requirements to IEEE 802.14. These objectives were used to develop the MAC and are as stated below:

- The MAC must support both connectionless and connection-oriented services.
- The MAC must support the concept of quality of service on a per-connection basis. Each connection should be able to specify its requirements for bandwidth, delay, and jitter.
- The MAC for bandwidth access must conform to application-dependent items such as CBR, VBR, UBR, and ABR.
- The MAC must have interoperability with ATM.
- The MAC must exhibit fair arbitration for shared access to the network within any level of service.
- The MAC must be simple and provide effective ranging.
- The MAC must be built with uncomplicated electronics leading to an inexpensive implementation (hardware and software).
- The MAC must have ease of installation and service activation.
- The MAC must exhibit simple service-provider provisioning, maintenance, operation, and administration.
- The MAC must allow accurate usage monitoring of the shared channels.
- The MAC must provide a level of protection against a faulty terminal transmitting in an uncontrolled fashion.
- The MAC must be designed for tree and branch architecture.
- The MAC must response in a fair manner regardless of where the station is positioned in the tree. Bandwidth allocation should be fair

across multiple connections of the same type. The MAC shall also be able to gain access to the media and satisfy the QOS of that specific application. Priority on sensitive delay applications must be accommodated so as not to reduce the value of the service.

■ The MAC must be responsive to traffic control and its behavior (e.g., flow control and congestion control) in order to maintain the number of transmitted packets within the network to below a level that degrades performance. Here are two examples that may be employed:
Calling admittance control (CAC), which may refuse a connection setup
Improving the contention mechanism under heavy load

■ The MAC must be robust and deal with failure or congestion, ideally without losing data or needing to reestablish the connection.

■ The MAC must react to prevent congestion collapse or overreact, resulting in an oscillating condition.

■ The MAC must minimize the cell delay variation of the CBR traffic (i.e., provide periodic access to the transmission channel). Synchronization to the network clock (common clock) is also required for circuit emulation.

■ The MAC protocol must balance bandwidth efficiency, robustness, fairness, simplicity of implementation, and efficient contention resolution.

6.4 Official MAC Proposals

In November 1995, all major cable vendors submitted their MAC proposal per the IEEE 802.14 request for standardization. Fourteen proposals were submitted in that meeting from cable modem vendors, universities, and others.

The MAC that was selected for standardization by IEEE 802.14 was derived from the proposals described below. During the selection process, it was recognized that each of the MAC proposals has strengths and weaknesses. These weaknesses have nothing to do with any technical flaws but more to do with the assumptions adopted by the IEEE 802.14 working group. A summary of the assumptions/objectives are stated above. A criteria document was also developed and was used extensively as the tool to fish out the best features of the submitted MAC proposals. Below are all the proposals made to the IEEE 802.14 committee for the MAC. MAC can be classified in two forms:

1. Management
2. Access

MAC management messages between the headend and stations are used for initialization, registration, time synchronization, ranging, and so on. This will be described further in Chap. 7. Access MAC deals with the data flow control structures. They are described below in random:

1. Cabletron	A simple and efficient multiple access protocol (SEMAP)
2. NEC Corporation	Frame pipeline polling (FPP)
3. Zenith Electronics	MAC layer protocol proposal: adaptive random access protocol for CATV networks
4. General Instrument Corp.	MAC for HFC
5. Scientific Atlanta (Georgia Institute of Technology)	Extended distributed queuing random access protocol (XDQRAP)
6. National Tsing Hua University	General multilayer collision resolution with reservation: a MAC protocol for broadband communication network
7. Philips Research Laboratories	General MAC
8. Lucent Technologies	Adaptive digital access protocol (ADAPt): a MAC protocol for multiservice broadband access networks
9. LANcity Corporation	A MAC protocol
10. Hybrid Network Inc.	MultiMedia MAC (M3)
11. Com21, Inc.	The UPSTREAMS protocol for HFC networks
12. Amati	MAC protocol to support both QPSK and SDMT

13. Panasonic Tech., Inc.   Collisionless MAC
14. IBM Corporation   MAC level access protocol (MLAP)

6.4.1 Cabletron Systems Inc.: A Simple and Efficient Multiple Access Protocol (SEMAP)

OVERVIEW. Cabletron proposed a MAC protocol that includes an implicit ranging mechanism feature and relies upon a number of control messages that are exchanged between the headend and the stations to perform the access control functions. Unlike the TDMA approach, SEMAP allows the use of upstream bandwidth for dynamic assignment of time slots. The protocol uses pipelining and implicit ranging, which define a nonintrusive method of propagation delay measurement complexity at the headend. Only the upstream protocol is addressed in this proposal. Management messages are included and also described.

FRAME STRUCTURE. The headend continually transmits frames at regular intervals. During the start-up phase, a station receives the frame and establishes synchronization. The upstream frame is divided into a contention period, reservation slots, and free slots.

The period between two frame markers as shown in Fig. 6.3 is referred to as the *frame*. A frame contains the contention window, reservation slots, and free slots. For timing purposes, the entire frame consists of fixed-length time slots.

CONTENTION WINDOW. The portion immediately following a frame marker is defined as the *contention window*. On power-up, a station may use the time immediately after a frame marker to send a control message to the headend. The contention window is long enough to ensure that the station farthest from the headend can send a control message without colliding with any of the reservation slots.

Figure 6.3
Upstream frame structure.

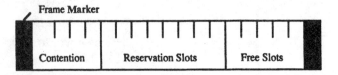

RESERVATION SLOTS.   The rest of the frame is divided into a *reservation slots window* and a *free slots window.* The headend is free to define the reservation window and the free window in any proportion. A reservation slot may only be used by a station when explicitly assigned to it by the headend.

FREE SLOTS.   The free slots may be used by all stations in a slotted ALOHA fashion. Free slots can be used for either control messages or for data packets.

MAC PACKET TYPES.   SEMAP assumes that the basic unit of data transmission is an ATM cell. Each time slot is able to accommodate one cell including the overhead. Data packets originating from LANs should be segmented using AAL5. The MAC frames shall be one of two types:

■ Data packets carrying payload
■ Control packets carrying protocol messages between the headend and the stations

CONTROL PACKETS.   The MAC functions are handled via a number of control messages that are exchanged between the headend and the stations. These messages provide mechanisms for ranging, slot assignment and termination, and for passing configuration parameters. This protocol also assumes that every station has a unique 48-bit IEEE MAC address assigned to its cable port. Initially, the MAC address is used to distinguish between individual stations. The following types of control messages are defined:

■ Initialization request
■ Initialization response
■ Terminate request
■ Terminate command
■ Special command

There are two basic types of control packets:

1. *Packets that are sent by the station to the headend.*   These are called the *upstream control messages* and include the initialization request and the terminate request messages.

2. *Packets that are sent by the headend to the station.* These are the *downstream control messages* and include the initialization response, the terminate command, and the special command.

THE UPSTREAM CONTROL MESSAGE. The upstream control message shown in Fig. 6.4 contains the following fields:

- *Destination address (optional).* The 48-bit unique address of the headend.

- *Source address.* The 48-bit IEEE address of the station.

- *Message type.*

- *Number of reservation slots requested.* An integer value 0 or greater.

- *Number of frames.* This integer value indicates the number of frames for which the reservation slot is requested. A value of all ones indicates a request for an indefinite period of time.

- Up to 22 octets are available for future definition.

It is assumed that initialization requests with respect to the frame marker are fixed. For instance, the time immediately following the frame marker is always used by a station to send the request message. The randomization of the request transmission takes place over multiple frames. In other words, when a request message experiences collision, the station waits a random number of frames before transmitting again.

Initialization Request. This message is sent by a station when it first wishes to transmit data on the 802.14 network. This is sent in the contention window at a specific time with respect to the frame marker. The timing of this message with respect to the frame marker is known a priori by both the headend and the station. In this protocol, a station would always send this message immediately following a frame marker. Alternately, the station could indicate the timing as an offset from the frame marker within the body of the message. In either case, the request message is used by the headend to calculate the range value of the station that is defined as the round trip propagation delay between the station and the headend.

Figure 6.4
Upstream control message.

					Bytes
6	6	2	2	2	
Destination Address	Source Address	Type	Number of Slots	Number of Frames	TBD

This message also contains requests for any reservation slots that the station may require. The request may be made for one or more slots. Further, the station may specify a fixed number of frames for which the slots are requested, or it may request the slots for an indefinite number of frames. This type of request would support a CBR-type application.

Terminate Request.   This message is sent by a station to the headend when the station has finished its transmission and no longer needs its assigned reservation slots. This indicates to the headend that these slots are no longer needed and may be assigned to another station.

DOWNSTREAM CONTROL MESSAGES.   Figure 6.5 shows the messages that are sent by the headend to the stations. They contain the following fields:

- *Destination address.*   This is the IEEE address of the targeted station.
- *Source address.*   The IEEE address of the headend.
- *Message type.*
- *Station ID.*   A temporary identification number assigned to the station by the headend.
- *Range.*   The round trip propagation delay value calculated by the headend.
- *First reservation slot.*   The number of the first reservation slot assigned to the station. A value of 0 indicates that no reservation slots are assigned.
- *Number of reservation slots.*   The number of contiguous reservation slots assigned to the station. A zero indicates that either a single slot is assigned to the station or no slots are assigned.
- *First free slot.*   The number of the first slot that is defined to be a free slot. A value of 0 indicates that there are no free slots defined in this system.
- *Number of free slots.*   The total number of free slots defined in the frame.
- Up to 14 octets are available for future definition.

Figure 6.5
Downstream control message.

6	6	2	2	2	2	2	2	2	Bytes
DA	SA	Type	ID#	Range	First Res. Slot	Last Res. Slot	First Free Slot	Last Free Slot	TBD

Initialization Response.   This message is sent by the headend to a station in response to its initialization request. It contains the basic parameters needed by the station to become operational. This includes the station's range value, its assigned reservation slots, and the location of the free slots within the frame, if any. Also included is an identification number assigned to the station for the duration of the frames.

Terminate Command.   This message may be sent by the headend if it needs to terminate a station's slot assignment in the middle of the service.

Special Command.   These messages may be used for special purposes such as obtaining status information and statistics gathering. A special command may also be used by the headend to deny the service to unauthorized users.

SYSTEM OPERATION.   In order for a station to be operational, it must first acquire synchronization by observing the frame markers. It must then send an initialization request frame to the headend. This is sent in the contention window and may contain a request for a slot assignment. This frame is also used for ranging. It is assumed that only the time immediately following the frame marker is used by a station to send this message. Therefore, the timing of this message with respect to the frame marker as observed by the station is known a priori by the headend. When the headend receives the initialization request frame, it can calculate the round-trip propagation delay from the station. The headend responds with an initialization response message. This contains the propagation delay value for the station and the slot or slots assigned to the station within the reservation slot window. This message may also indicate the free slots or slotted ALOHA slots that are defined in the frame. Upon receiving the initialization response message, the station is ready to transmit its data in its assigned reservation slot. If the station requested a reservation slot and no reservation slots are available, the headend may respond with a message that assigns no slots to the station. In this case, the station may use the free slots only, it may wait and request again to obtain a reservation time slot, or it may do both. If no reservation slots are available and no free slots have been defined, the station must wait and send a request message again to obtain service.

Since the initialization request message is transmitted during the contention window, it may experience collision and be corrupted before reaching the headend. After sending an initialization request, the station waits a predetermined length of time; if an initialization response is not received from the headend, it assumes that the request failed to reach the headend. The station then waits for a random number of frames within a prespecified range and retransmits its request.

TERMINATION PROCEDURE.   This protocol is based upon the station voluntarily giving up its reservation slots when it has no more data to send. Once the initialization is complete, a station may transmit its data in its assigned slots. If it has been assigned a fixed number of frames, it may transmit data until the last slot but one. The last slot must contain a terminate request message. If the station has been assigned an indefinite number of frames, it may transmit data for as long as it needs. When the station has no more data to send, it must give up its slot assignment by sending a terminate request message.

This method ensures that the reservation time slots are shared fairly between the active stations. For additional robustness, the headend must implement a time-out algorithm that automatically terminates the reservation slots assigned to a station if no transmission is detected from the station for a long period of time.

6.4.2 NEC Corporation: Frame Pipeline Polling (FPP)

OVERVIEW.   NEC proposed framed pipeline polling (FPP). It is based on polling protocol and pipeline processing. The FPP supports dynamic bandwidth allocation according to various traffic types including CBR, VBR, and ABR. This protocol achieves multimedia services with high channel efficiency. Polling protocol provides dynamic channel allocation. Because bandwidth scheduling is centralized, polling can be easily expanded to flexible bandwidth allocation for isochronous traffic and bursty traffic. The proposed protocol is developed based on the polling protocol. The polling protocol was modified for a large transmission delay environment and hence was called *pipeline polling protocol.*

MAC PROTOCOL ALGORITHM
PIPELINE POLLING PROTOCOL.   The proposal tackled the problem of polling by introducing an advanced polling mechanism. Traditional polling is not adequate for CATV topology because of the round-trip delay. Station to headend and back to the station can be as much as 160 km or 800 μsec. This time is wasted if advanced polling is not introduced. It is referred to as *pipeline polling.*

Pipeline polling protocol accesses the polling sequence in advance. The headend receives bandwidth requests from all stations from the previously polled period. Based upon these requests, the headend schedules bandwidth allocation to each station. The headend then polls each station in which each sends its data and bandwidth request information for the next polling. This request is used at the next bandwidth scheduling. After polling out to a station, the headend polls out to the next terminal without waiting for the end of the previous terminal's data transfer. With this pipeline process of polling, round-trip delay loss is no longer deadtime, and the bursts from all terminals are multiplexed without any gap in the upstream channel.

Framed Pipeline Polling.   Pipeline polling is suitable for bursty traffic such as ABR. However, it does not have any bandwidth allocation mechanism for CBR-types of traffic. For multiple traffic support, some additional techniques should be introduced in pipeline polling. In order to support isochronous traffic such as CBR or real-time VBR, transmission cell delay variation (CDV) must be bounded. To limit this jitter, a periodic frame structure is therefore introduced. The upstream and downstream will have a fixed-length frame structure in which, for each frame, polling starts from the first station and circulates through all stations at least one time. Hence all stations can send their data in every frame, and the delay variance is fixed within that one frame. This jitter can be further improved by the use of a buffer.

With this frame structure, polling circulates in the order of traffic types in each frame. As shown in Fig. 6.6, at the beginning of each frame,

Figure 6.6
Frame structure
of FPP.

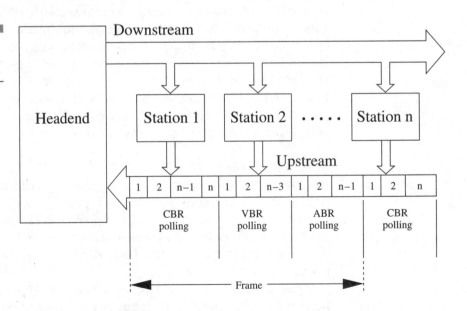

the headend polls all stations and allocates a fixed amount of bandwidth for CBR traffic. After CBR polling, the headend starts VBR polling. It allocates bandwidth for all VBR traffic, where the amount of allocated bandwidth is controlled in order to guarantee the needed QOS. After allocating bandwidth for both CBR and VBR traffic, the remaining bandwidth in the frame is used for ABR traffic. The headend polls stations for ABR traffic. The number of polls for ABR traffic depends on the remaining bandwidth. If the remaining bandwidth is large, polling circulates all stations many times. If bandwidth is small, the headend polls m stations. In the next cycle, it starts polling ABR at station $m + 1$ to maintain fairness.

FRAME FORMAT

Downstream Format. The downstream frame format is shown in Fig. 6.7. Downstream has a periodic frame structure. Each frame is composed of a starting frame synch pattern and a number of slots. Each slot has the same size and is used for transmission of poll or user data. Polling slots are placed according to bandwidth scheduling, and the other slots are used as data slots. For the indication of slot type, each slot has a poll/data indication field. The polling slot contains the following fields.

ID   *Terminal identifier.* ID indicates a destination terminal of this polling slot. This address is a short logical address assigned by the headend.

TYP   *Polling type.* This field indicates the type of data a destination station can send: CBR, VBR, or ABR.

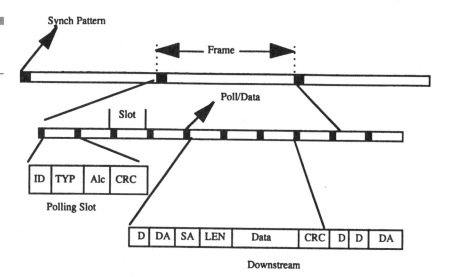

Figure 6.7
Downstream frame format.

Synch Pattern

Frame

Poll/Data

Slot

ID | TYP | Alc | CRC

Polling Slot

D | DA | SA | LEN | Data | CRC | D | D | DA

Downstream

Alc *Allocated slots.* This field indicates the number of slots that a destination station can transfer.

CRC *Cyclic redundancy check.* Error detection through ID to Alc.

Data slots are concatenated to make up a data stream. In this data stream, a number of packets are multiplexed. Each packet has the following fields:

D   *Delimiter.* Separator of packets.

DA *Destination address.* Indicates the destination station ID of this packet.

SA *Source address.* Indicates the source address of this packet.

LEN *Data length.*

DATA *User data.* LLC frame, PCM, and ATM cells are mapped according to data types.

CRC *Cyclic redundancy check.* Error detection through DA to DATA.

Upstream Format. Figure 6.8 shows the upstream frame format. The upstream frame also has a periodic structure. Each frame is composed of a number of bursts from all terminals. Each burst is composed of one or more slots. The first slot is the request slot, and it is used for transmission of bandwidth request information. The following slots are named data slots, and they are used for data transmission.

The request slot contains the following fields:

GT *Guard time.* For collision avoidance between neighboring bursts due to transmission delay fluctuation.

PR *Preamble.* For clock recovery.

Figure 6.8
Upstream format.

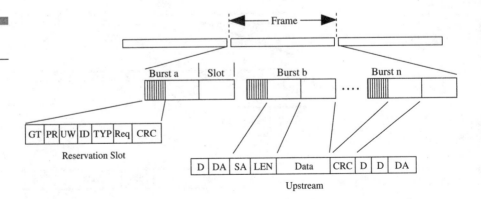

UW   *Unique word.*   For burst synchronization.

ID   *Terminal identified.*   Indicates the source station of this request slot. Definition of ID is the same with downstream case.

TYP   *Request traffic type.*   CBR, VBR, or ABR.

Req   *The number of request slots.*

CRC   *Cyclic redundancy check.*   Error detection through ID to Req.

Data slots belonging to the same station and the same traffic type, are concatenated to make up a data stream. This data stream has the same structure as the downstream.

D   *Delimiter.*   Separator of packets.

DA   *Destination address.*   This definition is the same as SA in downstream.

SA   *Source address.*   This field indicates the source station ID of this packet.

LEN   *Data length.*

DATA   *User data.*   LLC frame, PCM data, and ATM cell are mapped.

CRC   *Cyclic redundancy check.*   Error detection through DA to DATA.

6.4.3 Zenith Electronics: MAC Layer Protocol Proposal: Adaptive Random Access Protocol for CATV Networks

OVERVIEW. The Zenith proposal for the MAC was based on the adaptive random access protocol. The protocol combines a set of optimized contention and reservation schemes. Contention and reservation activities take place on a frame that contains a number of data slots (DSs) and minislots (MSs). The station uses the MSs to contend for the bandwidth and make a reservation request. It can also use a DS for immediate access. A successful reservation request will be allocated with a number of DSs based on the availability.

FRAME STRUCTURE. The frame structure consists of a pair of upstream and downstream channels. The upstream and downstream channels are divided into frames. The stations use the upstream channel for transmission and the downstream channel for receiving. A frame is a fixed size and should be at least the sum of total processing time and twice of the round-trip delay of a cable network. A frame consists of *n* number of slots, and each slot is a fixed size, containing a number of DSs and MSs.

DOWNSTREAM FRAME STRUCTURE. The downstream frames are used by stations for frame synchronization. Each frame consists of four sections:

1. The first section contains global parameters.

2. The second section is used for slot boundary variables; these variables indicate the status of each slot, which can be normal-priority minislot (Mc), high-priority minislot (Mp), contention data slot (DS), or reservation data slot (RS). The structure of the slot boundary variables contains three variables; each variable is one byte in size, and the variable indicates the boundary between the different types of slots.

3. The third section contains acknowledgment fields; each acknowledgment field includes a station ID and acknowledgment.

4. The fourth section contains the data slots. These are shown in Fig. 6.9.

Rc	Range of random number for normal contention
Rp	Range of random number for high-priority contention
DQL	DS reservation queue length
MB	Minislot priority boundary
CB	Contention boundary

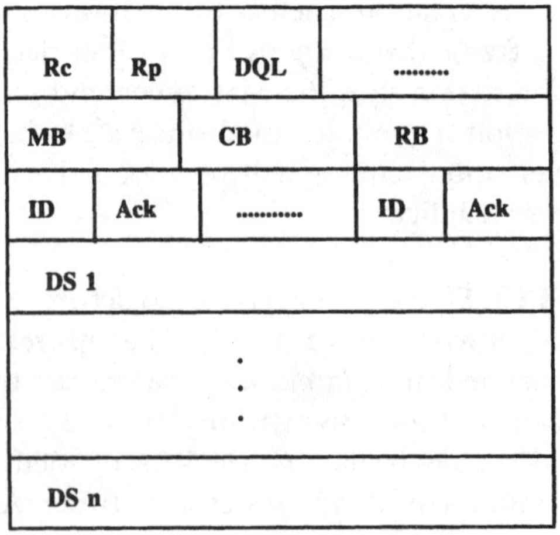

Figure 6.9
Frame structure of
downstream channel.

RB   Reservation boundary

ID   Station identification number

Ack   Acknowledgment

UPSTREAM CHANNEL FRAME STRUCTURE. The upstream frame consists of two types of data slots (DSs):

1. Contention DS
2. Reservations DS

Contention DS. Each contention DS can be further divided into a fixed number of minislots (MSs). The number of MSs and DSs is set dynamically. The number of MSs is increased as the number of backlogged contentions is increased and only bounded by the available bandwidth within the frame. The MSs can be further subdivided into Mc and Mp. Mc is a minislot with normal priority, while Mp is a minislot with high priority.

Initially, as a system starts up, the downstream frame provides a 1/1 ratio of MS and DS available for the stations to perform contention and reservation. This is illustrated in Fig. 6.10a, where equal numbers of Mcs and DSs are provided. The number of MSs (Mc and Mp) is incremented when more traffic is generated as shown in Fig. 6.10b.

On heavy load, the number of available Mcs will decrease as shown in Fig. 6.10c. This number reaches zero when blocking is invoked as shown in Fig. 6.10d. In this case, no Mps are processed.

DESCRIPTION OF THE MAC PROTOCOL. The operations of a station can be represented by using a state transition diagram as shown in Fig. 6.11.

The contention and reservation schemes consist of three sets of rules:

1. Transmission rules
2. Queuing rules
3. Reservation rules

TRANSMISSION RULES. The stations can access the MSs and contention DSs at any time, unless the system is in a blocking state. Blocking is experienced when there are excess collisions that spread in most of the MSs. In general, a station can transmit immediately onto a MS and DS under light traffic load, and it will only transmit onto a MS when there is no DS available under heavy traffic load. The selection of the MS and DS is on a

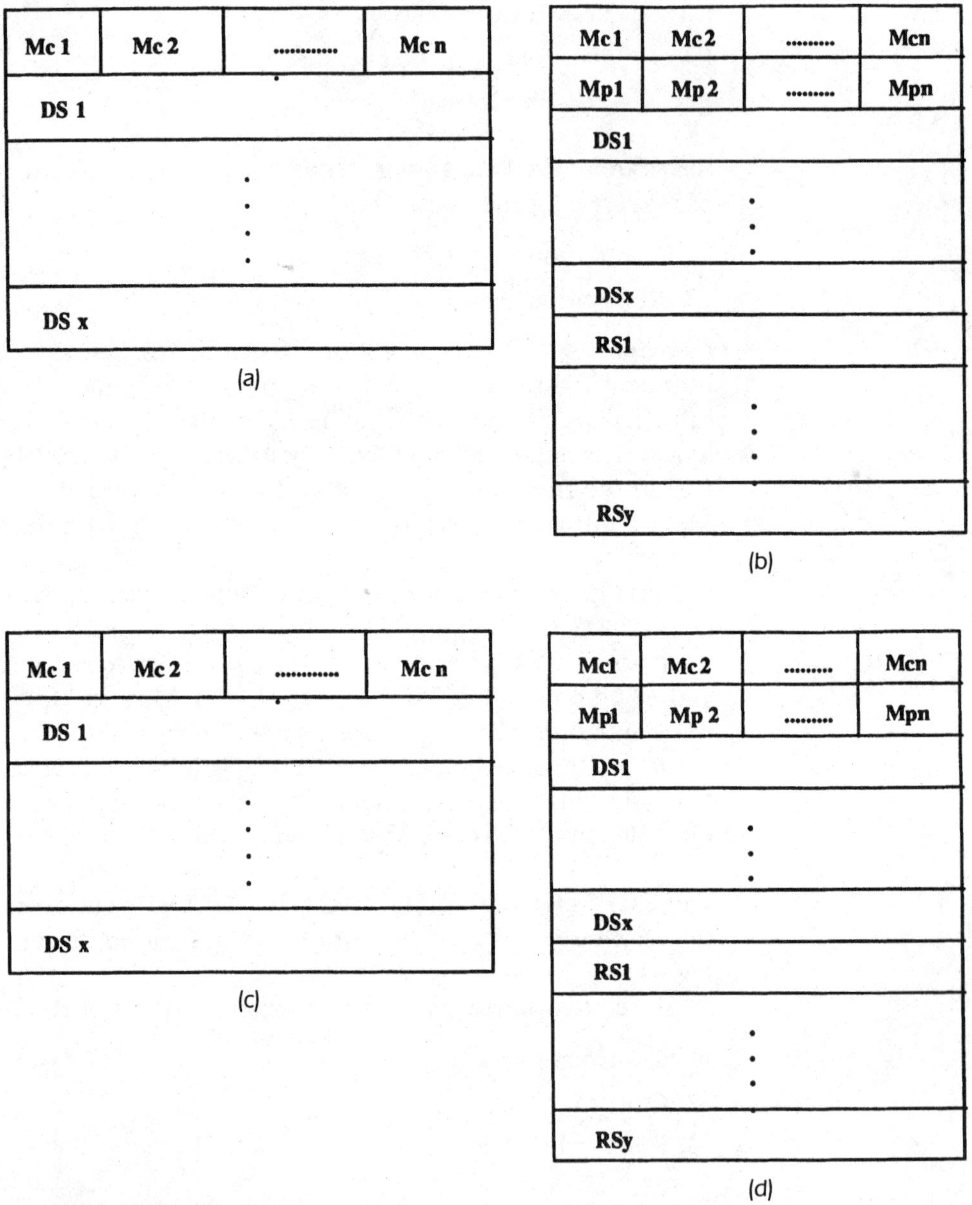

(a)

Mc 1	Mc 2	....	Mc n

DS 1

.
.
.

DS x

(b)

Mc1	Mc2	....	Mcn
Mp1	Mp2	....	Mpn

DS1

.
.
.

DSx

RS1

.
.
.

RSy

(c)

Mc 1	Mc 2	....	Mc n

DS 1

.
.
.
.

DS x

(d)

Mc1	Mc2	....	Mcn
Mp1	Mp 2	....	Mpn

DS1

.
.
.

DSx

RS1

.
.
.

RSy

Figure 6.10
Frame structure: a, on start-up; b, after contention and reservation; c, under heavy traffic load; d, after blocking.

Figure 6.11
State transmission
diagram of a station.

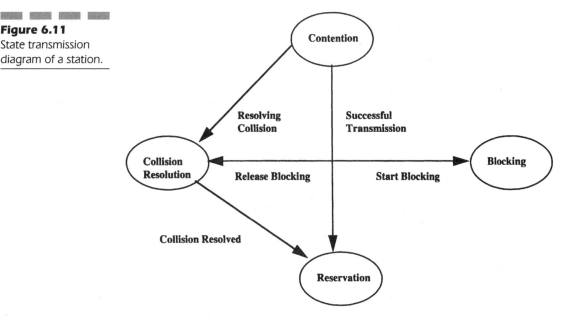

random basis, and a random number is generated that is based on the number of Rcs and Rps. The Rc represents the range of random numbers based on collisions in Mcs, and the Rp represents the range of random number based on collisions in Mps.

If the contention is successful, then the station will be acknowledged and granted the requested slots. If the number of reservations exceeds the number of available DSs, the reservation requests will be stored in a DS reservation queue (DQ). A headend controller will assign the requested slots later based on a first come, first serve basis. If the contention fails, no acknowledgment will be sent from the headend to the station.

QUEUING RULES. The queuing rules control the priority and ratio of the MS and DS. The headend maintains a DQ, which contains the number of slots allocated to each station. There are two transmission priority levels: normal and high. A contending station becomes a high-priority station after it experiences its first collision. Once the collisions spread out in all of the MSs, the ratio of the MS and DS will be increased by one unit per increment. After the ratio of the MS and DS is increased, only those stations with high priority can contend in the MS. The new arrivals of normal contention transmissions can contend the original Mcs only. Each station, after experiencing collisions, maintains its own priority level and randomly selects one of the Mcs and Mps, as it retransmits. The two variables, Rc and Rp, are used by the stations to access Mc and Mp.

After each frame arrives at the headend, the headend detects collisions and records the number of collisions. If there are many collisions in Mcs, then the headend will add another unit of MS called Mp. The limit on the number of Mps that can be added is based on the size of DQ. As the number of reservations in the DQ increases, the system starts to decrease the number of MSs (Mc and Mp). Once the number of Mcs reaches zero, the system starts to block any new transmissions, which is indicated by the slot boundary variables showing that no Mc is available. The blocking can prevent the system from congestion. At this time, only those pending numbers of Mps and Mcs can continue their transmissions until most of the collisions are resolved. In order to increase the successful rate of reservation, the stations contending the Mps will start to increase the range of their random numbers. As the range of the random numbers is increased, the probability of transmissions colliding in the same slot is decreased.

The blocking will be released (unblocking) after most of the previous collisions are resolved and the DQ is near empty; then the system opens a fresh number of MSs consisting mostly of Mcs. Note that it is possible that the system may have a good number of backlogged transmissions due to the previous blocking. Therefore, it is necessary to control these stations as they are transmitting by using a large range of random numbers. Also, the Rc is set to the same level as when the system entered the previous blocking. The flood of the new arrivals is moderated.

RESERVATION RULES. In addition to the transmission rules and queuing rules, the reservation rules are designed to meet the operation requirements of multiple types of traffic, such as constant bit rate (CBR), variable bit rate (VBR), and available bit rate (ABR) traffic. The classification of CBR, VBR, and ABR is based on the service requirements of voice, video, and data traffic. Each type of traffic has different packet delay and packet loss requirements. These different types of traffic can also be described to as being bursty, synchronous, and isochronous based on their transmission and receive characteristics. Bursty data traffic is mostly connectionless, ranging from bursty to sporadic. The synchronous and isochronous traffic is mostly connection-oriented, requiring a session to be set up between the source and destination nodes. This type of traffic is loss-tolerant and delay-sensitive, while bursty data traffic is delay-tolerant and loss-sensitive. The design of the reservation rules focuses on the support of the three types of traffic: ABR, CBR, and VBR.

PERFORMANCE CONSIDERATION. Based on the design of the adaptive random access protocol, the protocol dynamically adjusts the number of minislots and contention data slots for different traffic load. Under the light traffic load, the system uses a small number of Mcs. As the load increases, the system increases the number of Mcs, and the stations also change their priority level after the first collision. As the collisions increase, the system allocates additional MSs called Mps. By dynamically allocating the MSs and adjusting the priority level, the system can optimize the delay experienced by the stations. That is, as the traffic load increases, the station can still maintain the delay at an optimized level and keep the throughput as high as possible. The blocking scheme can help the MSs from being congested. As the blocking is released, the system uses a large range of Rc to prevent an impulse of transmissions. The large range of Rc decreases the probability of collisions.

As a result, the system consistently maintains its throughput and access delay at an optimal level. The system can be enhanced to expedite data by including a second data queue called DQx and adding an additional level of transmission priority.

6.4.4 General Instrument Corporation: MAC for HFC

OVERVIEW. GI proposed a MAC protocol to be used between the head-end processor (HEP) and the unit at the customer premises known as the subscriber client unit (SCU) or station. This MAC protocol is used to perform various functions related to administering the resources over the hybrid fiber coax (HFC) link between these two network entities. The MAC proposal addressed:

- Message structure
- Message addressing
- Upstream access to the shared medium
- HFC command message functions and protocols

HFC COMMAND MESSAGE DOWNSTREAM REQUIREMENTS. This section defines the requirements for the downstream MAC protocol. Figure 6.12 below shows a generic MAC reference model.

MAC MANAGEMENT ENTITY. The MAC management entity in the head-end controls the initialization, provisioning, and registration to enable

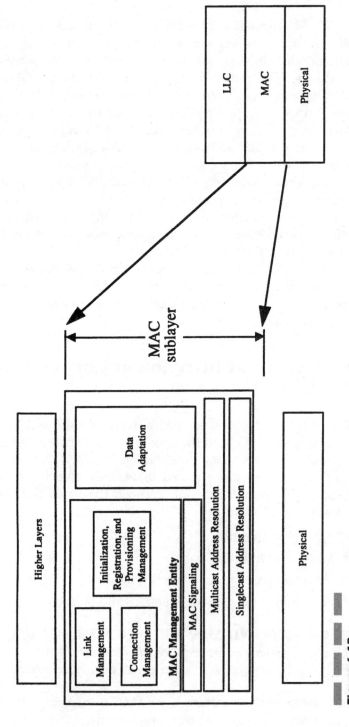

Figure 6.12
MAC reference model.

entry into the network. It controls connection management, which allows establishment and release of network bandwidth resources. It controls link management, which provides control and calibration of link-related and power and ranging parameters. It also maintains databases, security, management, diagnostics, and network configuration.

DATA ADAPTATION. The data adaptation entity provides adaptation services, specifically the ATM adaptation layer 5, between the MAC signaling (HFC command messages) and the ATM layer.

PHYSICAL LAYER INTERFACE. The MAC protocol is independent of the physical modulation scheme and data rate used. This provides the greatest flexibility to implement a common MAC signaling protocol across a variety of physical channel types. The physical modulation schemes include, but are not limited to, 256 QAM, 64 QAM, and QPSK.

UPSTREAM MAC REQUIREMENTS

UPSTREAM HFC COMMAND MESSAGE ENCAPSULATION. This section defines the upstream sublayers of the SCU. The SCU utilizes three layers to provide services for upstream transmission of HFC command messages. Messages are forwarded to the ATM adaptation layer 5 (AAL5). This layer is divided into the AAL5 common part convergence sublayer (AAL5-CPCS) and the AAL5 segmentation and reassemble (AAL5-SAR) sublayer. These sublayers provide segmentation of the higher-layer SDUs. The AAL5-PDU is forwarded to the ATM layer. The ATM layer generates a 53-octet cell by appending the 5-octet header to the ATM-SDU. The ATM-SDU next is forwarded to the MAC layer. The MAC sublayer receives the ATM-SDU and appends additional fields to support transmission on the upstream channel.

AAL5 Common Part Convergence Sublayer (AAL5-CPCS) PDU. The AAL5-CPCS sublayer is responsible for insertion of a trailer that indicates the message length. The CPCS-SDU is forwarded to the CPCS sublayer from the higher layers. The AAL5 trailer is appended to the end of the SDU, forming the CPCS protocol data unit (PDU). The fields associated with the AAL5-CPCS PDU are described in Table 6.1.

AAL5-SAR and ATM Layer. The SAR-SDU is passed to the AAL5-SAR sublayer via the AAL5-CPCS. The SAR sublayer provides the upstream fragmentation of the SDU. The fragmentation process consists of the partitioning of the SDU into 48-octet payloads of an ATM cell. AAL5-SAR and ATM encapsulation is described further in Chap. 3.

TABLE 6.1

AAL5-CPCS PDU
Format

Field	Number of bytes	Description
CPCS-PDU payload	Dependent on higher protocol	Contains the higher-layer protocol data unit.
PAD	(0–47 bytes)	Aligns the AAL5 trailer to be the last bytes in an ATM cell.
CPCS-UU	1	No function under multiprotocol adaptation.
CPI	1	One of the functions of the CPI field is to align the CPCS-PDU trailer to 64 bits. Other functions are for further study. When the 64-bit alignment function is used, this field is coded as all zeroes.
Length	2	Indicates the length of the CPCS-PDU. A length field of 0×00 is used as the abort function.
CRC	4	Cyclic redundancy on CPCS-PDU.

FRAME STRUCTURE.   Message structure for the downstream channels and for the upstream channels are as described below:

DOWNSTREAM FRAME STRUCTURE.   In the downstream direction, the transport at the data link layer is the ATM cell. The MAC layer adds no additional fields or bytes to the 53-byte ATM cell in the downstream. The headend resource manager specifies the VPI to be used in cells sent and received from a particular station. The VPI serves as a unique identifier for a particular station. The VCI value is assigned during the call setup signaling procedure. At the headend, the cell is passed down by the ATM level to the transmission convergence sublayer of the physical layer. At the transmission convergence layer, the overhead required for the downstream forward error correction (FEC) code is inserted. The type of FEC and the resulting physical-layer framing structure is described in Chap. 3.

UPSTREAM MESSAGE STRUCTURE.   In the upstream direction, the packet transported at the data link layer is composed of N ATM cells with one or more additional bytes. These additional bytes are sequence number fields that are used for the implementation of an automatic repeat request (ARQ) protocol. ARQ is used as a retransmission scheme for error correction. The bits in the sequence field are used to represent the sequence number of the upstream packet composed of one or more ATM cells. The sequence number is used so that the headend can request retransmission of an uncorrectably errored packet or a missed detection. In this case, the head-end sends an HFC command message to the station that acts as a negative acknowledgment (NAK) with the sequence number of the upstream packet that is expected next by the HEP from the station. Upon receiving a NAK MAC message, the station will retransmit the packet with the sequence number contained in the NAK message and will continue to transmit the subsequently sequenced packets. Thus, this is a go-back-N type of ARQ scheme, not a selective repeat type of ARQ.

MESSAGE ADDRESSING

UPSTREAM DATA MESSAGE ADDRESSING. In the upstream, source address information is contained within the VPI of the ATM cells. The VPI associated with a station is assigned during the power-up sequence. The VCI is assigned during signaling.

DOWNSTREAM DATA MESSAGE ADDRESSING. In the downstream direction, data is addressed also using the VPI. A station will filter the received downstream ATM cells for the assigned VPI that is assigned to the station during the power-up procedure.

HFC COMMAND MESSAGE ADDRESSING

DOWNSTREAM BROADCAST. Before a station is assigned a VPI identifier MAC management messages are sent with a well-known VPI used for broadcasting to all stations. For example, an "invitation" message for uninitialized units is a broadcast MAC message.

DOWNSTREAM SINGLECAST WITHIN A BROADCAST. This message uses an assigned VPI for broadcasting but is intended for a specific station. This occurs before a VPI is assigned to a station. The message that assigns the unique VPI to a station is an example of this singlecast message that is carried in a cell with the broadcast VPI. Each station will examine the address within the MAC message and determine if it matches the "burned-in" station address.

DOWNSTREAM SINGLECAST. HFC command messages intended for a particular station will have the VPI equal to the VPI that was uniquely assigned to that particular station by the headend. A reserved VCI is used to identify an HFC command message.

UPSTREAM MAC MESSAGES. Upstream MAC messages sent from the station contain the VPI that was assigned to the station. During the power-up sequence before the unique VPI identifier is assigned, a station uses the well-known "broadcast" VPI/VCI for upstream MAC messages. In this case, the station includes its unique "burned-in" address within an upstream HFC command message.

MAC UPSTREAM MULTIPLE ACCESS SCHEME. The MAC controls the upstream transmission from a station in both frequency and time using frequency and time division multiple access (FTDMA). The MAC also supports unsynchronized access to the network as required. The list below summarizes the access methodologies supported by the MAC:

- Fixed-assigned TDMA
- Dynamically assigned TDMA
- Slotted contention-based access
- Unslotted contention-based access

Multiple access modes can exist on the same channel. This is accomplished by resource management in the headend.

TDMA SCHEME. TDMA is the baseline access scheme for upstream transmission. The scheme supports a high degree of programmability and easy reconfigurability and allows variable data rates. The headend MAC management entity controls the agile upstream frequencies. At a given frequency, the headend controller MAC management entity relays TDMA configuration parameters to the station. The upstream TDMA, as shown in Fig. 6.13a, is fully configurable and programmable in terms of the parameters as described below.

Slot Size. The slot size parameter of the TDMA scheme is defined in terms of the number of timebase reference counts or *ticks* the slot occupies. The reference tick is used to measure the passage of time for upstream transmission for a bit or group of bits. The headend controller MAC management entity provides the station the slot size information during the initialization and provisioning state.

Frame Size. The headend controller MAC management entity shall specify the frame size associated with an upstream channel. Frame size is expressed in the number of slots per frame. The frame size value is configurable across different frequency channels, allowing a multitude of upstream data rates.

Super-Frame Size. The headend controller MAC management entity shall specify the super-frame size associated with the upstream TDMA

■■■ ■■ ■■ ■■
Figure 6.13
TDMA: a, structure;
b, frame format.

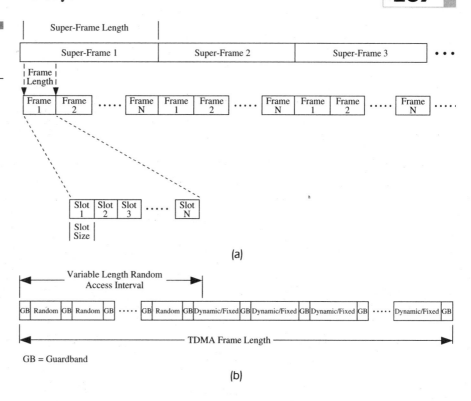

(a)

(b)

GB = Guardband

channels. The super-frame size is expressed as the number of frames per super-frame.

Frame Format. The MAC supports a dynamically configurable TDMA frame format shown in Fig. 6.13*b*. The components of a frame consist of a variable number of slots allocated to contention access. The remainder of the frame can be utilized for either fixed-assigned or demand-assigned TDMA access. The entire frame can be configured for contention-based access (or fixed access) or dynamically assigned access, or any combination of both. The MAC connection management entity shall determine dynamically the required frame composition based on user traffic requirements.

6.4.5 Scientific Atlanta (Georgia Institute Of Technology): XDQRAP

OVERVIEW. Scientific Atlanta, in partnership with Georgia Institute of Technology, proposed the extended distributed queuing random access protocol (XDQRAP) as the MAC for HFC. XDQRAP is an adaptive

MAC protocol for multimedia traffic over hybrid fiber/coax networks. The protocol features immediate medium access at low-offered traffic loads, and transitions to a reservation-based access at high-offered loads. It is assumed that synchronization and guard band between slots are performed properly. A station must know frame timing relative to the headend so it can send data on time, and guard band is also necessary to provide for slot-timing uncertainty due to inaccuracy of the ranging.

GENERAL DESCRIPTION. In the XDQRAP, the channel time is divided into a number of fixed-size time slots. Each slot contains the information field (payload) and the control field. The control field is used to negotiate with the headend a request permission to transmit onto the shared medium. This is done through contention resolution. The control field has two control minislots—more than two minislots is not precluded.

A station with data to transmit writes the transmission request in one of the control minislots. Because of the nature of the shared medium topology, contention will, no doubt, occur. The headend will monitor and manipulate the packet and send the status of the control minislots and the data slots back to the stations through the downstream channel. Each station, then, adjusts its local state machine accordingly: stations that experience no contention are queued in the global transmission sequence; while stations that experience contention will back off for a random length of time and retry in the next time frame. Figure 6.14 illustrates the upstream and downstream flow structure through the headend.

Figure 6.14
Upstream and downstream flow structure.

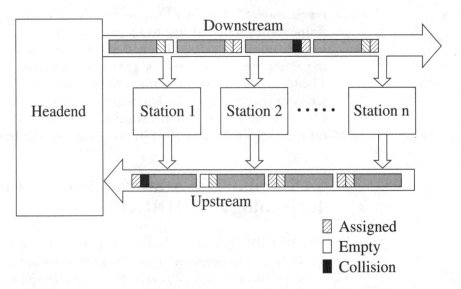

A control minislot contains a request for a certain number of data slots R_k, where k is the number of slots requested. The requested station is S_j, where j is the station number. Given this, a request for k slots from station j is formatted $(R_k S_j)$.

PROTOCOL DESCRIPTION. When traffic is generated at a station, it writes its request in the minislot to the headend. The request can be written in one of the two control minislots randomly. The headend provides feedback to the stations via a set of distributed queues. If one request collides with another request, the contention is resolved by a collision resolution algorithm. If the station received a grant request (no collision experienced), it transmits an appropriate slot as instructed by the headend.

FRAME FORMAT. The upstream and downstream pipes have different functions. Therefore, their frame formats are different.

Upstream Frame Format. The upstream frame, as shown in Fig. 6.15, is used for signaling and data transmission to the headend. It consists of a data slot and two control minislots (CMSs).

The following is a description of the upstream fields:

Preamble. Mainly used for synchronization of the frame.

ID. The ID is significant only for the XDQRAP protocol. Will be used to match this protocol with a grant received by the headend.

Length. Length is used by requesting the station and writing the number of data slots requested for transmission.

CRC. The CRC is used to check the integrity of the data.

FEC. FEC at the data slot is used for single-bit error correction.

Downstream Frame Format. The downstream frame, as shown in Fig. 6.16, carries the feedback information and user data from the headend to the

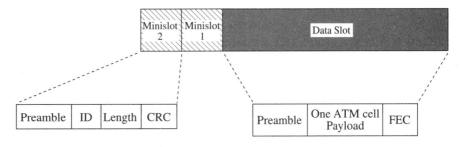

Figure 6.15
Upstream frame format.

Upstream Frame Format

stations. The feedback information includes both control minislot Ack and the status of the data slot. The feedback information field is followed by a data slot containing a number of ATM. The number of ATM cells in the data slot is determined by the ratio of the transmission speeds of the downstream channel and the upstream channel and the number of the upstream channels sharing the downstream channel.

Here is a description of the downstream fields:

Station ID.   The station ID is significant only for the XDQRAP protocol. It will be used to match this protocol with the station number requesting access.

Length.   The number of data slots granted by the headend for that particular station request.

Feedback.   Three states are provided in this field:
- Empty
- Single request
- Collision experienced

TQ feedback.   This signifies the current value of the transmission queue maintained by the headend.

RQ feedback.   This signifies the current value of the collision resolution queue maintained by the headend.

CRC.   The CRC is used to check the integrity of the data.

QUEUES IN XDQRAP.   There are three main queues, two of which are global and one of which is local to each station. The two global queues used by each station to monitor the global state of the system are:

Figure 6.16
Downstream frame format.

Downstream Frame Format

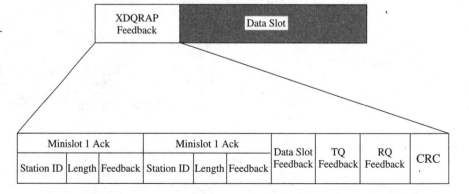

1. Data transmission queue (TQ).

2. Collision resolution queue (RQ). Arriving packets are stored in the arrival queue (AQ), which is local to each station.

The TQ ensures a collision-free transmission in the data slot, whereas the RQ allows a station to retry a request that collided in a previous attempt. The only time the TQ allows data slot collisions is during immediate mode transmissions. Requests that collide will go into a contention resolution algorithm. This algorithm in XDQRAP is similar to the tree resolution algorithm proposed as an enhancement to Ethernet.

To send data, stations write into the control minislot (CMS) randomly. Any or all minislots may experience collision with two or more stations trying to write in the same control minislot. When this happens, the system enters into a collision resolution phase. If both minislots have collisions, the stations writing in the second and higher minislots must wait until those stations writing in the first minislot have resolved their collisions. All stations monitor the status of this resolution algorithm while collisions are being resolved. New stations are not permitted from entering the reservation process during the resolution phase.

TRANSMISSION RULES. Data transmission rules (DTRs) and request transmission rules (RTRs) are based on the values of the TQ and RQ given by the headend. The TQ and RQ values are updated and maintained by the headend using the queueing discipline rules.

DATA TRANSMISSION RULES (DTRs). There are two data transmission rules:

1. Immediate medium access for single data slot messages

2. Reservation access

Immediate Medium Access. In the immediate access, a station can write information in the data portion of the current slot if one of the following situations applies:

■ The global reservation system is empty and only one slot is required to send data. This preemptive send (without request) can be used when there is no danger of affecting previous reserved slots. If collision is experienced, then no additional control reservation is needed, since the reservation process has already been requested and is being processed.

■ The station's turn to transmit data comes up according to the reservation process.

Reservation Access. There are two request transmission rules that dictate when a station can send a request in one of the control minislots:

1. *Initial request,* indicating that the first request from a station that has a message to transmit can be made if and only if resolution queue is empty.

2. *Retry request,* applied when a station has already made an initial request. Dictates when a repeat request can be made.

FEEDBACK BUFFERING. Buffering is utilized at each station to ensure that the station at the farthest end of the network has had an opportunity to receive the request. Additional buffering is added so that a station can apply the feedback information to the correct slot. Stations closest to the headend require the greatest buffering, while the station farthest from the headend applies the feedback without extra delay. The quantity of the buffering is expressed in slots.

PERFORMANCE. XDQRAP simulation performance results were the outcome of simulations performed using two independently developed models. These results characterize the behavior of XDQRAP under a variety of traffic loads and network configurations. In all cases, the network interleave was taken to be three slots. Figure 6.17 shows the average delay, in slots, versus offered load for 10, 30, and 50 stations. Traffic was uniformly distributed across the stations with all requests made for two data slots. Since there were no single-slot requests, there were no DTR1 transmissions made during the course of the simulations. Under light loads, with data rule 1 active, the throughput and delay characteristics are very similar to slotted ALOHA. However, it is under heavier loads that XDQRAP distinguishes itself, remaining stable, for most traffic patterns, with offered loads of up to 95 percent. With additional minislots, throughput figures can be approaching 100 percent.

6.4.6 National Tsing Hua University: General Multilayer Collision Resolution with Reservation

OVERVIEW. National Tsing Hua University proposed a MAC with collision resolution algorithm and reservation that utilizes the long propagation characteristic of the cable network. In the centralized operations, the need for complicated time synchronization procedures between headend

Performance chart of XDQRAP.

Figure 6.17

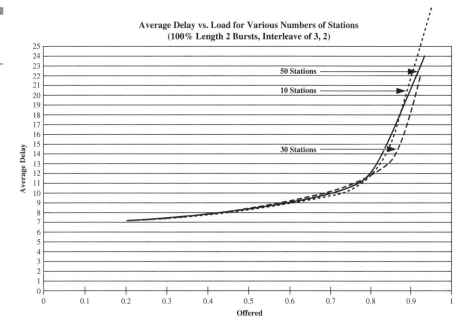

Performance chart of XDQRAP.

and users will not be needed. All that is needed is to provide the ranging function that calculates the round-trip delay of a node to headend. The following description assumes that each station has the information on its own round-trip propagation delay to headend.

A typical tree-and-branch CATV network is shown in Fig. 6.18. There may be hundreds to thousands of stations connected to a headend via the fiber/coaxial links. Large numbers of users together with long propagation delay (up to 800-μs round-trip delay) make the design of the MAC protocol a great challenge, especially when we consider real-time multimedia transmission over such a network.

PROPOSED PROTOCOL.   The concept of multilayer collision resolution is to reduce the probability of collisions by using numbers of independent resolution levels. As in the CATV network, we could have the following levels for resolution:

- *Frequency components* (orthogonal signaling), in which nodes with packets to transmit send a random number to the headend via orthogonal signaling.

- *Grouping resolution* (grouping by power levels and propagation delay). In the CATV network, nodes are grouped within the same contention slot, with similar power levels and propagation delays, in order to use

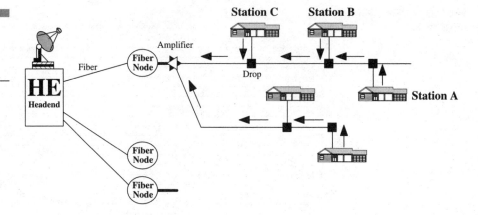

Figure 6.18
Traditional tree-and-branch CATV topology.

the channel more efficiently. Another reason for grouping is to make user numbers within each group roughly the same.

■ *Spatial domain resolution.* Because we divide users into groups by their propagation delays to the headend, we may use the pipelining method to improve the protocol efficiency.

■ *Time-domain resolution.*

Because of the flexibility provided by this protocol, we could accommodate different numbers of users in different CATV network. This could avoid traffic jamming in heavy-load situations by dynamically multiplexing and distributing users into different contention slots. The contention frame structure is illustrated in Fig. 6.19.

For a CATV network with hundreds to thousands of users, one time period is enough to prevent the network from jamming. Assume the following:

N	Number of subscribers in a CATV network.
M	Number of megagroups.
S	Number of supergroups per megagroup.
G	Number of groups in each supergroup.
p	Number of orthogonal signals used.
(i,j,k)	The group ID of this user; i is the megagroup, j is the supergroup, k is group number of this user. Note that this ID does not have any physical meaning but could help us to explain the detailed operations of this protocol.

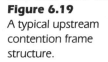

Figure 6.19

A typical upstream contention frame structure.

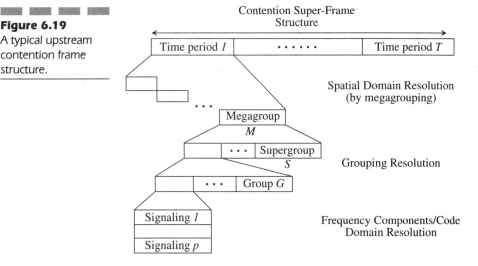

Within this multilayer structure, each node has a fixed group ID given by the headend when it first attaches to the network. Thus, we could equivalently have $M \times S \times G \times p$ random addresses for resolving N users in a CATV network, since the number of random addresses the system could provide depends on the complexity and reliability of the addresses detector. Thus, the value of p is limited, while other parameters could be optimized independently to provide maximum efficiency.

UPSTREAM TRANSMISSION. The upstream transmission could be divided into contention and reservation periods. Control and QOS signaling messages would use the random addresses transmission period to access the channel while time-bound multimedia traffic must go through a reservation slot. The general multilayer collision resolution algorithm provides an efficient resolution mechanism in the contention period, while the reservation period needs a bandwidth allocator to designate the bandwidth to users while maintaining the required QOS of the connection. The design of the bandwidth allocator is independent of the contention protocol and could fit well in this super-frame structure.

CONTENTION PERIOD. The operations of the general multilayer collision resolution protocol is specified as follows.

The headend receives upstream traffic when previous transmission is over. Nodes with packets to transmit send a random address (this could be done prior to receiving the broadcasting messages) when receiving the downstream ready signal broadcasted by the headend. This signal could be a special end-of-file message piggy-backing at the end of previous messages. Those nodes with the same group ID transmit their random addresses in the same contention slot. If the orthogonal addresses are properly designed, then simultaneous detection could be done.

Nodes transmit their addresses in the contention slot whenever they receive the ready signal (the signal belongs to their group). Since all nodes hear this signal at different times, the resulting contention slot would look like the one shown in Fig. 6.20. But this does not have any influence on the multiple addresses detector of the system. Furthermore, those nonoverlapping addresses, on the other hand, could be more easily detected. So that is why we said that the system does not need complicated synchronization among users.

The headend listens to the upstream channel, detecting any random addresses transmitted by active users. After listening to all groups, the headend starts to poll users according to the addresses it received. In this period, the headend only uses the random address information to poll active users and does not necessarily need to know the addresses of these users. Because this is a centralized polling scheme, users transmit according to the timing of the polling signal. At the end of the contention period, the headend sends a signal, indicating the beginning of the reservation period.

RESERVATION PERIOD. This period needs a bandwidth allocator to allocate bandwidth to different kinds of traffic. When a call is granted, the headend assigns slots to this user according to the request and broadcasts

Figure 6.20
Random address transmissions.

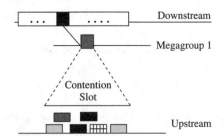

the assigned slot numbers to this user. When all nodes receive the starting message of the reservation period, they could calculate the time to transmit their data into the correct time slot.

In the upstream transmission, both fixed or variable super-frame lengths could fit in the protocol. Fixed super-frame size has the advantage of easy operation and bounded (at least smaller than variable-case) delay jitters of each reserved user. On the other hand, variable super-frame length, although it would result in large delay jitters, could still have smaller call/transmit (request) delay under heavy load conditions. But in order to satisfy the QOS requirements, the variable super-frame size should have lower and upper bounds.

In a CATV network, numbers of messages can be transmitted at the same time without collision, because users are divided into megagroups according to the propagation delay. Since the headend has the information of each user's propagation delay, it can coordinate the transmission time in the upstream channel so that the messages will not collide with each other. This is illustrated in Fig. 6.21.

6.4.7 Philips Research Laboratories: General MAC

OVERVIEW. Philips proposed an overall system approach. It was suggested that the MAC must accommodate the general structure of an interactive system as shown in Fig. 6.22. The MAC must be scalable, and it must support simple interactive applications, telephony (STM traffic), and advanced data communications (e.g., advanced ATM services). The proposed MAC, therefore, should accommodate multiple access protocols, including DQRAP, bitmap, ALOHA, polling, and so on.

Figure 6.21
Methods for improving efficiency in a CATV network.

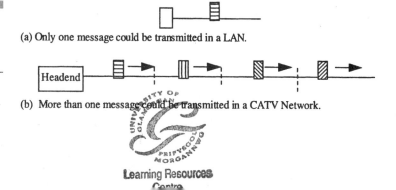

(a) Only one message could be transmitted in a LAN.

(b) More than one message could be transmitted in a CATV Network.

In the future, communications systems will probably converge toward the architecture shown in Fig. 6.22, in which all the communications equipment at the customer's premises will be connected through a single network interface module to the rest of the world. There could, of course, be several network operators for both the core and local access networks that would be in competition with each other. One of the concerns is the types of services that these systems need to support. The type of service that will be implemented in a certain network will depend on both technical and nontechnical factors. Whatever system is introduced in the return path today will have a lifetime of several years. This installed base will need to be supported for quite some time. Therefore, it needs to be future-proof. The system discussed in this section offers the flexibility to cope with different physical layer specifications, and, at the same time, it opens the possibility of implementing a variety of access protocols so that a variety of services can be supported.

INTERACTIVE AND INTERACTION CHANNEL. The different types of services that can potentially be delivered on a two-way CATV network are described by ITU, the ATM Forum, and other bodies. The ATM Forum, in the UNI specification version 3.1, defined four QOS (Quality of Service) classes. These classes are described in Chap. 3. In summary:

Class 1 is connection-oriented in support of CBR services.

Class 2 is connection-oriented in support of VBR services.

Class 3 is variable bit services without time dependencies.

Class 4 is variable bit rate using connectionless connection and no time dependencies.

Table 6.2 gives the overview of the services and their characteristics. The table also indicates the required bit rates. The value of these bit rates is a maximum value. The average value can be much smaller. If, for instance, a home worker transmits a burst of 1 Mbyte once an hour on a 10-Mb/s network interface, this burst is transmitted within 800 ms, but the average bit rate is approximately 2 Kb/s. In addition, every network needs some extra bit rate or bandwidth for management purposes.

The HFC network represents a tree-and-branch topology and therefore will always have a single point in the root of the tree (e.g., the headend). This means that the central point can also be used for processing the protocols centrally instead of in a distributed way. This simple observation

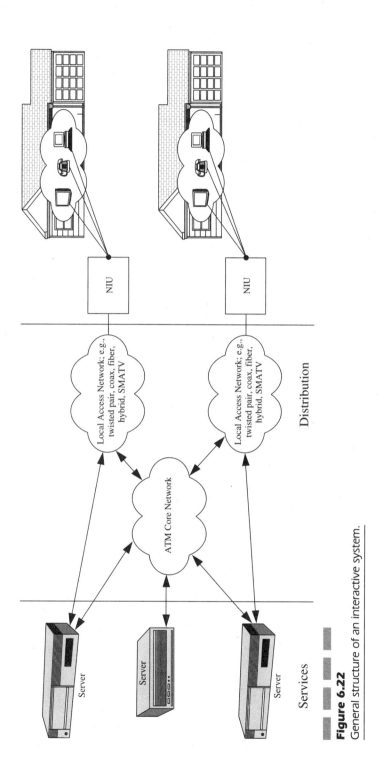

Figure 6.22

General structure of an interactive system.

TABLE 6.2

Characteristics of
Digital Services on
a Two-Way Cable
Network per User

Downstream		Upstream		
Peak bit rate/user	Spec. QOS class	Peak bit rate/user	Spec. QOS class	Application
1.5–2 Mb/s	1	1.5–2 Mb/s	1	Permanent leased line
1.5–2 Mb/s (few Kb/s)	1 3 or 4	1.5–2 Mb/s (few Kb/s)	1 3 or 4	Managed leased line Management data
12–64 Kb/s (few Kb/s)	1 3 or 4	12–64 Kb/s (few Kb/s)	1 3 or 4	Switched telephony; cordless signaling and management
64–384 Kb/s (few Kb/s)	1 3 or 4	64–384 Kb/s (few Kb/s)	1 3 or 4	Video telephony signaling and management
(few Kb/s)	3 or 4	(few Kb/s)	3 or 4	Telemetry, alarm, games
1.5–10 Mb/s	1 or 2	—	—	Digital TV distribution, pay-per-segment, simple pay-per-view, and NVOD
1.5–10 Mb/s (few Kb/s)	1 or 2 3 or 4	(few Kb/s)	3 or 4	Pay-per-view, NVOD, VOD, games, retrieval services, home shopping, alarm, teleeducation, teleadvertising, video mail and mail services, teleworking, telemetry, telegambling, end-to-end data communication, and signaling and management
		384 Kb/s— 10 Mb/s	1 or 2	Monitoring (traffic and security), surveillance
(few Kb/s)	(3 or 4)	(few Kb/s)	(3 or 4)	End-to-end data communication
(few Mb/s)	3 or 4	(few Mb/s)	3 or 4	Database application, teleworking
(few Mb/s)	unspec.	(few Mb/s)	unspec.	Data file transfer

has important consequences for the system. The main advantage of this
approach is the increased flexibility of the system. The nodes in the network have to signal to the central point that they require some bandwidth,
but the consequent processing of those requests and the allocation of
bandwidth to a node is done centrally. So the terminals only have to know

how the requests are transmitted but not how to deal with the requests. The focus then will be on the frame structure and the functionality that are necessary for standardization, so that an open system is defined that can be installed on present-operational and future-operational networks. The proposed system can support a variety of MAC protocols so that, depending on the class of services that the operator wants to offer, an optimal MAC protocol can be chosen.

FRAMING STRUCTURE. Figure 6.23 illustrates a general layout of the frame structure. It is assumed that the general frame structure is similar in the upstream and downstream directions.

At the bottom of the frame hierarchy, minicells are defined, which are built up from X bytes. A cluster of M minicells forms a basic frame (BF). At the top of the frame structure, we have the so-called multiframes. A multiframe contains B basic frames. There are two different classes of data:

1. System data
2. User data

SYSTEM DATA. In system data, minicells are used for synchronization, ranging sequences, housekeeping messages, and the MAC protocol. These MAC cells are used in the upstream direction for requests from the stations. In the downstream direction, these cells are used for acknowledgments on these requests and for bandwidth assignment to a particular station.

Figure 6.23
General frame structure.

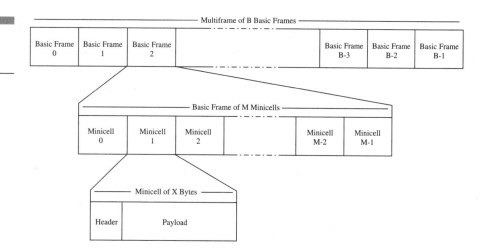

USER DATA. In the user data, minicells are used for STM services and all other sorts of services. These other services will all be based on ATM. Since in general an ATM cell is larger than a minicell, an ATM cell will be mapped into a sequence of consecutive minicells.

The distribution of the various cell types can differ per basic frame and is fully controlled by the headend. The amount of minicells allocated for housekeeping, the MAC layer, and user data is adapted to actual needs and can even be zero for some cell types. Therefore, the frame structure must not be thought of as a rigid one but more as a virtual structure to facilitate an easy implementation of multiple services. A multiframe with a repetition rate of B basic frames makes it easier to address a specific basic frame. This addressing facilitates implementation of ranging, very-low-bit-rate STM services, and the flexible use of MAC cells in relation to delay requirements, among other things.

The system can operate in two modes: a mode where there is no jitter on the STM data, and a mode where a certain amount of jitter on STM data is allowed. The latter mode allows a more efficient mapping of minicells for other, non-STM services within a basic frame. In the no-jitter mode, it is possible that, for a certain STM capacity, an integer number of ATM cells does not fit in the remaining interval between two successive STM blocks. This will cost a certain amount of ATM bandwidth. By allowing an amount of jitter on the position of the STM cells in the frame, we can pack the frame as tightly as possible. The general frame structure can be described by a few parameters as shown in Table 6.3.

UPSTREAM FRAME. The frame structure shown in Fig. 6.24 for both the upstream and downstream directions is similar. The cell mapping for an upstream frame is broadcast to all subscribers such that all subscribers know which minicells they may use and for which purpose. This information concerning the mapping of cell types per frame (the parameters N, K, M_{ATM}, etc.) is transmitted downstream via the HK cells. If there are some minicell locations not allocated due to improper matching of ATM/MAC clusters between successive STM blocks, they will be used for single ATM or MAC cells. The rule for filling these gaps in the frames can be, for instance, to first place as many single ATM cells as possible and fill the remaining part with MAC cells.

This rule might increase the ATM capacity in the upstream direction. The upstream frame needs no synch information, but it can have some space reserved for ranging purposes. Furthermore, there can be minicells allocated for housekeeping that can be used in this case for acknowledg-

TABLE 6.3

Frame Description
Parameters

M	Number of minicells per basic frame
K	Number of STM blocks per basic frame (M/K must be an integer)
M_{ATM}	Number of minicells used to store an ATM cell
N_{ATM}	Number of ATM cells per ATM/MAC-cluster
M_{MAC}	Number of minicells for MAC per ATM/MAC-cluster
M_{Sys}	Number of minicells for system purposes (HK, ranging, sync) per basic frame
M_{STM}	Number of minicells per STM block
STM Jitter Priority	STM jitter priority—this has two modes: no-jitter on STM, jitter allowed on STM

ments of received housekeeping commands, monitoring functions, alarms, and so on. A part of the frame is reserved for the MAC layer (i.e., the space where requests for bandwidth can be placed). The remaining part of a frame can be used for user data of various services. Again, as in the downstream direction, the amount of minicells allocated for a certain functionality can differ per basic frame and might be zero for some cell types.

DOWNSTREAM FRAME. Referring to Fig. 6.24, the downstream basic frames will start with an STM data block followed by a number of ATM/MAC clusters. An ATM/MAC cluster is a cluster of M_{clus} minicells formed by $M_{ATM} \times N_{ATM}$ minicells containing N_{ATM} ATM cells followed by M_{MAC} minicells with MAC information. This pattern of STM blocks and a number of ATM/MAC clusters is repeated up to the end of the basic frame. The MAC part is used to broadcast minicell allocation information for upstream traffic to all subscribers and to give acknowledgments on bandwidth requests from the NIUs. The downstream basic frames will end with the system data such as synch words and housekeeping. Synch information is present at regular intervals in a multiframe, so it is not necessarily in every basic frame. Housekeeping contains all physical layer data (ranging offsets, power settings, alarms, etc.) to keep the system running.

In the no-jitter mode, there are *K* near-identical subframes in a basic frame. A subframe is the interval between successive STM blocks in a basic frame. Each subframe starts with an STM block of M_{STM} minicells

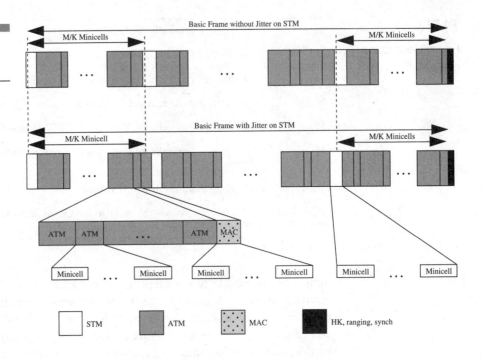

followed by a number of ATM/MAC clusters. The remaining minicells in the subframes are used for MAC, except for the last subframe, where a part is also used for system purposes (HK, ranging, etc.). The total amount of ATM cells per basic frame (ATM$_{cap}$) in this no-jitter mode is given by:

$$\text{ATM}_{cap} = N_{\text{ATM}} \times (K - 1) \times \left(\frac{M/K - M_{\text{STM}}}{M_{\text{ATM}} \times N_{\text{ATM}} + M_{\text{MAC}}} \right)$$

$$+ N_{\text{ATM}} \times \left(\frac{M/K - M_{\text{sys}} - M_{\text{STM}}}{M_{\text{ATM}} \times N_{\text{ATM}} + M_{\text{MAC}}} \right)$$

with the restriction of $M/K - M_{\text{sys}} - M_{\text{STM}} \geq 0$.

In the jitter mode, a basic frame starts with an STM block. This block is followed by a number of ATM/MAC clusters, where at intervals of about M/K minicells, an STM block is put in between. This means that the subframes do not have a fixed length in this situation. The STM blocks can be placed within an ATM/MAC cluster, but only at boundaries of ATM packets. The STM blocks are positioned such that there is a maximum jitter compared with the start of a basic frame of $\pm(M_{\text{ATM}}/2)$

minicells. The basic frame ends with the system data. The total amount of ATM cells per basic frame (ATM_{cap}) in this jitter mode is given by:

$$ATM_{cap} = N_{ATM} \times \frac{M - M_{sys} - K \times M_{STM}}{M_{ATM} \times N_{ATM} + M_{MAC}}$$

COMBINATION OF MINICELLS AND ATM CELLS. The ATM cells will be mapped into a consecutive train of minicells. For an efficient mapping scheme, it is preferred that an integer multiple of the minicell size is slightly greater than 53 bytes (the size of an ATM cell). Consider minicells of nine bytes each. In principle, a one-byte header including guard space and reference symbol for DQPSK is needed in the upstream direction for every minicell. However, if we map an ATM cell in a train of six consecutive minicells, we only need this header byte once. This is illustrated in Fig. 6.25.

Suppose we have two interfaces attached to the station: an ATM interface and an STM interface. In the packetizer, the incoming data from the STM interface is packed together in payload packets of eight bytes each to which the controller adds a one-byte header. The corresponding minicell is stored in the buffer, waiting for transmittal. The data from the ATM interface is mapped onto six consecutive minicells. The controller adds a one-byte header only to the first minicell. The 53 bytes of the ATM cell are mapped in the remaining eight bytes of this minicell and the header plus payload positions of the next five minicells. The six consecutive minicells are stored in the buffer, waiting for transmission. At the headend, the controller is aware whether it is a minicell or a train of six consecutive minicells containing an ATM cell. The reassembler in the headend strips the header from the minicell or from the first minicell of a train of six mini-

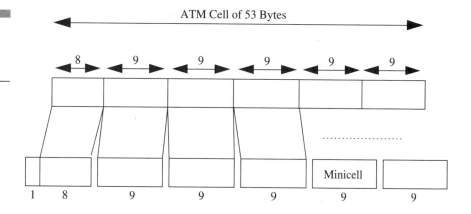

Figure 6.25
The segmentation and reassembly of the ATM cell.

cells. The payload of a single minicell is stored in the buffer, waiting to be passed to the STM interface. The header of the train of minicells is stripped, and the following 53 bytes from the minicell train are stored in the buffer, waiting to be passed to the ATM interface.

GENERALIZED MAC PROTOCOL SUPPORT.   Systems that have centralized control are positioned to handle various MAC protocols within the headend. With little added complexity, the same station can operate using different MAC protocols (bitmap, DQRAP, polling slotted ALOHA, etc.), depending on the particular headend implementation, operating conditions, network operator choices, and so on. The infrastructure described above enables the headend to support various MAC mechanisms so it can be classified as future-safe.

6.4.8 Lucent Technologies: Adaptive Digital Access Protocol (ADAPt)

OVERVIEW.   Lucent proposed the adaptive digital access protocol (ADAPt) that supports multiservice (STM and ATM) applications. The protocol adapts to changing demands for a mix of circuit and cell-mode applications and efficiently allocates upstream and downstream bandwidth to a variety of bursty and isochronous traffic sources.

In a hybrid fiber-coaxial (HFC) network, the protocol resides in the station and the headend controller. MAC processor provides for dividing the time domain for a given digital bitstream into successive frames, each with multiple STM and ATM time slots. Within the STM region of a frame, variable-length time slots are allocated to calls (e.g., telephony, video telephony), requiring different amounts of bandwidth. A contention access signaling channel is also provided in this region for call control and set-up requests. Within the ATM region, fixed-length time slots accommodate one individual ATM cell. These ATM time slots may be reserved for a user for the duration of a call or burst of successive ATM cells or shared via a contention process. At least one contention time slot is available for signaling messages related to ATM call control and setup requests. The range of services considered include voice and video telephony, video on demand, video games, and data applications. The protocol is also operable in the STM-only as well as ATM-only traffic scenarios. The high-bandwidth efficiency of the protocol is demonstrated by results based on performance and capacity analysis.

The protocol is described here generally in the context of a hybrid fiber coax (HFC) as shown in Fig. 6.26. It is also applicable to the fiber-to-the-curb (FTTC) architecture as well.

Many new applications such as high-speed data and MPEG2 video are likely to be transported using ATM techniques from a home or office. At the same time, applications such as voice and video telephony are likely to continue being transported using STM for reasons of delay, error sensitivity, and so on. Lucent proposal concluded that an access protocol to serve the needs of both STM and ATM applications was essential. The protocol should adapt to the changing demands of a mix of STM and ATM applications, and it should efficiently allocate bandwidth to a variety of bursty and isochronous traffic sources.

The protocol includes fixed time slot assignments (as in conventional TDMA) and ATM transport using a combination of contention mode and dynamic reservation mode access. It allows for dynamic sharing of the bandwidth within each radio frequency (RF) channel as well as across multiple RF channels.

PRINCIPLES OF ADAPt PROTOCOL OPERATION. In an HFC environment, the protocol is carried out both at stations and in the headend. A medium-access control (MAC) processor provided in each of the stations and in the bandwidth controller divides the time domain for a given digital bit stream channel into a series of successive frames, each having multiple time slots. There are two types of time slots: STM and ATM. The frame is divided into STM and ATM regions as shown in Fig. 6.27. The boundary between the two regions can be changed dynamically.

Figure 6.26
Headend/station
configuration.

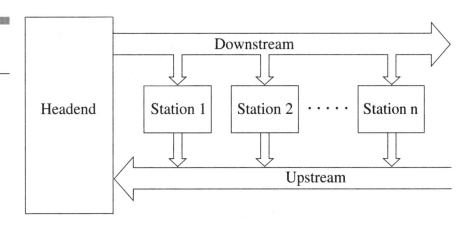

A view of an upstream provides an example of time slot allocations for the different types of STM and ATM services that are supported by ADAPt. The STM region contains an STM signaling time slot that is shared by contention access by all stations for STM call control and signaling. The STM signaling time slot as shown at the beginning of the frame in Fig. 6.27, requiring 384 Kb/s bandwidth, uses a relatively long time slot, while three separate voice calls, each requiring 64 Kb/s, use somewhat narrower time slots.

The ATM region of the frame is partitioned into several ATM time slots, each capable of accommodating one ATM cell. These cells are labeled as *R* (reservation) or *C* (contention), corresponding to the mode in which they are being used in this frame and as dictated by the bandwidth controller. An ATM/CBR data call requiring a guaranteed fixed bandwidth uses a reservation time slot. This time slot is made available to the call in a periodic manner to provide a fixed bandwidth for the duration of the ATM/CBR call. An ATM/VBR delay-sensitive call also uses a reservation time slot. However, this time slot may not be reserved for the entire duration of the call. Instead, it may be reserved only for a burst of data (i.e., group of ATM cells) that is currently being generated by the ATM/VBR source. Time slot allocations for this burst are requested in an upstream ATM contention time slot, and the allocation acknowledgment is received in the next downstream ATM map.

Figure 6.27
Upstream frame.

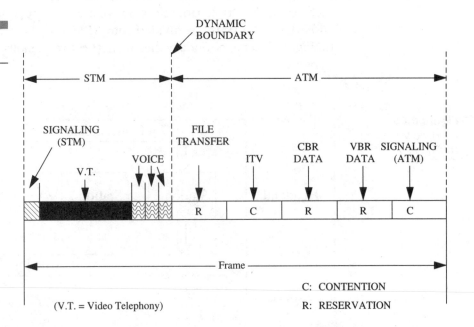

(V.T. = Video Telephony)

UPSTREAM/DOWNSTREAM FRAME STRUCTURE. Refer to Figs. 6.28 and 6.29. In the upstream direction, stations transmit in a burst transmission mode, and in the downstream direction, the headend transmits in a continuous transmission mode. The stations go through a ranging process and transmit, or *burst*, based on timing information received from the headend. Each station uses guard times to avoid interference with adjacent times. Further, stations use preambles for clock and level recovery.

In the downstream direction, the transmission channel is controlled and used only by the headend, and hence there is no need for guard times or preambles, and the transmission takes place in a continuous way. Figure 6.28 shows an upstream frame and the overheads associated with STM and ATM time slots including burst header (guard time and preamble).

The boundary between the STM and ATM regions is dynamic, depending on arrivals and departures of STM and ATM calls. Figure 6.29 shows an example of the downstream frame. The STM region (shown to the left of the STM/ATM boundary) consists of time slots corresponding to STM signaling, DS0 (e.g., voice), and nxDS0 (e.g., video telephony), and the ATM region (to the right of boundary) consists of one or more ATM cells that include headers. The MAP field carries control and status information from the headend to the stations. It indicates the status of contention time slots in the previous upstream frame and also informs stations of the bandwidth (or time slot) allocations in the next upstream frame.

Figure 6.28
Upstream frame.

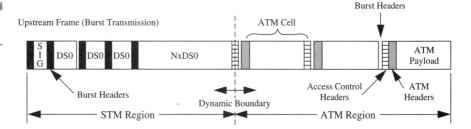

Figure 6.29
Downstream frame.

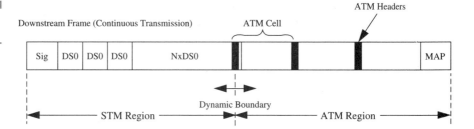

PERFORMANCE AND CAPACITY OF THE ADAPt PROTOCOL.
Lucent made a study of ADAPt performance. The result is self-explanatory
and as shown in Fig. 6.30.

Considering a neighborhood of 500 homes served on a coax leg, it is
possible to support in excess of 100 percent take rates for voice, video gaming, and work-at-home (VBR/ATM) data applications, and, in addition, a 20
percent take rate for VT can be supported. The VOD upstream traffic is
not shown. The upstream VOD messages (from 500 simultaneous users)
require only a very small percentage of the overall coax bandwidth and
can be accommodated in either the STM signaling channels or in ATM
contention slots.

6.4.9 LANcity Corporation: A MAC Protocol

OVERVIEW. LANcity MAC protocol supports a mix of contention,
reservation, and CBR bandwidth concurrently. It dynamically adjusts
the mix as offered load changes. Bandwidth assignment is done through
an allocator at the headend/central office. The protocol uses variable-
length transmission as a means of improving efficiencies. Concatenation allows multiple packets or cells to be put together in a single

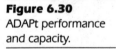

Figure 6.30
ADAPt performance
and capacity.

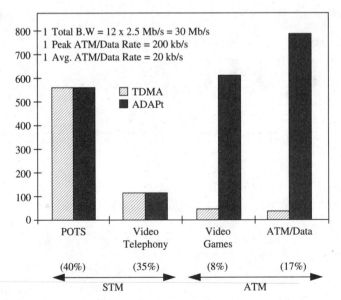

transmission. The protocol will support packet, cell, or both types of traffic simultaneously. The MAC protocol provides for asymmetric operation. The upstream and downstream channels can support different data rates. The protocol also provides support for many downstream channels to share a single upstream channel that meet the frequency topology of cable TV systems.

TRANSMIT TIMING AND CONTROL. The HFC environment poses a challenging problem to the network designer since propagation delays can be much larger than the transmission time for packets or cells. To overcome this, every node needs to have a global timing reference and knowledge of its round-trip delay. Once this is known, every node can then precisely transmit in a given time window and the allocator at the headend can assign windows or bandwidth as needed.

BLOCK SYNCH REFERENCE. The headend must transmit in the downstream channel a block synch reference at a fixed periodic rate. This reference will be tied into the downstream format. The exact interval is noncritical but is assumed to be on the order of several to tens of milliseconds. The time from block synch reference to block synch reference is known as a block synch interval (BSI). The BSI is used by the allocator in identifying transmit timing and assigning bandwidth. The BSI needs to have small jitter, since this will add to the guard-time budget for the IPG between the two nodes transmitting.

During its initialization sequence, each node must perform a ranging operation to calculate its round-trip delay from the headend. Once known, a node can now transmit in a precise time window in the upstream channel. The allocator controls when nodes can transmit and passes this information on the downstream-channel BSI time slots. Every node on the network can now identify a particular transmit time on the upstream channel relative to its time from the start of a BSI. The allocator can assign these transmit-time windows to an individual node (referred to as *dedicated access*), or these transmit times may be allocated to multiple nodes and be subject to collisions (a.k.a. contention access).

In order to minimize overhead, the allocator assigns bandwidth (i.e., transmit times) in fixed-sized units called transmit opportunity (TXOP) slots rather than assigning bandwidth at the bit level. An ATXOP slot size of 16 bytes was chosen. Typically, the size of the TXOP slot is chosen to equal the smallest possible transmission. This would map to a REQ packet in LANcity's UniLINK-2 protocol or a minislot in the XDQRAP protocol. An ATM cell transmission would consist of multiple TXOP slots. A variable-length transmission such as a packet or a concatenation could be accomplished by assigning an appropriate number of contiguous TXOP slots.

DYNAMIC MIX OF CONTENTION, RESERVATION, AND CBR. The headend controls the assignment of bandwidth on the upstream channel by reserving TXOP slots for individual nodes or allowing TXOP slots to be available to all in a contention mode. The protocol supports three distinct types of access:

1. Contention
2. Fixed dedicated assignments (i.e., CBR)
3. Variable dedicated assignments (reservation mode)

Contention access. Allows multiple nodes to contend for a single transmit opportunity and is subject to collisions. This provides low latency access to the network under light loads.

Fixed dedicated access. Implies that a single node has been assigned TXOP slots at a predefined interval. This is suitable for CBR-type traffic. This assignment is done through a signaling mechanism.

Variable dedicated assignments/reservation mode. Allows nodes to request and obtain bandwidth dynamically. The node starts by issuing a bandwidth request. The allocator assigns bandwidth to the node based on its request and the QOS required. The request may have been transmitted as either a contention or a previous dedicated assignment.

CONTENTION ACCESS MODE. Contention mode is the underlying access method that is always present. When the network is completely idle, all of the upstream bandwidth is set to contention mode. Thus, when a node has data to transmit, it grabs the first available contention opportunity and gets low latency access to the network. The MAC protocol restricts contention access to start only at specific times. This causes all collisions to completely overlap, thus increasing collision detection robustness. These contention start opportunities repeat at a fixed interval known as the contention resolution interval.

Contention Resolution Interval (CRI). A contention resolution interval is defined as the maximum time required for all nodes on the network to detect if a contention transmission was successful or unsuccessful. This means that nodes have transmitted on the upstream and heard the contention results in the downstream channel before the start of the next CRI. The size of the interval is determined by several factors, including maximum round-trip delay in the network, collision detection and processing time at the headend, downstream format restrictions, user node processing time to get ready for the next CRI, and efficiency concerns such as adjusting the CRI to accommodate an exact multiple of ATM cells or to align with the downstream format. In order to guarantee ordering of packets and/or cells, a node is restricted to having only one contention transmission outstanding for a given VC or service class.

The MAC protocol will use an XDQRAP-like contention resolution scheme. In XDQRAP, there is the concept of two distributed queues. The first queue is used in conjunction with small minislot transmissions for contention resolution. Once a collision is resolved, the data is placed on the second data queue for later transmission. Note that the data queue is contention-free. If enough minislot requests are allotted, then the data pipe can support utilization close to 100 percent. XDQRAP has several advantages that should be noted. The tree-based back-off algorithm resolves itself quicker and in a deterministic amount of time compared to the binary exponential back-off. Also, the use of minislots for contention area limits the bandwidth wasted during a collision to a very small amount of time. For this MAC protocol, the number of contention REQ minislots and the size of the data portion within a CRI is variable and is determined by the allocator. There will also be multiple data queues associated with different priority levels. The MAC protocol will *not* use XDQRAP's method for compensating for network delay, but stay with the approach used in LANcity's UniLINK-2 protocol.

The allocator monitors the collision status on the REQ slots and broadcasts the results in the downstream control for all nodes to receive before the start of the next CRI. In addition to collision resolution, the headend also informs the stations of the status of the data queue and whether the data region of the next CRI is being used.

Variable-Length Transmission over Several CRIs. One of the significant features of the protocol is its ability to support variable-length transmissions. This could come in the form of a variable-sized packet, or it could be a concatenation with multiple cells or packets put together.

The headend controls the bandwidth assignment and decides when it is appropriate to allow variable-length transmission. When the headend returns the collision results downstream, it indicates whether the current node transmitting is allowed to continue throughout the next CRI. If the transmission continues, then all other nodes would have to refrain from transmitting in both the REQ and data regions of the next CRI. For example, a node could have a large transmission pending, such as a max-sized Ethernet packet or a video frame consisting of several dozen ATM cells concatenated together, and request an appropriate amount of bandwidth. The allocator then processes this REQ along with the current status of the collision resolution. If no other contentions are pending (i.e., the REQ slots are not being used), the allocator can assign the next several CRIs to be a continuous transmission for this node.

FIXED DEDICATED ACCESS MODE. The fixed dedicated access mode allows stations to be assigned bandwidth at predefined intervals. This access mode is intended to support CBR applications. The assignment is done through signaling between the subscriber station and the headend. During the signaling, the station and the headend agree as to when the station is assigned transmit opportunities. When the time to transmit arrives, the headend broadcasts in the downstream control with the contention resolution results; a portion of the following CRI's data region is reserved and no one else is allowed to transmit at that time. The remaining portion of the data region could contain other queued data traffic.

RESERVATION ACCESS MODE. The reservation access mode provides the ability to assign bandwidth in a dedicated, dynamic manner. Thus, nodes only use what they need, and the remainder is available to the rest of the network. As its name implies, a node must reserve bandwidth by making a request for the number of TXOP slots it needs. The allocator then queues the transmit as appropriate. Every transmission has a bandwidth request field in its MAC layer header. Thus, the initial reservation request may come from either a contention or a fixed dedicated access. Once the data is queued, each transmission allows the subsequent transmission to be queued in a dedicated, contention-free mode. If a node has no more user data to transmit, it may still request bandwidth for a dedicated REQ to be assigned later by the headend.

Once in reservation access mode, a node will continue to have dedicated, contention-free access to the network. All stations in reservation mode at a particular priority/service level will share the bandwidth fairly.

Under heavy network loads, almost all network accesses will be dedicated, and only a handful of contention-based REQs will be needed to handle nodes just entering reservation mode. Once a node becomes idle for a sufficient amount of time, it may drop out of reservation mode by simply not requesting any more bandwidth. In both reservation mode and fixed dedicated mode, the node transmits in a collision-free environment. Thus, the node considers its transmissions complete once it is sent since there are no retransmissions due to collisions. This allows nodes to have multiple dedicated transmits on the cable at the same time. This allows higher data throughput from a single node than can be achieved in contention mode where only a single transmission may be outstanding at any time.

MULTIPLE PRIORITY LEVELS. In order to support the various types of quality of service (QOS), the MAC protocol supports multiple priority levels. This means that there are multiple data queues, with the allocator determining when each data queue is serviced. We estimate that there should be support for at least eight priority levels and that sixteen may be defined to allow for future expansion. If a node is transmitting data for a particular priority level and the allocator receives a request for a higher priority level, then the current transmit is preempted at the end of the current CRI. The higher priority is allowed to transmit starting in the next CRI. When a REQ for bandwidth is made, the node must supply the appropriate priority level for the request. This allows the data to be put in the proper data queue. The highest priority level should be reserved for urgent transmissions. These are network-critical events that must be sent immediately. The next highest priority level would be designated for real-time traffic that needs to be sent ASAP. Several priority levels could be assigned for REQ-only transmission. These could then be serviced at fixed intervals. For example, there could be three levels serviced at a rate of once per 6 ms, once per 30 ms, and once per 100 ms and used for applications like packetized voice, video frame generation, and transaction processing, respectively. Lower priority levels would allow nodes to share remaining bandwidth in a round-robin fashion. The lowest priority level would be used for bulk background data when the network is otherwise idle.

A node can intermix its use of contention and reservation traffic. If a node has traffic to transmit and has a dedicated REQ scheduled for a later time, it can try to use currently available contention bandwidth. If the node is still trying to resolve its contention when the dedicated REQ becomes available, then the node uses the dedicated REQ and drops from the contention resolution process.

VBR (REAL-TIME) SUPPORT. The MAC protocol can support real-time VBR by using a mix of fixed, dedicated, and reservation access modes. Using signaling, a node can obtain a small amount of CBR equal to the minimum amount of data that is to be sent periodically. If the minimum data amount is zero, then the CBR assignment becomes simply a dedicated REQ at periodic intervals. At these fixed intervals, the node can transmit its minimum amount of reserved data and, at the same time, request in a contention-free manner any additional bandwidth it may need. The allocator can then queue this additional data request at the appropriate priority level to insure its QOS.

VBR (NON-REAL-TIME) SUPPORT. For non-real-time VBR, the protocol exclusively uses reservation mode. As part of the signaling process, the node would receive a dedicated REQ assignment at the appropriate priority level. It would retain the dedicated assignment as long as the connection is established. Even if there is no data to send, the node would still request enough bandwidth to have its dedicated REQ put on the data queue. In this manner, the VC will always have dedicated access to maintain its QOS.

ABR AND UBR SUPPORT. The priority level structure allows the protocol to support ABR and UBR traffic. ABR traffic can use reservation mode to obtain a minimum cell rate by being assigned to an appropriate priority level. If additional bandwidth is needed, then the ABR traffic can use contention mode and/or reservation mode at a lower priority to get the additional bandwidth as needed. UBR traffic could use the lowest priority available. This would prevent an unrestrained UBR source from grabbing an unfair share of the bandwidth from ABR sources.

BANDWIDTH ALLOCATOR. The allocator controls the assignment of bandwidth. As such, it is responsible for enforcing the QOS requirements. It is also responsible for maintaining maximum efficiency by choosing network parameters and deciding when functions such as continuous transmission across CRIs or concatenation may be used. The allocator can be simply implemented, or it could be made more complex to take advantage of some network efficiencies. This would be transparent to end-user devices and allow companies to provide value-added services at the headend. The control interface between the allocator and end nodes is what needs to be standardized.

MAC HEADER. The MAC header is the most basic unit of transfer on the upstream link. It consists of a PHY overhead portion and a MAC control function. When transmitted by itself, it may be used as a REQ frame. This may be transmitted in either contention or dedicated modes. The format for a generic MAC header is:

Guard + preamble (PHY)	6 bytes
MAC control	8–12 bits
BW REQ	4–8 bits
TYPE specific	4 bytes
CRC/FEC	4 bytes
Total:	16 bytes

Guard time. The guard time between modem burst transmission plus the preamble and start of frame for the transmission is needed.

MAC control. A MAC-type field is used to distinguish what kind of frame format follows. This includes the following: MAC header only, 802 LAN packet, ATM cell, transparent user data, or concatenation header. A priority group field identifies the priority level, or group, associated with this frame. The intent is that these groups would include: VBR (real-time), VBR (non-real time), interactive, background, and maintenance traffic. The MAC control field would also indicate whether the transmission was sent in contention or dedicated mode.

BW REQ. The BW REQ field allows a node to request additional bandwidth. The units are either in ATM cells or TXOP slots, depending on the MAC header type.

TYPE specific. The TYPE specific field is unique for each type and is defined below.

CRC/FEC. The CRC/FEC field provides verification for the data in the MAC header. The actual polynomial used is TBD. However, since this field is used in collision resolution, it should be at least four bytes in length to sufficiently reduce the possibility of a false detection.

A TXOP slot size of 16 bytes is assumed. Therefore, a MAC-header-only transmission (i.e., REQ) would require a single TXOP slot.

The ATM cell format consists of a MAC header followed by an ATM cell. The format is:

MAC header	16 bytes
PHY overhead	6 bytes
Control + REQ	2 bytes
ATM header	4 bytes
CRC/FEC	4 bytes
ATM payload data	48 bytes
FEC (optional)	(4) bytes
Total:	64–68 bytes

The ATM header includes the GFC, VPI, VCI, and PTI/CLP fields. The HEC field is not included since these fields are already covered by the MAC header. Depending on the PHY selection, it may be necessary to include an FEC field covering the payload data to obtain the required BER. The ATM cell would fit into 4 TXOP slots.

TRANSPARENT USER DATA. The transparent user data format is a way of passing user data directly over the network with a minimal overhead. An example of this may be STM (synchronous transfer mode) data used in supporting a DS0-type service. This could require 16 bytes of data being transmitted every 2 ms. The format for transparent user data is:

MAC header	16 bytes
PHY overhead	6 bytes
Control + REQ	2 bytes
Reserved	4 bytes
CRC/FEC	4 bytes
User data	16 bytes
Total:	32 bytes

The 802 LAN packet format contains:

MAC header	16 bytes
PHY overhead	6 bytes
Control + REQ	2 bytes
SID	12 bits
PKT length	12 bits
Reserved	8 bits
CRC/FEC	4 bytes

DA	6 bytes
SA	6 bytes
Ether TYPE	2 bytes
User data	14–4000 bytes
CRC/FEC	4/+ bytes (TBD)
Total:	48–4096 bytes

SID. The TYPE specific field of the MAC header would contain an SID field, which is a unique identifier assigned as part of the initialization process. There would also be a length field that defines the total length of the packet in bytes.

DA/SA ether type. The SA, DA, and ether TYPE fields are defined to be identical to 802.3's packet format. This provides for ease of bridging between these networks.

User data. The user data field is a superset of 802.3. The particular range was chosen for discussion purposes only. A minimum-sized Ethernet packet (64 bytes) would fit into 5 TXOP slots.

CRC/FEC. The choice of simple CRC or FEC is strongly dependent on the PHY layer selection. It is assumed that a minimal configuration would support the 32-bit CRC defined in 802.3. A more powerful FEC could possibly replace or supplement the CRC. This choice is beyond the scope of this book.

It is noted here that, with multiple priority levels and supporting the various QOSs, a variable-length packet transmission may be fragmented if a higher-priority transmission becomes pending and the packet transmission is preempted. To ease implementation issues, the protocol limits only one packet outstanding at any time for any given priority level. When a preempted packet resumes transmission, the control field in the MAC header indicates that this is a continuation of a previous packet as well as the associated priority level. The number of priority levels that supports packets may also be limited.

REQ FRAME. REQ frames are MAC-header-only transmissions used to request bandwidth from the allocator. They come in two formats: ATM-style or PKT-style. Both formats are shown below.

REQ format (ATM-style) can be broken down as follows:

MAC header	16 bytes
PHY overhead	6 bytes

Control + REQ	2 bytes
ATM header	4 bytes
CRC/FEC	4 bytes
Total:	16 bytes

The ATM-style REQ frame allows requests to be associated with a particular VC. The unit of the request field is the number of ATM cells needed to transmit.

The following information applies to REQ format (PKT-style):

MAC header	16 bytes
PHY overhead	6 bytes
Control + REQ	2 bytes
SID	12 bits
PKT Length	12 bits
Reserved	8 bits
CRC/FEC	4 bytes
Total:	16 bytes

The packet-style REQ frame is associated with nodes that are transferring 802 LAN packets over the network. The SID identifies the source node making the request. The bandwidth request field is the number of TXOP slots required. The PKT length field is not used for the REQ-only frame.

CONCATENATION. A significant feature to improve network efficiency is concatenation. Concatenation is the ability to group multiple packets or cells into a single transmission with the overhead of a single packet or cell. The MAC header by itself represents over 25 percent overhead on an ATM cell or a 64-byte LAN packet. Nodes that have multiple packets or cells to transmit can concatenate them together into a single transmission called a concatenated frame. These concatenated packets or cells are treated as a single entity and may be transmitted in either dedicated or contention modes. Each packet or cell within a concatenated frame still contains all of its addressing information. Thus, each may have its own unique destination. The concatenation remains completely transparent to all upper layers.

The first major benefit of concatenation is the reduction of transmission overhead. Only one MAC header is needed for many packets or cells that are put together in the concatenation. The second benefit is increased efficiencies from reduced collisions. For example, if ten cells are concatenated together, it appears to the network as a single transmission that is ten times as large with only one opportunity for collision instead of ten. There are different types of concatenations for each of the formats previously described as well as a special compacted ATM cell format.

CONCATENATED ATM CELLS.   The concatenated ATM cell format consists of a MAC header followed by multiple ATM cells. The format is:

MAC header	16 bytes
PHY overhead	6 bytes
Control + REQ	2 bytes
ATM[1] header	4 bytes
CRC/FEC	4 bytes
ATM[1] payload data	48 bytes
ATM[2] cell	53 bytes
.	.
.	.
.	.
ATM[n] cell	53 bytes
Total:	$16 + n \times 53$ bytes

The MACTYPE field would indicate this is an ATM concatenation. The control field in the MAC header now contains concatenation-specific information such as the total number of ATM cells included. Every additional three ATM cells would require ten TXOP slots instead of four TXOP slots each if transmitted separately.

COMPACTED ATM CELLS.   Many times, a node is given a burst of data that is to be sent over the same VC. In this case, the ATM header is virtually identical for all the ATM cells. The compacted ATM cell format consists of a MAC header followed by multiple ATM cells all destined to the same VC. The format is:

MAC header	16 bytes
PHY overhead	6 bytes
Control + REQ	2 bytes
ATM header	4 bytes
CRC/FEC	4 bytes

ATM[1] payload data   48 bytes

.   .

.   .

.   .

ATM[n] payload data   48 bytes

Total: $16 + n \times 48$ bytes

The MACTYPE field would indicate this is a compacted ATM concatenation. Part of the MAC control field would indicate the total number of ATM cells that follow and whether the last cell indicator in the PTI should be set. In this mode, each additional ATM cell would occupy three TXOP slots instead of requiring four TXOP slots if transmitted separately.

CONCATENATED TRANSPARENT USER DATA. The concatenated transparent user data format would allow support for a multiple DS0 service. The format is:

MAC header   16 bytes

PHY overhead   6 bytes

Control + REQ   2 bytes

Total length   1 byte

Reserved   3 bytes

CRC/FEC   4 bytes

User data $n \times 16$ bytes

Total: $16 + n \times 16$ bytes

The MACTYPE field would indicate this is a transparent user data concatenation. The total length field would indicate the number of TXOP slots filled with user data that follow.

DOWNSTREAM SIGNALING. Any upstream access control mechanism requires that control information, such as bandwidth assignments and contention resolution status, be transmitted downstream in a timely fashion. Here is a summary of the information that needs to be passed downstream and its interval:

- Information every BSI
- Block synch reference

- CRI format
- Global network parameters (e.g., TX Freq)
- Information every CRI
- Empty/successful/collision status for each REQ
- Bandwidth assignment for next CRI
- Transmit inhibit on REQ in next CRI
- Transmit inhibit on all or part of next data region
- Information as needed
- Network authorization
- Ranging
- Power adjust
- TX equalization (as needed)
- Operations and management (OAM) traffic
- Efficiency comparisons

The following example is given where a node is trying to transfer 576 bytes of user data across the network. This data is assumed to be packaged into 12 ATM cells. It starts with the simple CRI case from above. There would be 2 REQs plus 1 ATM cell, or 6 TXOP slots, per CRI. To transfer all of the user data would require 72 TXOP slots (12 CRI × 6 slots/CRI). This case is similar to XDQRAP with 2 minislots and a fixed data region of 1 ATM cell. If the CRI is made larger as in the second case and/or the allocator allows the node continuous transmission, then all 12 ATM cells could be sent consecutively and only require 48 TXOP slots (12 cells × 4 slots/cell). If the 12 ATM cells are also concatenated together, this further reduces the transmit time to 41 TXOP slots (4 for the first cell + 10 slots for every 3 cells). Finally, if compacted ATM concatenation is used, only 37 TXOP slots are needed (4 for the first cell + 11 cells × 3 slots each).

Thus, the data-carrying capacity of the upstream has almost doubled by reducing the transmit time from 72 TXOP slots to 37 slots. The theoretical limit would be 36 slots.

ASYMMETRIC SUPPORT. The protocol makes no assumptions on the data rate other then the downstream rate being equal to or greater than the upstream. The rates may be a multiple of each other to ease some implementation issues such as feeding control information back downstream. Another basic asymmetry exists on HFC cable plants that needs to be exploited. Most modern cable plants are being built to support 450–750 MHz downstream but only 5–42 MHz upstream. Thus, there are more downstream channels than upstream. The MAC protocol supports a single upstream channel being shared by multiple downstream channels. To accomplish this, the block synch reference and the downstream control information (e.g., collision resolution and bandwidth assignments) must be broadcast across all downstream channels. The total delay incurred is not critical, since the ranging process compensates for this. However, the jitter introduced into the block synch reference must be carefully controlled, since this adds to the guard time required as part of the PHY overhead.

6.4.10 Hybrid Network Inc.: MultiMedia MAC (M3)

OVERVIEW. Hybrid Network MAC protocol proposal supports features and characteristics of a media access protocol (MAC) for a hybrid fiber and coax metropolitan area network that operates on CATV cable plants. The MAC layer offers an adaptation service to several other standard MAC layers that include asynchronous transfer mode (ATM), 802 packets (802.3), MPEG packets, and time division multiplexed data (TDM). The MAC is not dependent on a particular physical layer and functions in an asymmetrical branch-and-tree network. The asymmetry comes from the availability of more data bandwidth in the downstream channel than is available in the upstream channel.

In most installations, both downstream and upstream channels are on one physical cable at the subscriber's home. In the cable operator's network equipment, the media access algorithm controls the subscribers' transmission on the upstream channels. Media access control is achieved through downstream grant messages, known as *credit packets*, and upstream relinquish/request messages, known as *done messages*. The upstream and downstream M3 MAC layers provide an adaptation function between 802-compliant data units and M3 data units.

M3 MAC FEATURES. Features of the M3 system are:

- A simple MAC adaptation layer for 802 packets, ATM cells, time slot data, and MPEG video streams. The adaptation has low data overhead and produces native data streams that can be processed by existing protocols.
- CBR, VBR, and ABR services are supported.

■ Multiple downstream channel data rates in a fixed bandwidth for cable systems with marginal, good, and very good cable characteristics.

■ Message-based upstream media access control. The message-based control permits diverse upstream technologies to be mixed on the same downstream without changing the downstream architecture. This allows 802.14 to define the downstream MAC independently of any upstream PHY like QPSK, QAM, DWMT, or DMT.

Features of the M3 upstream channel architecture are:

■ *Three upstream channels bandwidths and data rates (132 Kbps, 536 Kbps, and 2.144 Mbps).* Small-bandwidth channels are used in noisy parts of the spectrum. Large-bandwidth channels are used for high-performance services and to lower the headend costs.

■ *Transmission slots.* Forward error correction (FEC) is used for low error rates on lines that have high carrier-to-noise ratio. (A BER of $<1E^{-9}$ at a C/N of 13 dB is desirable.)

■ *Delineation of the start and end of a packet.* RF preamble for burst mode transmissions and guard bits for TDM slots.

■ *Encryption.* The MAC supports the selection of an encryption key based on a source address. The MAC also tags what data is encrypted and what is not.

■ *Support of low and high data rates.* Low-speed video gaming and telenet applications, <19.2 Kbps; high speed for video conferencing, 384 Kbps; high speed for business page Web servers, up to 2 Mbps.

■ *Support for low latency applications like video gaming.* Access and transmission latencies <20 msec.

■ *Large number of channels.* Based on 500 to 2000 homes passed, a take rate of 30 percent and active/idle ratio of 30 percent, the system design needs to service between 55 and 220 active subscribers.

UPSTREAM CHANNEL ARCHITECTURE. The upstream channel speed is configurable at the headend and at the terminal equipment unit. The demodulator port is configured for either 132 Kbps, 536 Kbps, and 2.144 Mbps. The frame architecture is the same for each speed, but the actual structure changes.

FDM AND TDM UPSTREAM CHANNEL STRUCTURES

2.144-Mbps Channel Structure. The M3 upstream architecture is based on a 20-msec frame that is split into 80 TxSlots. See Fig. 6.31.

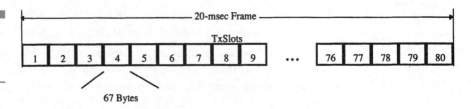

Figure 6.31
2.144 Mbps
Upstream Frame
Structure.

The 20-msec frame is selected so that the minimum transmit rate for the payload of a TxSlot is 21.2 Kbps. This number is based on a 53-byte payload in a 67-byte TxSlot. The raw bit rate of a TxSlot is 26.4 Kbps. Eighty TxSlots are used so that the aggregate bandwidth of the upstream channel is close to 2 Mbps and a multiple of lower bandwidths near 512 Kbps and 128 Kbps. The raw transmit bit rate works out to be 2.144 Mbps.

536-Kbps Channel Structure. The 536-Kbps channel has a 20-msec frame with 20 TxSlots. This gives it the same data rates as the 2.144-Mbps channel up to the maximum of 536 Kbps.

132-Kbps Channel Structure. The 134-Kbps channel has a 20-msec frame with 5 TxSlots. This gives it the same data rates as the 2.144-Mbps channel up to the maximum of 134 Kbps. The TxSlot data structure is the same for any channel and given in Table 6.4.

The *RF preamble* length is based on similar systems that have been proposed by demodulator suppliers.

The *virtual channel ID* enables the upstream demux to have a simple function for extracting virtual channels. Without the ID, there has to be a way of phase-aligning the upstream demux's 20-msec frame to the downstream so that the upstream knows when TxSlot 1, 2, and so on, can be extracted. In addition to this, each upstream demux would have to maintain a TxSlot counter and slot assignment memory that contained virtual channel assignments.

The *type field* assists the upstream receiver in processing the TxSlot data. This field controls the distribution of data to packet, ATM, and control streams. It also marks which data is encrypted, and it marks when the payload contains an end-of-message (EOM). The ranging message type is a special message designed to time the clients delay relative to the TxSlot boundary.

The *Payload field* contains either an ATM cell, a packet, or a portion of a packet. The ATM cell is carried without modification in the payload. If encryption is turned on, a two-byte sourceID field is appended to the ATM cell. If encryption is turned off, the two-byte sourceID field is ignored.

The *TxSlot payload* is as shown in Table 6.5.

TABLE 6.4

TxSlot Data
Structure

Length in bytes	Field	Function
3	RF preamble	This field is used by the receiver to demodulate the signal.
1	Virtual channel ID	The virtual channel number ID for this upstream channel is assigned by the headend and is used for all the TxSlots that are a part of the virtual channel.
1	Type	The type field identifies the contents of the TxSlot payload.
	1 bit	Identifies a channel control message (1) or client data message (0).
	2 bits	0×1 EOM "end-of-message" (note: EOM infers an EOP state) 0×2 EOP "end-of-packet" 0×0 No EOP or EOM and not a ranging message 0×3 Ranging message
	1 bit	Encrypted (1) or not encrypted (0)
	4 bits	Data type: 0×0 802.3 packet 0×2 TDM data 0×1 ATM cell 0×3 MPEG packet
	TxSlot Payload	This field contains either an ATM cell or part or all of a packet.
6	FEC	Reed-Solomon forward error correction bytes. The FEC protects all the data in the TxSlot except the RF preamble and guard bytes.
1	Guard	Guard time expressed in bits to allow for timing variations between clients transmitting in time slots.

TABLE 6.5

TxSlot Payload ATM
Cell Format

Type control fields	Packet payload field	Field size (bytes)	Function
Client data	sourceID	2	Source ID of the client
ATM	Payload	53	Client ATM cell
Encrypted			
Client data	Pad	2	Padding bytes (set to 0×00)
ATM	Payload	53	Client ATM cell
Not encrypted			

The *sourceID* field is used to select the decryption key. The sourceID is established either at startup or when an encryption key is exchanged in a session. The sourceID is reassigned after the client leaves the network. The small, non-802 address, sourceID is selected to facilitate using lookup tables to recover encryption keys. Since the sourceID is selected by the system, there is the implementation option to restrict its range to reduce the size of the lookup tables needed. System requirements may call for a maximum of 2000 homes passed per fiber node with a take rate of 30 percent and an active percentage of 40. In these systems, 8 bits of address could be used. The format of the 55-byte TxSlot payload format for packet data depends on the control bits in the type field. Table 6.6 lists the format options.

All packet transmissions start at the beginning of a TxSlot. This eliminates the need for preamble and start symbols like in 802.3. The EOP flag and length field are used to determine the end of the packet. Depending on the length, the final TxSlot may contain a fragment of a packet. In this case, the remainder of the TxSlot is padded. Traffic statistics show that 92 percent of the upstream 802.3 data packets are 64 bytes long. This means that the 46-byte 802.3 payload fits into the TxSlot payload field. The packets sent upstream in the M3 system are not true 802.3 packets. Most of the overhead of an 802.3 packet is not needed for the cable upstream and has been removed.

The end-of-message flag (EOM) signals the headend when the end of the transmission occurs that is associated with a credit packet. A message may contain several packets and end-of-message "done" information. This

TABLE 6.6

TxSlot Payload
Packet Format

Type control fields	Packet payload field	Field size (bytes)	Function
Client data packet, EOM encrypted	sourceID	2	Source ID of the client
	Length	1	Length of client data payload in bytes
	Payload	1—52	Client data
	Pad	0—51	Padding bytes (set to 0×00)
Client data packet, No EOM encrypted	Payload	53	Client data
Client data packet, EOM not encrypted	Length	1	Length of client data payload in bytes
	Payload	1—54	Client data
	Pad	0—53	Padding bytes (set to 0×00)
Client data packet, no EOM; not encrypted	Payload	55	Client data

information contains the state of the terminal equipment's transmit queue. The guard byte allows for timing variations among geographically dispersed clients for when they insert data into TxSlots.

VIRTUAL CHANNELS.   Multiple TxSlots in an upstream channel are combined to form a virtual channel. In Fig. 6.32, an example shows four virtual channels (A, B, C, and D) on a 2.144-Mbps upstream channel. VC-A uses four TxSlots and has a raw bandwidth of 107.2 Kbps. The TxSlots are spaced to give the channel an access delay of about 5 msec. VC-B uses 15 TxSlots for a total bandwidth of 402 Kbps. Its TxSlots are randomly selected from those available, so there is no specified access delay. VC-C and VC-D are two other virtual channels with the bandwidths of 643.2 Kbps and 402 Kbps, respectively. The remaining TxSlots are unused in this example and are available to form other virtual channels.

For a dedicated virtual channel, a terminal equipment unit is told which upstream channel to transmit on, which TxSlots to use, and the assigned virtual channel ID (VC-ID). The terminal equipment unit uses the assigned TxSlots whenever it has data to send. If it has no data, then it does not transmit, and the assigned TxSlots are left empty. Terminal equipment units lose the dedicated channel when the headend sends a new assignment message to a shared channel. For shared virtual channels, a terminal equipment unit is told how much data to send, which upstream channel to use, and which TxSlots to transmit on. Depending on how may TxSlots are assigned and how much data can be sent, a unit may use TxSlots from several frames. At the completion of a transmission, the terminal equipment unit signals the relinquishing of the shared channel. The TxSlots that make up a virtual channel may not be contiguous. This means that a terminal equipment unit may transmit several bursts of data within a frame. Transmission bandwidth is assigned in increments of TxSlots. This allows varying

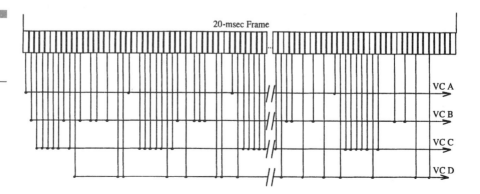

Figure 6.32
Example of a 2.144-Mbps upstream virtual channel.

amounts of bandwidth to be assigned to a dedicated or shared virtual channel.

TIME RANGING. Each terminal equipment unit's upstream transmitter places its data into fixed-length TxSlots. To keep one unit's transmission from interfering with another on an adjacent TxSlot, there is a guard time between TxSlots that is equal to 8-bit times. The natural variation in transmit times is greater than 8-bit times. This variation comes from propagation length variations between near-end and far-end units and component propagation differences in each unit's circuitry. Because of the unit-to-client variations, it is necessary to adjust each unit's frame-period phase delay that is used to time the start of transmission. These adjustments align the data within the guard time of the TxSlot. The process of aligning terminal equipment units is called *time ranging*.

The propagation length variations dominate the total transmit time differences between units. A modern cable plant can have 4.5 miles of coax at the end of a fiber run. If units are restricted to sharing an upstream channel with only other units on the same fiber node, then the total time variation between them is $(2 \times 4.5 \text{ miles}) \times 5.4 \text{ }\mu\text{sec/mile } (C) \times 1.6 \text{ } (v/C) = 77.76 \text{ }\mu\text{sec}$.

Eight bits of guard translates to:

- 3.73 µsec @ 2.144 Mbps
- 14.93 µsec @ 536 Kbps
- 60.61 µsec @ 132 Kbps

Therefore, a ranging function is needed for the upstream channel speeds. The accuracy of the ranging function needs to be greater than half the guard time to ensure that adjacent channels do not interfere. The accuracy selected is 1.5 µsec. This leaves a 365-ηsec margin for inaccuracies in the terminal-equipment-unit circuitry. Therefore, the requirement placed on the unit is that it shall be able to adjust its transmit time within a TxSlot relative to the downstream 20-msec frame in increments of 1.5 µsec and to an accuracy of ±365 ηsec.

For the 2.144 Mbps channel, the adjustments could be made by delaying the transmission on a bit-time boundary. For the slower channel (536 and 132 Kbps), the bit timing is too coarse, and a fractional bit clock has to be used. The system has to perform an initialization algorithm for each terminal equipment unit to determine its transmit timing adjustment. To do this, the system reserves a time period equal to the round-trip time from the closest client to the farthest client plus the transmission time for a simple message. For 20 miles of coax, this equals 346 µsec + the message time. The ranging message can be small. All it has to do is uniquely identify the client's machine using its 802 source MAC address. Therefore, the time ranging can be done within two TxSlots on a 2.144-Mbps channel and in one TxSlot on the 536- and 132-Kbps channels with a ranging message that has about 30 bytes of payload.

UPSTREAM PROTOCOL STACK. The protocol stack example depicted in Fig. 6.33 shows four types of software interfaces. They are:

1. An IP interface directly to the IP to M3 adaptation layer (AL)
2. An IP interface to an intermediate, IP compression layer
3. An ATM cell interface to the M3 upstream MAC
4. A MAC control-layer interface for maintenance and control

Figure 6.34 is an illustrative example. Actual stacks can have a smaller number of interfaces like ATM and MAC control or direct IP and MAC control. Stacks can also be expanded to include other data stream types such as MPEG and time slots. A diagram of the encapsulation and fragmentation process of compressed IP frames is depicted in Fig. 6.34.

The compressed IP stack layer takes standard IP packets and compresses them using a CSLIP-like compression algorithm. The compressed IP PDU is then passed to the "IP to M3 AL" layer, where it is encapsulated into an upstream packet PDU. The upstream packet PDU has Up_Length, M3_PDU_Type, and Up_CRC information. It also has an optional field that contains feedback information that is used to influence the bandwidth allocation policy and adjust the terminal equipment unit transmitter's power. This information is usually appended to the last packet transmitted upstream in a message. If there is no data ready to send in response to a credit packet, the terminal equipment unit sends a discrete done packet.

The following are fields that are defined by option 1:

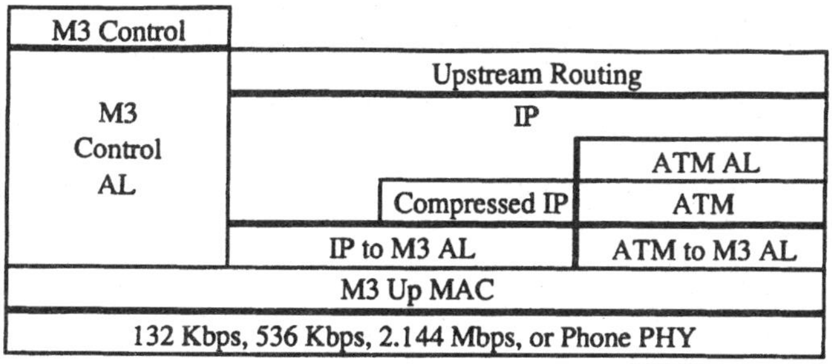

Figure 6.33

Terminal equipment unit upstream protocol stack.

M3 Control				
M3 Control AL	Upstream Routing			
	IP			
			ATM AL	
		Compressed IP	ATM	
	IP to M3 AL		ATM to M3 AL	
M3 Up MAC				
132 Kbps, 536 Kbps, 2.144 Mbps, or Phone PHY				

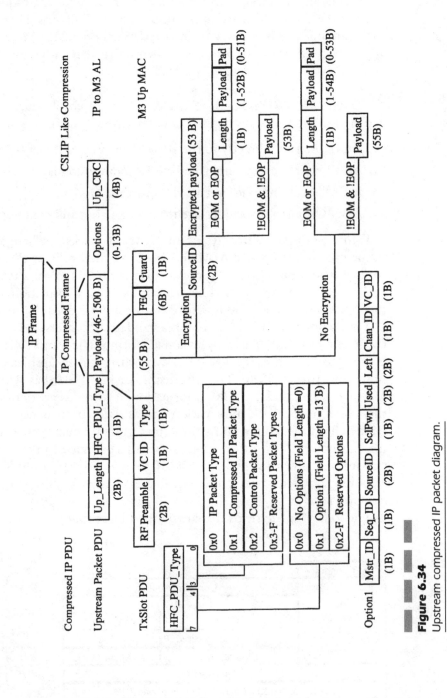

Figure 6.34
Upstream compressed IP packet diagram.

■ *Mstr_ID and Seq_ID* are used to identify the downstream transmission grant message (credit) that generated the upstream transmission. These values are reflected from the downstream credit packet, unless the channel is a dedicated one and the upstream message is a heartbeat packet not controlled by a credit packet.

■ The *sourceID* is the identification number given to the client at power-on or when the client logs on to the cable network. The ID is used to select an encryption key and identify a client during the network session.

■ *SclPwr* is the credit scale (4 bits) and client transmit power level (4 bits). The scale indicates the credit scale: packet count, byte count, word count, and long word count.

■ The *Used field* reports how much of the credit was used to transmit. The scale of the Used field is determined by the SclPwr scale.

■ The *Left field* indicates how much data is left to send after the credit is used. The scale of the Left field is determined by the SclPwr scale.

■ *Chan_ID* identifies the upstream channel the PDU came in on.

■ *VC_ID* identifies the virtual channel the PDU came in on. The VC_ID is a number assigned by the POP equipment that generates the credit message. From the VC_ID, the POP equipment can look up the individual TxSlots that make up the virtual channel.

The upstream packet PDU is then fragmented into blocks that range in size from 52 to 55 bytes by the IP to M3 AL. The size depends on whether encryption is being used or not. The fragments are then placed in a 67-byte TxSlot. The M3 Up MAC then transmits the TxSlots on the assigned virtual channel.

Uncompressed IP frames can also be sent upstream. These frames go directly to the IP-to-M3-AL layer, where they are encapsulated into upstream packet PDUs as shown in Fig. 6.35. The IP to ATM addresses and connection setup are handled by protocols like RFC 1577, and the IP frame is encapsulated in a standard ATM AL PDU; most frequently, AAL5 is used. This is all depicted as one ATM AL layer. The ATM AL PDU is then segmented into standard ATM cells. The ATM cells are placed into TxSlot payloads by the ATM-to-M3-AL layer. The TxSlot PDUs are transmitted by the M3 Up MAC on the assigned virtual channel.

The upstream M3 control packets are not networked and do not make use of the IP layer.

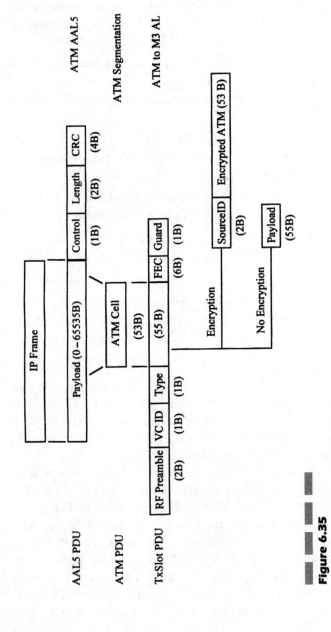

Figure 6.35
Upstream IP packet over ATM diagram.

314

DOWNSTREAM PROTOCOL STACK. The downstream stack depicted in Fig. 6.36 shows an example IP/802.3 and IP/ATM adaptations to a M3 network.

For direct IP frames, the IP message is encapsulated in 802 packets by the IP to the 802.3 adaptation link layer. For this adaptation, we can choose either 802.2 standard or the widely used RFC 894 protocol. The 802.3 packet is then encapsulated in an HDLC-like frame by the 802.3-to-M3 AL (this is illustrated in Fig. 6.37). The M3 Down MAC then multiplexes the 16-bit words of the IP-to-M3 PDU onto the physical channel. Unlike the upstream, the M3 control packets in the downstream use UDP/IP so that they can be routed over a network in the POP. IP frames may also be sent via ATM cells on the downstream channel (as shown in Fig. 6.38). IP-to-ATM address assignments and ATM connection establishment tasks are managed by protocols like RFC 1577. The IP frames are then encapsulated in a standard ATM AALX (again AAL5 is the most common adaptation layer). The combination of RFC 1577 and AAL5 are shown as one layer in the protocol stack diagram. The ATM AAL5 PDU is then segmented into 53-byte ATM cells by the ATM layer. The ATM cells are placed in a 55-byte ATM/M3 PDU by the ATM-to-M3 adaptation layer and then fragmented into downstream-frame PDU words. The M3 Down MAC multiplexes the words onto the downstream channel. ATM cell delineation is determined by scanning for the word boundary that gives good ATM CRC checks. This is the same method used for ATM on Sonet and T1/E1 serial lines. The ATM-to-M3 adaptation link layer generates idle ATM cells at a prescribed rate when there are none to send.

END-TO-END DATA PATH. Figure 6.39 shows the round trip data path for an FTP application running between a client PC and an FTP server at the POP. The client PC is connected to a terminal equipment unit via an 802.3, 10BaseT LAN. The terminal equipment unit is connected to the upstream headend equipment (UHE) on either a shared or

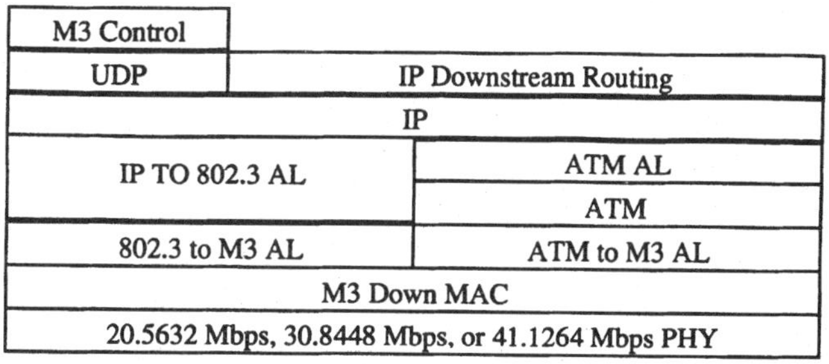

Figure 6.36
Terminal equipment
unit downstream
protocol stack.

M3 Control		
UDP	IP Downstream Routing	
IP		
IP TO 802.3 AL	ATM AL	
	ATM	
802.3 to M3 AL	ATM to M3 AL	
M3 Down MAC		
20.5632 Mbps, 30.8448 Mbps, or 41.1264 Mbps PHY		

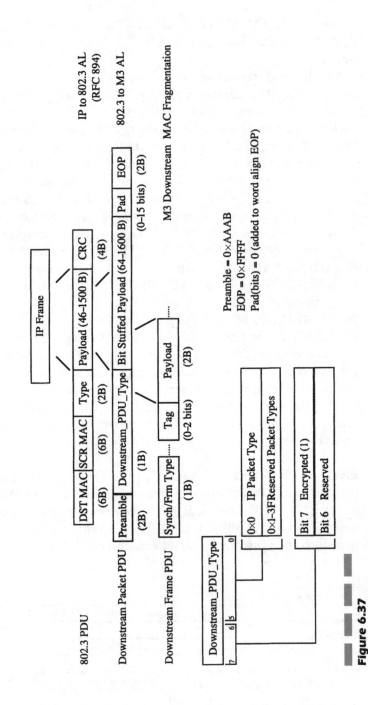

Figure 6.37
Downstream IP packet diagram.

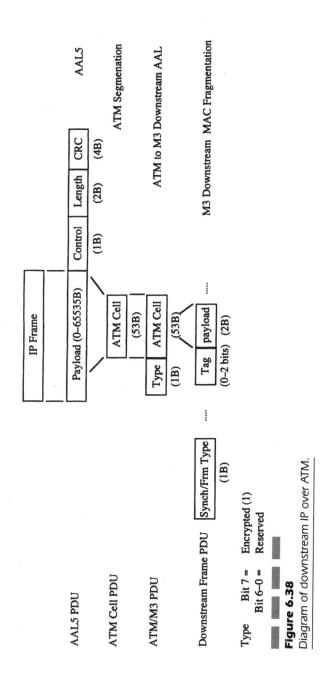

Figure 6.38
Diagram of downstream IP over ATM.

317

dedicated M3 upstream virtual channel. The UHE is connected to the headend FTP server through an 802.3 switch or an ATM switch. The headend server is also connected to the downstream headend equipment (DHE) via the 802.3/ATM switch. The DHE connects to the terminal equipment unit via an M3 downstream channel. The 10BaseT return connection from the terminal equipment unit to the client PC is the same physical 10BaseT connection used in the upstream direction. They are depicted as separate to keep the drawing simple. Data packets generated by the client PC are received by the terminal equipment unit. For a single PC connected to a terminal equipment unit, the upstream IP routing is a simple IP bridging function that relays the IP packet to the upstream and intercepts the ARP packets.

The terminal equipment unit configures the upstream transmitter according to the UDP M3 control packets it receives on the M3 downstream channel. Upstream M3 control information is sent either in the option field of the upstream packet PDU or in separate upstream messages. The upstream M3 control information tells the UHE about the state of the terminal equipment unit's transmit queue, power setting, ranging setting, and so on. The DHE receives these messages and, if required, passes the information on to the headend network management function in IP packets.

The UHE routs the client's IP packets to the 10/100BaseT port. This routing function is simple and more like an IP bridging function. The networking to and from the FTP server uses standard stack protocols. The FTP server sends response packets back to the client. The DHE receives these packets and routes them to the downstream channel that is servicing the user. As with the upstream routing, this is a simple IP bridging operation. The terminal equipment unit receives the FTP packets and M3 control packets. The terminal equipment unit routes the FTP packets to the client and uses the M3 control packets to maintain the upstream channel.

UPSTREAM CHANNEL MANAGEMENT

UPSTREAM CHANNEL TYPES. The headend equipment makes two types of upstream channel assignments:

1. Dedicated
2. Shared

The upstream channel assignments are made with credit packets generated by the headend. All terminal equipment units receive credit messages that are for either dedicated or shared channels.

Dedicated Channels. Dedicated channels are given to a terminal equipment unit on a semipermanent basis. When the unit receives a dedicated

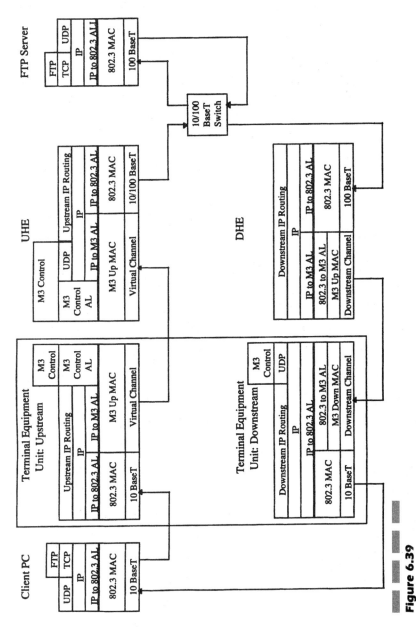

Figure 6.39
Example of an FTP communication path.

channel, it transmits data whenever a data packet is ready. The terminal equipment does not have to wait to gain access to the channel. If no data is available to send, the channel remains inactive.

The terminal equipment generates periodic heartbeat messages on a dedicated channel so that the unit can be monitored by the headend. The headend also monitors the utilization of the channel.

Subscribers may be given a dedicated channel if:

- They have subscribed to a dedicated service
- Their measured traffic utilization needs a dedicated channel
- An application like voice or video requests one
- The system utilization is low and channels are readily available

Shared Channels. Shared channels are used to discover terminal equipment units that have entered the network and to provide transmission service at moderate rates and access times. Shared channels allow the M3 system to expand automatically the number of terminal equipment units serviced by a fixed number of upstream channels. The basic queuing model for shared channels is a pool of users serviced by a pool of channels. The shared channel model can use a variety of algorithms to implement the service queue. The simplest is a first-in-first-out (FIFO) queue where there is no attempt to distinguish between busy and idle terminal equipment units. More sophisticated queuing protocols can be implemented in the headend that move units up in the queue based on their recent data transmission rate or on the size of their transmit data queues. Queue protocols can also be designed that give time-based access to the upstream channel pool so that CBR, VBR, and minimum ABR services can be guaranteed.

The M3 system implements more than one pool of terminal equipment units and channels. There can be pools for the following:

- *Nonresponding (powered-off) units.* This is usually a separate pool because credits are sent to them to detect when there is a powered-on unit, usually time out and lead, to slow access times and low throughput for the other units sharing the channels.

- *Video game users.* These users need low-access delay times but low throughputs. The ratio of units to channels is kept low, and the bandwidth of the channels selected can be low.

- *Regular users.* These users require moderate amounts of upstream bandwidth, mainly to send downstream transfer acknowledgments. Traffic profiles of a large majority of existing M3 users are serviced by the performance characteristics of shared channels.

Terminal equipment units that receive a credit packet for a shared channel use as much of the credit as they can or immediately return it to the headend. Terminal equipment units cannot hold on to shared channel credit like they do with a dedicated channel. Credit packets for a shared channel are addressed to a single terminal equipment unit or are sent to multicast addresses of a group of units. Multicast credits are used to implement upstream contention channels. Successful transmissions are acknowledged by downstream messages. Contention-based channel assignment is only used for low-utilization functions like looking for powered-on units. Because contention-based channels are only used for low-utilization functions, a simple back-off algorithm can be used.

UPSTREAM CHANNEL CONTROL. The headend equipment controls the assignments of upstream channels through in-band messages. These messages perform a poll-and-response access function that avoids embedding control-specific information in the downstream and upstream physical frame formats. By making the access control all message bases, the M3 system achieves a high degree of flexibility that allows it to incorporate a variety of underlying physical layers, each of which requires different parameters to be controlled.

CLASS OF SERVICE SUPPORT. The M3 MAC protocol supports constant bit rate (CBR), variable bit rate (VBR), and available bit rate (ABR) traffic classes defined by the ATM Forum. The headend has to be told when to set up a specific class of service for a subscriber's application.

CBR downstream traffic is carried in either the ATM data stream marked by the ATM-tagged MuxSlots or in the time slot data stream marked by the time-slot-tagged MuxSlots. In both cases, there are mechanisms for reserving bandwidth for particular end-to-end connections.

CBR upstream traffic can use a dedicated virtual channel or a shared virtual channel that the headend assigns using a CBR upstream virtual channel allocation algorithm. The CBR allocation algorithm periodically sends a credit message to the terminal equipment unit that is used to service the CBR connection.

VBR downstream traffic uses either the ATM data stream or the TDM data stream. For the ATM data stream, VBR traffic is handled by the standard ATM mechanisms. For the TDM data stream, VBR traffic, when present, is sent in the assigned MuxSlots tagged as TDM data. When VBR data is not available in a downstream frame, the reserved MuxSlots are tagged as packet or ATM data. This relinquishes the VBR bandwidth for that frame.

VBR upstream traffic uses a shared virtual channel. The headend uses a VBR upstream virtual channel allocation algorithm. VBR allocations are made periodically to a terminal equipment unit. If the unit has no data to send, then it returns the unused credit, which makes upstream bandwidth available to others.

ABR downstream traffic can be transported on either the ATM data stream or the packet data stream. Existing ATM mechanisms provide the ABR minimum service rate. For the packet data stream, a headend downstream bandwidth management function has to provide the minimum service rate guarantee.

ABR upstream traffic uses a shared virtual channel. The upstream virtual channel allocation algorithm guarantees the minimum service rate.

6.4.11 Com21, Inc.: The UPSTREAMS Protocol for HFC Networks

OVERVIEW. Com21 proposed the UPSTREAMS (upstream protocol sharing transmission resources among entities using an ATM-based messaging system) protocol for HFC networks. The UPSTREAMS proposal includes all aspects of the HFC system and stations, including the physical layer, architecture, management, MAC layer, service aspects, and so on. Only the access MAC layer will be described in this section.

FEATURES OF THE UPSTREAM MAC. The following MAC-related features are part of the UPSTREAM protocol:

- UPSTREAMS is an ATM-over-HFC media access control (MAC) protocol.

- The UPSTREAMS protocol itself is not a traffic management scheduler, but rather a protocol messaging system that can support one of several bandwidth management models; e.g., a best-effort Ethernet-like management model and/or a scheduling model with constant and best-effort bit rate.

- UPSTREAMS resource management support is capable of integrating with an ATM layer to provide the full suite of ATM traffic management classes: CBR, variable bit rate (VBR), available bit rate (ABR), and UBR.

- The headend controller is responsible for all bandwidth management and all resource management of the P802.14 segment, including profile assignment, modulation, frequency, bandwidth, and power assignment.

■ The protocol is designed to support all hybrid fiber-coax (HFC) models, i.e., combined returned and separate return systems over a homes-passed size of 500 up to 1200 homes, and then possibly up to a maximum of 4093 homes.

■ Both the downstream and upstream are framed and divided into slots.

■ A request-and-grant-style of bandwidth allocation is used to allocate slots in the upstream. A contention protocol slot is used to reduce polling requirements and improve upstream latency for making requests.

■ An ATM network-to-network interface (NNI) cell format is used as the basic information exchange data unit for both the downstream and upstream traffic flows. The larger, 12-bit virtual path identifier (VPI) of the NNI cell maps directly to a P802.14 station address with 4093 active unicast station addresses maximum per segment.

■ Station addresses are issued during the initial registration and acquisition process.

■ Unicast, multicast, and broadcast addressing is supported in the downstream.

■ Single-cell ATM messages are used for all downstream traffic, including but not limited to normal user data traffic; upstream resource management traffic; encryption management traffic; operations, administration, and management (OAM) traffic; etc.

■ Per-cell overhead is limited to 1 octet of security-envelope overhead in the downstream and 1 octet of both security-envelope and station-request information in the upstream.

■ The UPSTREAMS MAC is independent of the physical (PHY) layer so long as the physical transmission convergence (TC) layer supports specific basic framing functionality, a one-transmitter-to-many-receivers model is supported in the downstream, and a many-transmitters-to-one-receiver model is supported in the upstream.

■ Downstream and upstream encryption key management for the ATM payload is supported, although the actual encryption is done in the PHY TC layer.

■ An Ethernet/802.3 LLC overlay is accomplished using an ATM AAL5 segmentation and reassembly (SAR) function.

- IEEE 48-bit link layer control (LLC) MAC addresses are used to identify all stations and are communicated between stations during registration acquisition and in all Ethernet/LLC frames.

- The unspecified bit rate (UBR) and constant bit rate (CBR) services are supported for networking needs of the 802.

- Forward error correction in the downstream and upstream is performed as part of the PHY-TC function.

BASIC METHOD OF OPERATION. The basic method of operation for UPSTREAM is that each station must request upstream resources using a resource request message. Resource requests are transmitted to the headend controller as a response to a contention grant. The headend controller accumulates resource requests from all stations and apportions upstream resources appropriate to meeting the fair needs of all stations.

Station responses to contention grants are processed using a truncated binary exponential backoff contention mechanism in the slots assigned in the grant. Direct grants are nonshared allocations, each to a specific station. An acquisition grant is used as part of the new station registration process. All grants assign resources based on the location of an upstream frame that is common knowledge to all active stations. The upstream frame is synchronized to a downstream frame. For each station, the relation of its notion of the upstream logical frame timing to the downstream frame reception is set as part of the acquisition process. A time division multiplexing (TDM) technique is used on the upstream channel; i.e., the upstream is slotted, an UPSTREAMS message is placed within a slot, and upstream slots are numbered. All grants then specifically assign a small window of resources based on the notion of slot number and number of slots to use for the grant. All grants are renewable in that they expire immediately after their grant window. All grants are sent as ATM messages in the downstream channel. One ATM message may contain many direct grants, each to a different station.

All stations on an UPSTREAMS segment are assigned a unique logical station identifier as part of the acquisition process. This is called the station unique identifier. This identifier is then used to send messages to that station. Station identifiers are not persistent, in that each time a station is powered up, it will be assigned its station identifier during the acquisition process. Stations are also assigned one or more station multicast addresses. The IEEE 48-bit MAC address of the station is discovered by the headend during the acquisition process. The headend controller maintains a mapping of MAC addresses to logical station addresses. Each UPSTREAMS message contains only the logical subnetwork station identifier.

BANDWIDTH MANAGEMENT FOR THE UPSTREAM

STATION RESPONSIBILITIES. Each station maintains a FIFO queue of grants it receives from the headend controller. Grants expire when their frame number and slot number have passed—in other words, use it or lose it. A station also has a contention-denied timer that is activated whenever the grant queue is empty. The period of this timer is adjustable from 0 milliseconds to 255 milliseconds. Each station has an upstream cell transmit queue, and a queue empty condition is known to the upstream scheduling function. Directed grants allow one station access to the upstream channel for the specified permit. The grant indicates an individual station identifier, the starting slot, and a length that indicates how many additional slots to use. The length is from zero to seven additional slots. Therefore, each directed grant may only give out up to eight slots on the upstream channel. The slot numbering mechanism allows the station to determine if the grant is for the current logical frame or the next. The grant queue allows many directed grants to be queued for the station, and the station may use these back to back, if necessary. When processing a directed grant, each station will either transmit a cell from its upstream transmit queue or it will transmit an idle cell if the queue is empty. When required to transmit an idle cell, an idle cell is transmitted for each allocated slot in the permit.

Referring to Fig. 6.40, the ATM layer of the station continually updates the request register of the PHY layer through a well-defined layer interface. The PHY layer communicates the contents of this register on each and every cell transmitted from the station to the headend controller, whether it be a real user data cell or an idle cell.

If the grant queue is not empty, the station does not queue contention grants. If the idle timer has not expired, the station also does not queue contention grants. If the grant queue is emptied, the station resets and starts the idle timer. Any reception of a directed grant prior to the expiration of the timer, resets the timer. The idle timer is halted as soon as the request queue is no longer empty. If the grant queue is empty and the timer expires, the station will queue up a contention grant, if it receives one. Contention grants, which are used for requesting bandwidth, are issued as an all-stations group identifier in the permit. Contention grants are for a one-slot opportunity to transmit an idle cell if and only if there is a cell in the upstream transmit queue; that is, the station only makes a request if it has something real to send. If a station transmits an idle cell at this time, the station enters the contention state. The station may only exit the contention state via a management reset or by the reception of a directed grant. The contention state executes a truncated binary exponential back-off based on the number of contention grants it receives.

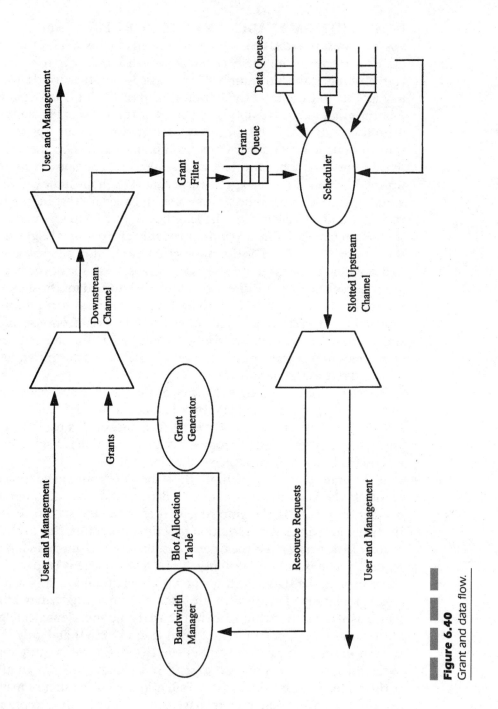

Figure 6.40
Grant and data flow.

Note that in this mechanism the contention result's feedback is inferred by the nonreception of a directed grant. Furthermore, since real data isn't being sent during the contention opportunity, the dropping of the idle cell or the duplication of the requested information is not critical. The mechanism is robust in the presence of upstream cell error.

HEADEND CONTROLLER RESPONSIBILITIES.   The messaging and allocation capability of the UPSTREAMS protocol places the responsibility for managing different traffic classes solely in the hands of the bandwidth manager located in the headend controller. For example, if the station has an established CBR service, such as voice at 64 Kbps, the bandwidth manager must allocate grants to that station at or above the rate required for the 64-Kbps service. Any additional UBR bandwidth would be sensed by the bandwidth manager by observing the request tokens sent on every transmission from that station on the upstream channel. The bandwidth manager calculates what additional allocations are necessary for the best-effort UBR traffic and allocates upstream permits sufficient for the CBR rate maintenance and the bursty, request-oriented nature of the UBR traffic. The cell rate allocated to a station on the upstream channel is referred to as the allocated cell rate (ACR) in this proposal. Note that a station with a CBR service is always active in that the bandwidth manager is always aware of the request status of the station. Stations may also be idle in that they might have not sent any allocated traffic upstream for some period of time. In this case, the bandwidth manager is not allocating any upstream resources, requiring the station to make use of the next available contention opportunity. The UPSTREAMS protocol uses the contention mechanism to communicate only requests from idle stations to the bandwidth manager. A successful reception of a request requires that the bandwidth manager schedule a directed grant. The station can then transmit its traffic.

PROVISION LAYER SERVICES.   The functions of the layer provide access control to the downstream and upstream channels. These functions are used by a range of convergence functions to create the P802.14 layer as shown in Fig. 6.41. This proposal overviews the convergence functions for the provision of the MAC sublayer service to the LLC sublayer and the provision of sublayer service to an Ethernet/802.1D service.

All P802.14-layer services obey ATM cell-ordering requirements. At no time do two convergence functions simultaneously use the same VCI for a given station. Hence, if a station is to receive a sequence of cells that encode an LLC frame, the VCI can be used to direct the cell flow to the appropriate convergence function at the station (or headend).

Figure 6.41
ATM structure rela-
tionship with 802.14
MAC.

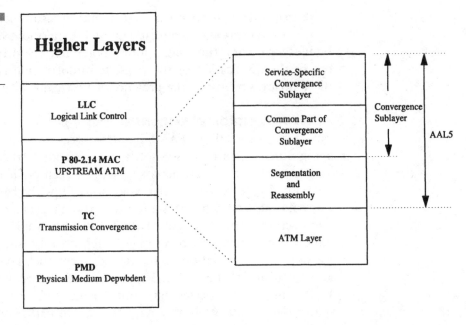

PROVISION OF MAC SERVICE TO LLC. The provision of a MAC service to the LLC consists of the encapsulation of the MAC service data unit (m_sdu) using ATM adaptation layer 5 (AAL5), the subsequent segmentation of the AAL5 PDU at the source into ATM cells, and then the transfer of these cells to the destination, which reassembles them into the m_sdu. The encapsulation of the LLC frame into AAL5 occurs in the service-specific convergence sublayer (SSCS) of the ATM model. For the purposes of this proposal, the specific encapsulation used will be the NULL encapsulation. The SSCS sublayer must also provide a mapping function of the 48-bit MAC destination address to the appropriate VPI (station identifier) for the destination. The segmentation and reassemble (SAR) function conforms to the ATM AAL5 common part convergence service sublayer. Chapter 3 described AAL5 segmentation procedures.

The encapsulation of the Ethernet/IEEE 802.1D frame into AAL5 occurs in the service-specific convergence sublayer (SSCS) of the ATM model. For the purposes of this proposal, the specific encapsulation used will be the NULL encapsulation. The SSCS sublayer must also provide a mapping function of the 48-bit MAC destination address to the appropriate VPI (station identifier) for the destination. An individual destination address maps to an individual station identifier. A group destination address maps to a group station identifier. The assignment and use of such group station identifiers is the function of the layer management entity. The segmentation and reassemble (SAR) function conforms to the ATM AAL5 common part convergence service sublayer. The maximum supported length of the frame is set at 1518 octets.

REASSEMBLE AT THE DESTINATION. To receive ATM cells, each station will monitor all cells being sent on the downstream channel. The station will copy cells that have been addressed to that station:

1. The VPI matches the individual station address

2. The VPI specifies an all-stations broadcast

3. The VPI matches a preassigned group address of which that station is a member

Upon receiving the ATM cells, the station will place the cells in queues, in arrival order, specific to the VCI specified in the cell header. Each queue is then associated with the specific convergence function, and the cell stream is reassembled appropriate to that function. In the case of the segmentm_sdu stream, the received cells are placed in a queue and then processed as an ATM AAL5 cell flow.

STATION FUNCTIONAL ARCHITECTURE. The fundamental architecture for an UPSTREAMS station is shown in Fig. 6.42. It consists of two layers: the physical (PHY) layer and the P802.14 access layer. The P802.14 layer uses the services of the PHY layer to provide a number of different services. Two of these services are the MAC sublayer service to the LLC sublayer and the sublayer service to the Ethernet/IEEE 802.1D sublayer.

UPSTREAMS STATION MAC LAYER FUNCTIONS. The general MAC layer contains the ATM layer, common and access functions, the ATM AAL5 common part convergence sublayer, the MAC/LLC service-specific convergence sublayer, the Ethernet/IEEE 802.1D service-specific convergence sublayer, the ATM Forum residential broadband (RBB) ATM interface unit (AIU) transmission convergence emulation service interface, and layer management functions. The P802.14 to LLC service is provided to the LLC service layer entity at a station controller via a service access point, or SAP. The P802.14 to Ethernet/802.1D service is provided to the Ethernet/802.1D service layer via a SAP. The P802.14 to ATM RBB services is provided to an ATM Forum RBB ATM AIU service layer via a SAP.

ATM Transmission Convergence Emulation Function. The UPSTREAM P802.14 identifies the interface at the station that interfaces to and sup-

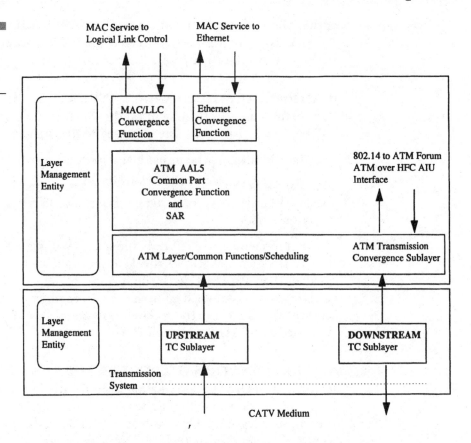

Figure 6.42
UPSTREAM station functional architecture.

ports the services to the ATM Forum's ATM Interface Unit. The AIU function and the nature of the interface is described in Chap. 3.

STATION MAC LAYER FUNCTIONS. Within the UPSTREAMS MAC layer, there are five principle functions:

1. The ATM common and scheduling functions
2. The ATM AAL5 convergence functions
3. The MAC/LLC service-specific convergence functions
4. The Ethernet/IEEE 802.1D service-specific convergence function
5. The layer management functions

ATM Common/Scheduling Functions. The ATM common, and scheduling function block acts as an ATM layer for the transfer of ATM cells between the PHY SAPS and the UPSTREAMS functions. It provides other MAC layer functions that may be common in both the downstream and upstream PHYTC sublayers. It also provides upstream transmission scheduling and queueing management control over the ATM cell streams on active VCIs. The common functions block supports the layer management entity for coordination of the configuration control and the virtual channel allocation of this function.

ATM AAL5 Convergence Functions. This is a standard ATM AAL5 common part convergence layer. See Chap. 3.

MAC/LLC Service-Specific Convergence Function. The MAC/LLC service-specific convergence function is responsible for adapting the ATM AAL5 PDU-based service provided by the ATM AAL5 CPCS function to the standard MAC service required by the LLC sublayer. The convergence function is realized primarily by the implementation of the m_sdu to CPCS-PDU encapsulation process.

Ethernet Service-Specific Convergence Function. The MAC Ethernet/IEEE 802.1D service-specific convergence function is responsible for adapting the ATM AAL5 PDU-based service provided by the ATM AAL5 CPCS function to the services required by the Ethernet/IEEE 802.1D sublayer. The convergence function is realized primarily by the implementation of the Ethernet/IEEE 802.1D protocol data unit frame to CPCS-PDU encapsulation process.

Layer Management Functions. The MAC-LME performs the management of the local MAC layer functions.

6.4.12 Amati: MAC Protocol to Support Both QPSK and SDMT

INTRODUCTION. Amati propose a MAC protocol that will support HFC transmission using SDMT.

Traditional protocols use controlled transmission (i.e., each station transmits only when permitted); they differ mainly in the method of requesting permission: sequenced (S), polled (P), and random (R) requests.

Amati observed that there is no MAC currently proposed for QPSK that is identical to that used by SDMT. For example, SDMT handles CBR by using FDM, while other systems handle CBR within the TDMA mechanism. VBR and ABR are handled as TDMA in both systems. In SDMT, all stations use the same number of bits in each frequency, although this may change as the ingress noise changes. The headend continually tells the stations what carriers are available for VBR/ABR use, sending out the information once per superframe.

BASIC PRINCIPLES OF SDMT MAC.   Three types of station SDMT transmitters are envisioned:

1. *Class 1.*   A simple one that can transmit one or two subcarriers. This would be very low-cost for use for telephony.

2. *Class 2.*   One that can transmit up to 32 subcarriers in a bandwidth of approximately 1.1 MHz; this is comparable to some of the wider QPSK modems. The maximum data rate would be greater than 8 Mbit/s, but this midrange transmitter would typically be used for data rates from any one station of up to 1.536 Mbit/s.

3. *Class 3.*   One that can transmit 128 subcarriers in a bandwidth of 4.4 MHz. This would be used for data rates up to 10 Mbit/s.

For maximum efficiency in the use of the upstream bandwidth, it is desirable that all station transmitters—whether QPSK or SDMT—should be able to operate anywhere in the 5- to 40-MHz band and that a single controller should oversee all stations connected to the network. Depending on the complexity, the MAC can operate in any one of the above transmit modes. An example of operating a medium-complexity (class 2) transmitter mode is described below.

DATA TRANSMISSION.   Each time a station is granted access, the subcarriers to be used and the number of bits to be modulated onto each should be defined again. Since the group (or subgroup) was subdivided and the subchannels allotted to the multiple transmitters, this was a form of FDMA.

CBR.   For CBR data, FDMA is still the simplest and most efficient method of multiplexing. The symbol rate is 32 Kbaud, so the throughput bit rate can be fixed in units of 32 kbit/s. Each station requesting CBR service is assigned full-time or part-time use of one or more subchannels (e.g., 64 kbit/s could be provided by use of every symbol of a 2-bit s-c; every other symbol of a 4-bit s-c; etc.). These subchannels are then not available for VBR or ABR.

VBR AND ABR.   The achievable data rates from all the stations are approximately equal, and therefore the times required to transmit a packet can be made equal without any significant loss of throughput. The bandwidth available for packetized data may, however, vary widely because:

- HFC systems will differ in attenuation, amplifier noise figures, amount of RFI and EMI, etc.
- Frequency groups will suffer from different amounts of RFI and EMI.
- The pool of subchannels available for packetized transmission will change as CBR users are added or removed.

The number of symbols required to transmit a single ATM cell could vary from 2 for the quietest system with no CBR (providing an average of 7.9 bits/s-c capability) to very large if much of the bandwidth of a noisy system were preempted by CBR.

The basic unit of SDMT transmission is a symbol. Therefore, packet sizes of about half of an ATM cell (approximately 30 bytes) could be handled with no loss of efficiency on the fastest channel by assigning a station just one symbol. With shorter packets, the efficiency would decrease on the very fast channels because of the one-symbol granularity.

ACCESS. Before a new station is registered, it must know the subband in which it should transmit. Therefore, the headend should continually balance the assigned load in each band against the capacity of that band and broadcast the number of the subband that the next-registered station should use. There are then three stages that a new station must go through:

1. Registration
2. Session sign-ons
3. Access

REGISTRATION. The following steps are required for station registration:

- Ranging
- Identification of modem type (QPSK, SDMT Class 1, 2, or 3) and limitations on frequency agility
- Assignment of RF band
- Exchange of IEEE device address for 6-bit network address

SESSION SIGN-ONS. These can occur only once per super-frame and are sequenced by the network address; they involve:

- Type of service requested: CBR, VBR, or ABR
- QOS negotiation
- Exchange of network address for 4-bit access address if VABR requested

- Preliminary definition of packet length
- Equalizer training

ACCESS.   Access features are as shown below:

- All requests for access are sequenced; that is, allowed only in pre-assigned, regularly occurring time slots.
- Even though most subgroups will be able to support multiple bits per subchannel, all requests for access use QPSK. Thus, the MAC protocol can remain invariant and secure across the whole range of capacities and is the same as for QPSK.
- The ability to reserve a small part of the frequency bandwidth for access requests and transmit efficiently very short request messages gives SDMT a big advantage over all single-carrier systems.
- Because all access requests are sequenced, one part of the access time is proportional to the number of possible requesters that must be cycled through. Therefore, the fastest access is achieved for VBR only if active stations have an assigned address, which is only valid for a session.

MAC DETAILS.   The system is locked to the downstream symbol and super-frame clocks. Access is effected by:

- Every super-frame period: downstream broadcasts
- Every super-frame period: downstream-addressed grants for either training upon installation or CBR service
- Every symbol period: upstream requests sequenced by the 5 LSBs of the network addresses
- Every symbol period: downstream-addressed grants for VBR or ABR service

DOWNSTREAM PER SUPER-FRAME PERIOD.   The control frame contains at least 18 bytes (36 nibbles). The first 17 bytes are broadcast; the rest are addressed to one station that has requested CBR service.

UPSTREAM PER-SYMBOL PERIOD (ACCESS REQUESTS).   Most subchannels in most HFC systems will be able to support multiple bits (up to 10), but in order for the MAC protocol to be independent of the quality of the system, all access requests must be limited to 2 bits/s-c.

CBR.   Each station uses one subchannel for three symbols to send 6 bits. The pattern 000000 should be reserved for a sign-off; that is, end of session—no more CBR needed.

VABR.   Each active VABR station is assigned a 4-bit access address that controls how and when the station may request a packet grant. A 0 (or 1) LSB indicates that the station should use the subchannel identified by the code 1100 (or 1101) in the downstream broadcast of subchannel usage. The 3 MSBs indicate which of eight symbols in sequence throughout the super-frame the station should use.

It is assumed that all QOS parameters, including whether the requests will be for VBR or ABR, are negotiated and defined at the beginning of a session. Then, since the source of a request is apparent to the HEC from the symbol and subchannel numbers used, the only information that need be transmitted in the two bits available is the requested length of the packet. The minimum time to transmit a packet (with the quietest channel that is not carrying any CBR) is 2 symbols. Therefore, since a station may send a request once every 8 symbols, the maximum number it need request at one time is 4, and all lengths <4 can be defined with 2 bits:

00 = 1 cell
01 = 2 cells
10 = 3 cells
11 = 4 cells

6.4.13 Panasonic Technologies, Inc.: Collisionless MAC

OVERVIEW.   Panasonic proposed a flexible scheme to support diverse applications. It provides a method of point-to-point signaling access and flexibility of variable bandwidth. The MAC facilitates digital services communication and multiple services like voice, video, and data servers. These diverse services have varying characteristics in terms of delay, error, and quality. The protocol covers connection-oriented and connectionless services. Quick access and bandwidth on demand are key features of this MAC layer. The MAC layer monitors and maintains a link between station and headend. This synchronization is needed for system broadcast information. The CATV telecommunication (CATEL) MAC layer also interfaces with the higher layer using predefined primitives. Using these primitives, the higher layer can ask services from the MAC layer, and the MAC layer can inform the status of the requested services to the higher layer. To perform services for the higher layer, the MAC layer needs to request a traffic channel from a line control unit

(LCU). At the end of the higher-layer service, the MAC layer needs to release the traffic channel.

FALLOUT OF MULTIPLE-CONTENTION ACCESS.   It was observed the following fallout of the multiple-contention access and collision resolution schemes:

- The average access delay among various protocols falls between 2.521 msec and 6.6 msec.
- Minislots contribute significant overhead in comparison to the signaling payload.
- Collisions increase the traffic load and result in wasted bandwidth. Collisions also contribute to broadband noise in adjacent channels. Efficient collision resolution and clipping management is therefore required.
- Station-initiated network management tasks are also required. This increases upstream access traffic.
- Fancy protocols add complexity in applications software (headend/station); for example, access via contention in minislots by way of variable/fixed or plain data slots.
- Network response is slow. Fast access by the modem does not buy much. Fast access is feasible only if the outgoing call (by the station) terminates locally, i.e., within the headend boundary.

MAC FEATURES.   Panasonic's MAC protocol attempts to resolve the above and focuses on:

- Defining bandwidth groups; e.g., RFG-1, RFG-2, ... to cover 5—45 MHz in 6-MHz chunks
- Defining each carrier in each radio frequency group; e.g., RFG-1 = RFC-1, RFC-2, RFC-3, RFC-4, ... to support a T1/E1 rate effectively
- Defining frame rate for each carrier
- Defining the number of slots in a frame
- Defining slot size

FRAME/SLOT SIZES.   The basic frame structure, as shown in Fig. 6.43, is 24 slots/frame, with frame size = 1 msec and slot size = 12 bytes with 8 bytes as the payload and 4 bytes as overhead (1 byte GT, 1 byte L2 header, and 2 bytes for CRC). Therefore, effective slot bandwidth is 64 Kbps.

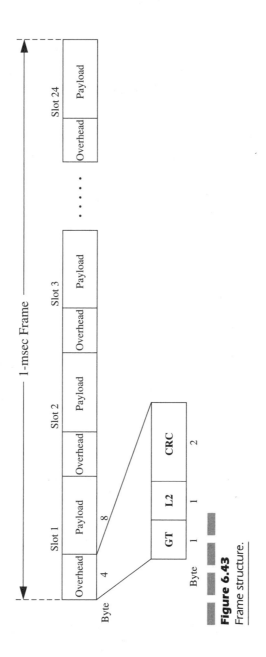

Figure 6.43
Frame structure.

Each carrier is divided into frames. Each frame is divided into slots. The slots are classified as signaling slots, user slots, and OAM (operation, administration, and maintenance) slots. All these slots are integer multiples of user slots.

For an ATM cell (53 bytes), five sequential slots are assigned by the headend. This effectively gives 60 bytes of bandwidth. We distribute 60 bytes as follows:

ATM cell	53 bytes
Guard time	1 byte
L2 header	4 bytes
CRC	2 bytes

The bandwidth manager can assign one of the following:

- Contiguous slots in the same frame to satisfy the entire bandwidth request
- Noncontiguous slots in the same frame to satisfy the entire bandwidth request
- A few slots per frame with 1 or 2 requiring multiple frames
- One slot per frame

For STM traffic assuming 64 Kbps bandwidth multiples, there is one slot with the following distribution:

Payload	8 bytes
Guard time	1 byte
L2 header	2 bytes
CRC	1 byte

UPSTREAM ACCESS PROTOCOL. In the upstream direction, the following dialog is proposed: the headend assigns slot 1 as a signaling slot. (This could be any fixed engineered number; e.g., the first slot in a frame or the last slot in frame.)

In the registration process, the headend assigns a signaling rate to stations. For example, the ith station will use every yth frame of slot 1. The value of y is headend-controlled. If there are more stations, y increases, making the signaling rate go down. The headend may dynamically configure uniform distribution of stations on each RFG and RFC. This distribution may be defined as less active, moderately active, and so on. The headend can change the assigned signaling rate if it finds a station generating more upstream requests. It can then place the station on a separate RFG and RFC with fewer active stations.

No user activity means network-related management activity. That is, the station needs to send periodic loop tests and device status (errors, counters, timers, etc.).

In downstream signaling, the frame number needs to be checked periodically. This can be done:

- Explicitly by indicating the frame number in the broadcast message
- Implicitly; the stations derive the frame number from the time marker broadcast (used for time synchronization)

SIGNALING SLOT.   Signaling slots are used for call establishment, call maintenance, and call tear-down. The outgoing call can either be voice/video or an LLC packet. Bandwidth for the signaling slots is kept small. This accommodates new stations. Providing a separate signaling bandwidth to each station (out-band signaling) provides a virtual point-to-point link with the headend. This simulates a direct link with the headend just like a telephone directly connected to the central office switch. This direct link is desired because it is more like a centralized control system. The complexity of the overall system is reduced by making the station provide stimulus-mode signaling as opposed to functional-mode signaling. This technique has several advantages:

- Ease in network management and updates
- A guaranteed allocated bandwidth as opposed to bandwidth stealing for signaling
- Comparatively less headend processing as it processes only the signaling slots
- User-allocated bandwidth can also be used for administration and maintenance (software downloads/updates, device activate/deactivate commands, etc.)

OAM SLOT.   These slots are used for joining the cable network and are mainly used for registration, ranging, device activation/deactivation, periodic status queries, and so on.

TRAFFIC MIX BOUNDARIES.   It is proposed that the bandwidth allocater in the application layer does the following reservations. This is similar to ADAPt as described in Sec. 6.4.8, with a modification that upper limits of STM and ATM be defined. This is desired for fairness among anticipated traffic mix.

6.4.14 IBM Corporation: MAC-Level Access Protocol (MLAP)

OVERVIEW. IBM made three proposals covering all aspects of the modem. The MAC-level access protocol (MLAP) proposal is described below. MLAP supports the shared-medium topology that addresses MAC functions such as message formats, bandwidth request/grant mechanisms, collision resolution, synchronization, and bandwidth allocation. MLAP is an optimum shared-medium protocol, providing the building blocks necessary to support ATM encapsulation and QOS guarantees.

KEY FEATURES SUPPORTED BY MLAP. Features supported by MLAP include:

- Variable-length slots
- Variable-length MAC messages supported with high utilization
- Multiple PHY options enabled (modulation rates/types/preambles)
- ATM encapsulation
- MAC management protocol (which is called MLMP)
- Reservation-mode and contention-mode data transfer (mix is programmable)
- Reservation-mode requests
- Reservation grants pipelined to optimize implementation cost
- High-throughput collision resolution
- n-ary stack resolution algorithm with $n = 3$ (START-3)
- Nonblocking characteristic
- Multiple traffic priorities
- Bandwidth allocation optimization (fairness within each priority)
- QOS enabler
- Error detection and recovery

MESSAGE STRUCTURE OVERVIEW. The upstream and downstream channels are delineated by a constant time interval, hereafter referred to as a *block*. Blocks are further divided into SLOTs of variable size as shown in Fig. 6.44.

Each slot contains one message, hereafter referred to as a *primitive*. The following terms are defined for the downstream primitives:

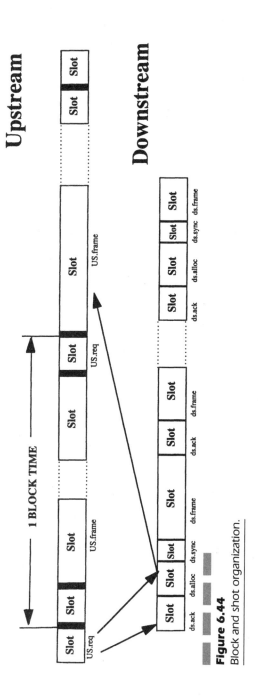

Figure 6.44
Block and shot organization.

DS.SYNC   Enables the station to maintain synchronization with the start of the downstream block .

DS.ALLOC   Carries upstream slot allocations

DS.ACK   Carries the results of contention-mode transmission attempts

DS.FRAME   Carries a payload (data or MAC management)

The following terms are defined for the upstream primitives:

US.REQ   Carries a bandwidth request

US.FRAME   Carries payload(s) (data or MAC management) and a bandwidth request

DESCRIPTION OF DOWNSTREAM PRIMITIVES

DS.SYNC. The SYNC primitive is sent in the downstream to provide a synchronization marker to allow the station to align its downstream and upstream block boundaries with the headend controller. The SYNC primitive consists of an 8-bit pattern '11011000'B.

DS.ALLOC. The ALLOC primitive, as shown in Fig. 6.45, is sent in the downstream to define a slot in the next upstream block. Each ALLOC primitive defines one slot in the next upstream block, and each slot can carry one upstream primitive. The following information is contained in the primitive:

Source address. The Source address field specifies the station or group of stations that may access the specified slot. If the slot is reserved for a particular station, the field will contain the individual address of that station. If the slot is a contention slot, the field will either contain a group address or the broadcast address.

Offset. The offset field specifies, in units of 1/1024th of the upstream block length, the starting point for the slot. A value of zero indicates that slot begins at the start of the upstream block.

Probability. The probability field specifies the random probability that all newly contending stations will transmit in the contention slot as follows: 00, 00%; 01, 25%; 10, 50%; 11, 100%. All stations that are already

Figure 6.45
Downstream ALLOC.

8 bit	16 bit	10 bit	2 bit	8 bit	8 bit	12 bit	16 bit
'D9' x	Source Address	Offset	Prob	Freq	PHY CNTL	Grant	CRC

participating in collision resolution for a contention slot do not use the probability field.

Frequency. The frequency field specifies the frequency to be used for transmitting in the slot. The frequency subfield is 8 bits and specifies the frequency in 150-kHz steps.

PHY control. The PHY control field specifies the modulation rate, modulation type, and preamble type to be used for transmitting in the slot. Two subfields of 3 bits each are used to specify modulation rate and modulation type. The remaining 2 bits specify the preamble type to be used in the upstream slot.

Grant. The grant field specifies the number of payloads that may be sent in the slot and the service class. The allocation count subfield is 4 bits and specifies up to fifteen 53-byte allocations. A value of zero indicates that the slot may contain only the bandwidth request primitive (US.REQ). The service class/contention ID subfield is 8 bits. If the allocated slot is a reservation slot, then this subfield specifies the service class for which the grant is given. If the slot is a contention slot, then this field contains a contention ID number that is correlated with the DS.ACK primitives that are sent to indicate collision status.

CRC. The CRC polynomial is $X^{16} + X^{12} + X^5 + 1$ and covers the entire primitive, including the first byte.

DS.ACK. The ACK primitive, as shown in Fig. 6.46, is sent in the downstream to indicate the status of a contention slot. The ACK contains the information necessary for contending stations to determine whether a collision occurred in a particular slot. ACK primitives are not sent for reserved slots because no collision resolution will be performed and because lost requests will be resent as part of future queue depths. ACK primitives are sent synchronously with respect to block timing; however, each primitive contains a contention ID number that may be used by the station to relate the acknowledgment to a particular upstream slot. Each ACK primitive begins with the pattern 'DA'X and is protected by a 16-bit CRC field. It is ignored by the station if the CRC is bad. Below is the format of the primitive:

Figure 6.46
Downstream ACK primitive.

8 bit	16 bit	8 bit	8 bit	8 bit	16 bit
'DA' x	Source Address	Freq	ID	STAT	CRC

Source address. The source address field specifies the group of stations that the acknowledgment is for. The field will either contain a group address or the broadcast address.

Frequency. The frequency field specifies the frequency of the channel carrying the contention slot. The frequency subfield is 8 bits and specifies the frequency in 150-kHz steps. The frequency range to which the step is applied is specified in an MLMP message. For example, in a low-split system, the range would be from 5 to 42 MHz.

ID. The ID field specifies which slot is being acknowledged. The contention ID field is 8 bits and specifies the contention ID number. This number is assigned by the headend controller in the ALLOC primitive and is unique per upstream channel.

Slot status. The slot status field contains two bits and specifies the status of a contention slot as follows:

00	Empty
01	No collision
10	Reserved
11	Collision

The remaining 6 bits are reserved.

CRC. The CRC polynomial is $X^{16} + X^{12} + X^5 + 1$ and covers the entire primitive including the first byte.

DS.FRAME. The DS.FRAME primitive, as shown in Fig. 6.47, defines an 802.14 downstream frame. All downstream frames are identical in size; however, they carry two different payload types:

1. Management (MLMP) payloads
2. Data (ATM cell) payloads

Frames that carry MLMP payloads are referred to as *MLMP frames*. Frames that carry ATM cell payloads are referred to as *data frames*. The term *MAC protocol data unit* (MPDU) can be used interchangeably with the term *frame*. Each DS.FRAME primitive begins with the pattern 'DB'X and is protected by a 16-bit CRC field. It is ignored by the station if the CRC is bad. Below is the format of the primitive:

Figure 6.47
Downstream frame primitive.

8 bit	16 bit	8 bit	53 Bytes	16 bit
'DB' x	Destination Address	Frame Control	Payload	CRC

Destination address. The destination address field specifies the station or group of stations that the frame is sent to. The field will either contain an individual, group, or broadcast address.

Frame control (FC). The FC field specifies information about the payload of the frame. The first bit of the FC field indicates the payload type as follows:

0   Data (ATM cell)
1   Management (MLMP)

The second bit of the FC field specifies encryption of the payload as follows:

0   No encryption
1   Encrypted

The remaining bits are reserved.

Payload. The payload field contains a 53-byte payload that is either a data (ATM cell) payload or a management (MLMP) payload.

CRC. The CRC polynomial is $X^{16} + X^{12} + X^5 + 1$ and covers the entire primitive, including the first byte.

DESCRIPTION OF UPSTREAM PRIMITIVES

US.FRAME. The US.FRAME primitive, as shown in Fig. 6.48, defines an 802.14 upstream frame. Upstream frames carry two different payload types:

1. Management (MLMP) payloads
2. Data (ATM cell) payloads

Frames that carry MLMP payloads are referred to as *MLMP frames.* Frames that carry ATM cell payloads are referred to as *data frames.* US.FRAME may be allocated (and therefore sent) in either a contention slot or a reservation slot. During normal operation, US.FRAMEs that are allocated as the result of a station request will always occur in a reservation slot. The headend may allocate US.FRAMEs in contention slots in order to allow payloads to accompany a station request. This primitive is assumed to be encapsulated in an upstream PHY frame with error detection and correction. Below is the format of the primitive:

Figure 6.48
Upstream frame primitive.

16 bit	8 bit	16 bit	$N \times 53$ Bytes
Source Address	Frame Control	RF	Payload

Source address. The source address field specifies the station from which the frame is sent. The field will always contain an individual address.

Frame control (FC). The FC field specifies information about the payload of the frame. The first bit of the FC field indicates the payload type as follows:

0   Data (ATM cell)

1   Management (MLMP)

The second bit of the FC field specifies encryption of the payload as follows: 0, no encryption; 1, encrypted. The remaining bits are reserved.

RF. This field provides the headend with the information needed to describe a request for bandwidth from the station. The queue depth subfield is 8 bits and specifies up to two hundred fifty-five 53-byte values. A value of zero indicates that no bandwidth is being requested. The service class subfield is 8 bits and specifies the service class for which the request is made.

Payload. The payload field contains either data (ATM cell) payloads or management (MLMP) payloads. The number of payloads is defined by the grant field of the corresponding DS.ALLOC primitive sent in the downstream direction. The number of 53-byte payloads may range from 1 to 15. A value of zero in the grant field indicates a US.REQ (instead of US.FRAME) primitive in the slot.

US.REQ. The REQ primitive, shown in Fig. 6.49, is sent in the upstream direction to indicate a request for reservation-mode slots. During normal operation, REQ is always allocated (and therefore sent) in a contention slot. This primitive is assumed to be encapsulated in an upstream PHY frame, with error detection and correction (if required) being the responsibility of the PHY. Below is the format of the primitive:

Source address. The source address field specifies the station from which the frame is sent. The field will always contain an individual address.

RF. This field provides the headend with the information needed to describe a request for bandwidth from the station. The queue depth subfield is 8 bits and specifies up to two hundred fifty-five 53-byte

Figure 6.49
Upstream REQ primitive.

16 bit	16 bit
Source Address	RF

values. A value of zero indicates that no bandwidth is being requested. The service class subfield is 8 bits and specifies the service class for which the request is made.

CONTENTION RESOLUTION.   Collisions that occur in the upstream contention slots are resolved using the n-ary stack resolution algorithm with $n = 3$ (START-3).

START-3 ALGORITHM CONCEPT.   The contending stations randomly select a number between zero and two for their count value. Any station that ends up with a zero count value will attempt to transmit again in the slot, with the same slot ID number, in the next upstream block. All other stations monitor the feedback to determine if another collision occurs. If no collision occurs, then all contending stations subtract one from their count values and those that have a zero count value will attempt to transmit again in the next upstream block. If a collision does occur, then the stations that were only monitoring add two to their count values. Those stations that were involved in the collision do the same as the initial collision by picking a new count value between 0 and 2. The process repeats until all count values go to zero and all transmissions occur without collisions. In order to determine when collision resolution is complete, the headend controller will maintain a counter for each contention ID in use. The initial value of the counter is one. When a collision is detected, the worst-case value of 2 is added to the counter. If no collision occurs, then 1 is subtracted from the count value. Contention is known to be resolved when the controller counter reaches a value of zero.

BANDWIDTH REQUEST PROTOCOL.   The station implements an algorithm for bandwidth requests that prioritizes requests based on service classes. Each cell queue that is implemented by the station is associated with a particular service class. The order of requests sent will be the order of priority of the service classes, from 0 (highest) to 255 (lowest).

The station MAC should limit requests to a maximum of one per service class per upstream block or the number of contention slots in the upstream block, whichever is less. The station will maintain an ALLOC timer for each service class in use such that failure to receive an ALLOC in response to a request (contained in either US.REQ or US.FRAME) for a service class within the allowed time (programmable) causes the request to be sent again. A request for a particular service class should indicate the queue depth at the time the request is made after being adjusted for any grants that were received. If a reservation-mode US.FRAME of that particular service class has been allocated, the request is placed in the US.FRAME's RF field. If a reservation-mode US.FRAME's RF field is not being used by the service class matching its payload, then requests from other service classes (in priority order) may use the RF field. If a request is unable to use the RF field of a reservation-mode US.FRAME, then it will use the RF field of either a US.REQ (always contention mode), or a contention-mode US.FRAME. Since a contention-mode US.FRAME always has at least one payload region, in this case the station may also send payloads matching the service class, in which case the queue depth sent will be adjusted by the number of payloads sent. The station chooses between US.REQ and contention-mode US.FRAME by using a "service class to request-type" mapping that is maintained at the MLMP level. This mapping will define if a service class is allowed to use a contention-mode US.FRAME. If the station is allowed, it will use contention-mode US.FRAME if available; otherwise, it will use US.REQ.

DETERMINING AVAILABLE CONTENTION IDs. The set of available contention IDs (and associated REQ or FRAME) is determined by using the probability fields contained in the ALLOCs received during the last block to do an independent calculation for each ID not already in use by the station to determine if the ID is available. The station then uses the available IDs in random order, conforming to the rules described in the previous paragraph to choose between IDs that correspond to US.REQ and IDs that correspond to a contention-mode US.FRAME. Each service class is allowed to use one ID (total, including any IDs already in use), assigned in priority order until the IDs are all used or each service class has obtained one.

BANDWIDTH ALLOCATION PROTOCOL

DOWNSTREAM. In the downstream direction, ATM cells that are received across the MAC service interface are transmitted in available slots after being encapsulated as 802.14 frames. There is no multiple access contention in the downstream direction.

UPSTREAM. Upstream channels are controlled by a time division multiple-access (TDMA) scheme based on slots. There is a mixture of slots that may be accessed on a contention basis and slots that are reserved for a particular station or group of stations at a particular service class. As part of the request mechanism, a station may request a permanent slot allocation by sending a special MLMP stream request frame.

After allocating any permanent slots, upstream slots are allocated fairly to station MACs within each service class. Service classes are allocated slots based on priority, in order from 0 (highest) to 255 (lowest). Each service class is fully satisfied, in order, before any slots are allocated to lower-service classes.

6.5 MAC Performance Evaluation

The evaluation process of the MAC proposals submitted to the 802.14 committee was a three-step process that led to the convergence to a unified MAC protocol approach. In the first step, there was the design of an evaluation model and the specification of test scenarios. Then there was building and implementing the models and obtaining the simulation results. In the third stage, there was the analysis of the data and results. Many iterations were required in this process before a converged MAC protocol started taking shape. In this section, we describe this process and give an overview of some of major performance evaluation results.

6.5.1 Common Evaluation Model

A performance evaluation process document was put together. Several MAC proponents took part in this effort led by Georgia Institute of Technology and the National Institute of Standards and Technology. The document attempted to define a set of test scenarios based on a well-defined evaluation criteria. The major goal was to ensure that a common procedure is used to evaluate the performance of all MAC protocols. Layers above the MAC layer were considered as much as possible outside the scope of the evaluation process, although it was impossible to evaluate the MAC layer in isolation. Certain assumptions were made about adjacent layers so that a general procedure could be adopted. The evaluation model consisted of three parts:

1. The traffic model
2. The parameters of the simulation
3. The output measurements

The traffic model is defined for the generation of packets of information or messages. These messages represent the information passed from the upper sublayer of the data-link layer to the MAC layer. They contain whatever headers the upper layers may have chosen to attach to the message. The MAC layer will add further fields to the message and may segment it into smaller *cells* for the purpose of transmission.

The evaluation procedure makes no assumptions on how the MAC layer might function. In order to generate realistic traffic, it would be possible to model all activities from the application itself down to the MAC layer. Such a procedure *might* give a very accurate model of the traffic generated by that application. However, we are concerned here about the impact the traffic pattern will have on the performance of the MAC layer alone. And it turns out that the performance of the MAC is sensitive only to the gross statistics of the traffic such as the burstiness of the source. Therefore, we propose models of the data source that capture the gross behavior of the traffic. The model should cover the extremes that are likely to occur in any anticipated use.

The evaluation parameters are of two types: those that define the system itself, such as the number of stations, and those that describe the simulation parameters, such as the duration of the simulation. The latter are rather arbitrary but should be chosen to exercise the extremes that are likely to be encountered. They should also present statistically stable results.

There are many output measurements that can be recorded, each determining the performance of the system for a particular type of desirable or undesirable behavior. We would like to know the delay under load (and hence the maximum throughput), the amount of delay variation, the fairness of the system, the ability to give priority to different types of traffic, the effect of transient loading, and the robustness of the protocol (and possibly other behavior as well). We would like to know all this under a range of conditions. The number of required measurements grows as the product of the number of variables and can very quickly become enormous. The challenge is to come up with as small a set as possible while at the same time giving reasonable assurance that the protocol has no undiscovered erratic behavior.

6.5.2 Component Evaluation Process

Although a common test procedure or platform for the evaluation has been defined, the task of comparing different proposals is still impossible at this point because of the different assumptions each MAC proposal uses. Simulation results produced by the different MAC-proposal proponents were put side by side but were often inconclusive. It was like comparing apples and oranges. NIST was solicited as a third unbiased party to conduct the evaluation of the MAC proposals. But what was needed was a consistent way of testing individual MAC protocol functions belonging to different proposals as opposed to testing different MAC protocols in their entirety. This very popular approach was proposed by NIST, which actually started implementing and comparing the different MAC protocols submitted to the 802.14. This approach made a lot of sense, since the standard may not be based on a single MAC protocol but rather on a collection of features or functions assembled together from different proposals.

The main idea was to think of a MAC protocol as a collection of components, each performing a certain number of functions. An HFC MAC protocol can be broken into the following sets of components: ranging or acquisition process, message format, support for higher-layer traffic classes, bandwidth request, bandwidth allocation, and contention resolution mechanism. Ranging is the phase during which the round-trip delay time to the headend is calculated and the station synchronization to the downstream timing is performed. The message format element of the MAC defines the upstream and downstream message timing and describes their contents. If the MAC needs to provide support for ATM, it also needs to differentiate between different classes of traffic supported by ATM, such as constant bit rate (CBR), variable bit rate (VBR), and available bit rate (ABR). Bandwidth allocation represents an essential part of the MAC and controls the granting of requests at the headend. Finally, the contention resolution mechanism, which may be the most important aspect of the MAC, consists of a backoff phase and retransmission phase.

This structure constituted the basis for the MAC evaluation and convergence process. New tests specific to each component were designed and performed in addition to the ones initially laid out in the evaluation model so that the best of each solution could be chosen per the criteria document.

CONTENTION RESOLUTION. A collision resolution algorithm needs to be implemented because request packets and possibly data packets are transmitted in a contention fashion. A wide variety of algorithms can be used. It must be noticed, however, that the characteristics of HFC networks impose a number of constraints. Since stations cannot monitor collisions, feedback information about contented requests must be provided by the headend. The algorithm must also take into account the delay before a station receives the feedback information. If the number of contention slots is small and if they are located at the beginning of the frame, it may be possible for a station to receive feedback before the end of the frame. In this case, the stations can attempt a retransmission in the next frame if its request collided. However, if the number of contention slots is large or variable, stations may have to skip a frame before retransmitting a collided request. There are basically two families of contention resolution algorithms:

1. p-persistence
2. Tree-based

One algorithm from each family is evaluated, and the results are compared.

TREE-BASED CONTENTION RESOLUTION ALGORITHM. The principle of the n-ary tree-based contention resolution algorithms is that, when a collision occurs, all the stations involved in this collision split into n subsets, and each of them randomly selects a number between 1 and n. The basic idea is to allow different subsets to retransmit first while the subsets from 2 to n wait for their turn. One can see the waiting subsets as a stack. The position in the stack represents the number of slots the station has to wait before it can retransmit its request. If a second collision occurs, the first subset splits again. The subsets that are already waiting in the stack must be shifted up to $n - 1$ positions in the stack to leave room for the new stations that collided. If no collision occurs, the stations occupying the lowest stack level can transmit.

In its original definition, the algorithm assumed that stations receive feedback immediately. However, in the HFC system, the stations must wait at least until the beginning of the next frame before they receive feedback from the headend and thus are able to retransmit. The algorithm is modified as follows in order to accommodate this delay in feedback. For example, let's consider frame $j - 1$ containing $c(j - 1)$ collided minislots. All stations involved in the first collided slots are dispatched in the first n slots; those involved in the second collided slots use the next n slots; and, more generally, the stations involved in the ith collided slots select a subset between $(i \times n + 1)$ and $[(i + 1) \times n]$ slots. If frame j contains p contention slots, the first p subsets are able to retransmit, the ith subset transmitting in the ith slot. The other subsets wait in the stack. If cj new collisions occur in frame j, the waiting subsets must be shifted by $n \times c(j) - p$ positions in the stack to make room for the new subsets; n is set to 3, and that is based on several studies in the literature that show optimal results with $n = 3$. New packets arrivals may be handled in two different ways. If the tree-based algorithm is nonblocking, new stations transmit without waiting in any slot selected randomly. On the other hand, if the algorithm is blocking, the newcomers are not allowed to use a slot reserved for collision resolution. In other words, they are directly put on top of the stack. When new stations are able to transmit, they randomly select a slot among the remaining available slots. Note that a variety of mechanisms could be used in order to limit the entry of new packets to the system. For example, new packets can occupy different levels in a stack based on their arrival time and thus could send their requests on a first come, first serve (FCFS) fashion. A *p*-persistence approach could even be used.

ADAPTIVE *P*-PERSISTENCE.   The adaptive *p*-persistence algorithm is an adaptation of a stabilized ALOHA protocol to frames with multiple contention slots. Newly active stations and stations resolving collisions have an equal probability of access *p* (*p*-persistence) to contention slots within a frame. *p* is determined by an estimate of the number of backlogged stations, computed by the headend and sent to the station in the downstream frames. The estimate $N(j+1)$ of the number of backlogged stations in the $(j+1)$th frame is determined by

$$N(j+1) = \max \left\{ \min\left[\frac{n, N(j-1) - ni(j-1) - ns(j-1) + nc(j-1)}{(e-2) + MS(j-1)/e} \right], MS(j+1) \right\}$$

where n is the number of stations, $MS(j)$ is the number of minislots in the jth frame, and $ni(j)$, $ns(j)$, and $nc(j)$ are the number of idle, successful, and collided minislots in the jth frame, respectively.

Here, the estimate for the $(j+1)$th frame is determined by the parameters of the $(j-1)$th frame. This is due to the feedback from the $(j-1)$th frame not being received at all stations before the beginning of the jth frame when frames consist of all minislots. However, the number of minislots allowed in a frame may be restricted so that feedback from a frame is received at all stations before the beginning of the next frame. If that is the case, then the estimate for the jth frame is determined by the parameters of the $(j-1)$th frame. When a station needs to make a request in frame j, it generates a random number ij uniformly distributed in the interval $[1, N(j)]$. If ij is less or equal to the number of minislots in the frame, it will make its request in the ijth minislot; otherwise, it will attempt to make its request in the next frame using the estimate for that particular frame.

TREE-BASED VS. *p*-PERSISTENCE CONTENTION RESOLUTION. Many simulations were performed in order to compare the ternary-tree and the *p*-persistence contention resolution algorithms. The comparison is conducted in a realistic MAC protocol environment for different types of traffic (bursty, long packets, short packets), network loads, and in steady state and transient conditions.

We report here some of the major results obtained. In an attempt to decouple the contention resolution component from the bandwidth allocation component (the number of contention minislots and data slots available on the upstream channel), we look at the time it takes a packet request to reach the headend from the time the packet is generated at the station, and we call it *request delay.* Note that access delay is the sum of the request delay, the queueing delay for the request at the headend, and the propagation and transmission delays. Figure 6.50 illustrates the mean request delay versus the offered load (or packets generated by all stations) for the tree-based blocking, nonblocking, and the adaptive p-persistence algorithms when the upstream request/acknowledgment cycle (including round-trip delay, headend processing delay, minislot transmission, etc.) is approximately equal to 3 ms on an a 3-Mbits/s upstream channel, with 32 minislots available for contention and a request size of 2. This test uses short packets of 64 bytes (48 bytes as payload and 16 bytes as MAC overhead) generated according to a Poisson distribution with a mean arrival rate lambda. Lambda is varied for different applied loads.

Figure 6.50
Mean request delay (ms) vs. offered load.

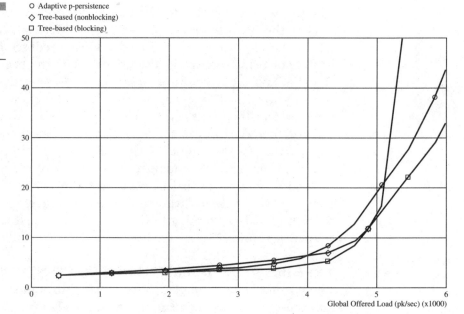

O Adaptive p-persistence
◇ Tree-based (nonblocking)
□ Tree-based (blocking)

Global Offered Load (pk/sec) (x1000)

We note that the tree-based blocking algorithm performance is superior to both the *p*-persistence and tree-based free-access algorithms in that order. This result was consistently observed throughout all simulations performed: the difference in terms of request access delay (or mean access delay) between the two contention resolution schemes is small in mid-load regions but always to the advantage of the blocking ternary-tree algorithm. The mean access delay is almost identical in low-load regions, where the number of collisions is relatively small.

Another test that probably determined the choice of the contention resolution scheme to adopt for the IEEE 802.14 consisted of the following. A bandwidth allocation strategy is applied at the headend and the probability density function of the access delay for a station is computed, which is the time a packet is generated until it is received at the headend. This measurement was thought to be very important, especially for issues related to cell delay variation (CDV) in ATM environments.

Figure 6.51 shows the cumulative density function for the tree-based blocking and *p*-persistence with 50 percent applied load using the same bursty traffic described above. We observe that, for the tree-based blocking, the probability the access delay is less than 50 ms is almost 95 percent, while it is close to 85 percent for the *p*-persistence. This result is expected considering the randomness nature of the *p*-persistence and the blocking feature of the tree-based scheme.

Figure 6.51
Probability that mean access delay < *x* ms, offered load = 50 percent of capacity.

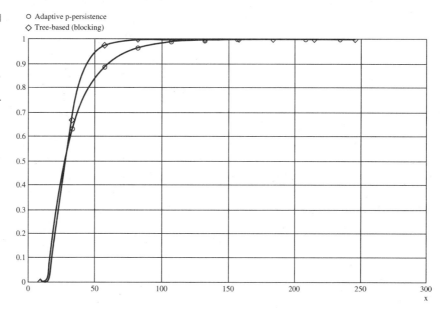

BANDWIDTH ALLOCATION. One of the main characteristics of bandwidth allocation in the MAC protocol is the updating of the contention slot to data slot (CS/DS) ratio on the upstream channel, and that function is controlled exclusively by the headend. Two main schemes are studied:

1. Fixed
2. Variable

In the fixed scheme, the number of CS and reservation DS in the request/acknowledgment cycle is fixed. In the variable scheme, the headend varies the ratio of CS to DS based on the traffic in the system.

FIXED CS/DS RATIO. The fixed allocation is simple; however, the optimal CS/DS ratio depends on the traffic pattern. Figure 6.52 shows the access delay when the load is varied for CS/DS ratios of 1:1, 2:1, and 3:1. The traffic pattern used is a bursty source with a batch Poisson arrival model. The message size distribution is defined as follows. Messages of sizes 64, 128, 356, 512, 1024, and 1518 bytes are generated at the stations with a probability of 0.6, 0.06, 0.04, 0.02, 0.25, and 0.03, respectively. The message interarrival time is exponentially distributed with mean T where T is varied according to the load. The cycle time is set to 3 ms on a 3 Mbits/s channel. The maximum number of stations used in the simulation was equal to 200 stations, and the request size was set to 32. The blocking ternary-tree algorithm is used for resolving collisions.

It is readily observed that a CS/DS ratio of 1:1 does well for this particular type of traffic. However, additional simulation results showed that the optimal CS/DS ratio is heavily dependent on the traffic type (i.e., message size and burstiness). In this particular example, due to the large average packet size that results in larger request sizes, less CSs are needed. Also from Fig. 6.52, we note that, when the ratio of CS/DS increases, the mean delays at higher applied loads become larger because less DSs are available to data transmission.

VARIABLE CS/DS RATIO. The variable allocation scheme adapts the CS/DS ratio according to the traffic. Two flavors are illustrated in Fig. 6.53. A DS priority allocation scheme computes the required number of DS and converts the remaining slots to CS. On the other hand, a different approach consists of computing the required number of CSs and leaves the remainder as DS. Several factors may be considered in the computation of the CS/DS ratio; for example, the number of collisions, the data queue request size, the traffic backlog, and so on. Figure 6.53 shows the mean access delay

Figure 6.52
Mean access delay vs. offered load, fixed CS/DS ratio.

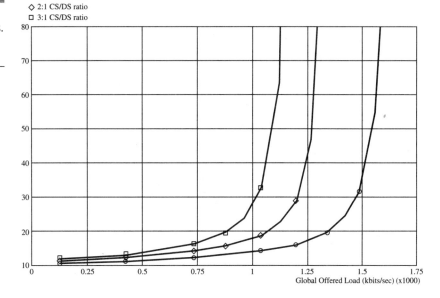

Figure 6.53
Mean access delay (ms) vs. offered load, variable CS/DS ratio.

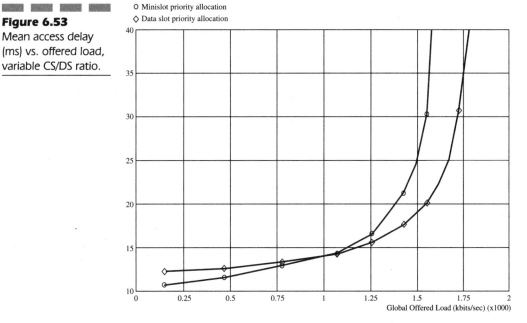

for both schemes using a blocking ternary-tree algorithm for contention resolution. The network conditions and traffic patterns used are the same as in the fixed CS/DS ratio experiment (Fig. 6.53). We immediately note that both schemes perform better than the fixed ratio allocation policy.

6.6 Overview of the IEEE 802.14 Selected MAC

The component performance evaluation conducted by NIST served as a basis for discussion and helped build consensus towards a converged MAC protocol that combines many useful features of the original MAC proposals submitted to the IEEE 802.14 committee. The convergence agreements in document IEEE 802.14-96-112R2 were used as a starting point for the MAC working group draft specifications.

The main features of the MAC protocol currently being specified by the IEEE 802.14 group are its ability to support both ATM cell transfers and variable-length packet transfers while maintaining the ability to provide high quality of service (QOS). The upstream channel is divided in time into basic units called minislots. The minislot is the smallest unit of transmission. There are several types of minislots. Their functions are defined by the headend and conveyed to each station by means of downstream control messages. Several minislots can be concatenated in order to form a single data PDU, ATM, or variable-length fragment. There are no fixed frame structures, and there is a variable number of minislots in any given time. Thus, the upstream channel is viewed as a stream of minislots. The contention resolution algorithm is based on a ternary-tree algorithm with variable new packet entry persistence.

While the station's behavior will be fully defined at the time the MAC specifications will be completed, most of the headend algorithms for computing bandwidth allocation and granting requests will be vendor-implementation-specific. The MAC protocol specifications will concentrate on the following issues:

- Station addressing that provides mechanisms to distinguish one station from another. The basic idea is that each station, in addition to a 48-bit MAC address administered by IEEE 802, will have a local 14-bit identifier used within the 802.14 network.

- Upstream bandwidth control formats define the request minislot types and structure.

- Upstream PDU formats specify the protocol data unit format for both ATM and variable-length PDUs segmented into variable-length fragments for an efficient transport of LLC traffic types.

- Downstream formats specify the downstream data flow that can be seen as a stream of allocation units each 6 bytes long. ATM cells and variable-length fragments can be sent by concatenating several basic cells together. It also defines the format for downstream bandwidth management cells carried in ATM cells. Each ATM cell can carry a number of information elements such as bandwidth information elements: grant information, allocation information (request minislot allocation), and feedback information (request minislot contention feedback). Downstream MAC-layer management messages are assumed to be carried in the same fashion.

- MAC operation describes the entry mechanism and steady-state operation of the station's behavior. That includes ranging and power leveling, contention resolution, time synchronization, queue transmission operation, and message segmentation.

- Encryption and decryption are done using the DES algorithm.

- MAC interfaces primitives to the PHY/TC layer are defined at both the headend controller and the station.

Chapter 7

Noise and the High-Speed Cable Modem

7.1 Overview

Cable network providers in the last four years have been deploying the HFC architecture to provide full-service networks that can deliver voice, video, and data to meet the demands of the increasingly competitive market. The upstream channel in an HFC network has been a source of great concern. The channel frequency in which it must operate positions it in a very hostile noise environment. Ingress noise in the upstream direction is the main cause of impairments in an HFC system. This noise comes in different flavors and severity. Egress noise hence plays a part in the overall network performance.

Technical solutions are, of course, always at hand. The challenge facing the cable operators is to build an HFC system that can be a viable business model and operate reliably in an environment that is prone to noise generated by all kinds of electrical interference. IEEE 802.14 developed channel models that mathematically defined the nature and physics of the noise. These models, described later in the chapter, were used to develop the physical and MAC layers for the cable modem.

This chapter will discuss in detail the types of noise characteristics that find their way into HFC systems, and ways to combat them. A channel model developed by IEEE 802.14 will be discussed. The channel models were designed to give some insight as to the root of the noise and were also used to evaluate the proposed physical layers.

7.2 Noise Roots

Unlike other noise phenomena, the environment in the cable network is very unique. The cable system acts as a giant antenna for various noise and impairments that are additive, especially in the 5- to 45-MHz band of the RF spectrum. Each type of noise must be addressed at its source before it propagates further into the network and mutates. What is just as challenging is that the noise phenomena in the cable network is time-dependent. What is measured in the morning is quite different from measurements made in the peak TV viewing hours. Moreover, these measurements are different from one region to the next. The age of the cable plant, drops in humidity of the region, the number of subscribers in the drop, the inside home wiring, and past maintenance practices all play a part in how the network behaves under different loads. To say that the system must be developed for a worst-case scenario is not the optimal solution. In most instances, it is likely to be less costly to clean an old neighborhood than to provide a complex design, hence penalizing most other cable systems. In some situations, a field technician can enhance video signal quality and reduce noise measurably simply by mechanically and electrically securing the cable plant. Therefore, an important aspect of cable plant installation or modernization is to ensure the system is both mechanically and electrically sealed. For it to be otherwise will invariably cause a significant contribution to ingress and impulse noise within the system. Electrically, all powered devices and the cable plant must be grounded appropriately. This may prove difficult in arid and/or rocky climates due to the inability to establish a good electrical ground.

Most if not all cable networks comply with FCC regulations; hence, all signals across it are at a level of 3 to 10 dBmV. However, in the real world, we find few networks that have been truly designed to comply with these requirements for upstream transmissions. It is recommended that the cable network be designed initially or modified to support unity gain in the upstream direction. This implies that, regardless of where in the system a signal of specific amplitude is input, it should arrive at the headend location at the same consistent level.

To design or modify the network, a loss plan must be determined for the downstream and upstream transmission paths. The loss plan must be calculated at the upper- and lower-band edge frequencies in both directions for the input and output of all network devices. Table 7.1 illustrates the importance of upstream engineering by showing the loss through a 1000-foot length of 0.5-inch coax cable.

As can be seen, the downstream gain variations are significant, about 12 dBmV. However, the upstream transmission has very little variation. For a network engineered only for downstream transmission, the upstream transmissions vary greatly in amplitude, as measured at the headend, dependent on the launch location.

TABLE 7.1

Downstream/
Upstream Typical
Loss

Direction	Frequency	Loss
Downstream	55 MHz	5.20 dB
	550 MHz	17.30 dB
Upstream	.5 MHz	1.50 dB
	45 MHz	4.70 dB

7.2.1 Noise General Description

In general, network problems come from three areas:

1. Within the subscriber's home
2. Drop plant
3. Rigid coaxial plant

Seventy percent of the problem comes from the home, 25 percent is generated from the drop portion of the network, and 5 percent comes from a rigid coaxial plant. Troubleshooting intermittent problems is costly and time-consuming. Finding the problem does not always mean it can be fixed.

HFC INGRESS NOISE. To understand and evaluate the effects of ingress noise, we have to examine both the sources of ingress noise as well as the weak points in the HFC network that are prone to ingress pickup. Among the sources of ingress noise in the upstream frequency region, we have:

1. FCC-regulated transmission sources such as shortwave, amateur, CB, and maritime radio.
2. Transient sources like lightning, corona noise from high-voltage lines, impulses synchronous with power systems, and sparks from ignition systems. Also noise generated within the home by appliances, bad electrical contacts, etc.
3. Harmonics of power system atmospheric noise.

Ingress noise can be classified either as wideband or narrowband based on its bandwidth relative to the data channel bandwidth. A system with 6-MHz channel spacing, for example, would consider a 100-kHz bandwidth interferer as narrowband. On the other hand, a system with 100-kHz channel spacing would have to consider such interference as wideband. In the time domain, we can classify the type of ingress based on duration in relation to the symbol period. For example, an ingress burst that lasts less than n symbol periods is considered short, while ingress that lasts more than n symbols is considered long. The wideband and narrowband classification helps to evaluate and measure ingress relating carrier-to-noise ratio in the wideband ingress cases and carrier-to-interferer ratio for the narrowband cases. Systems that operate at a BER of

10^{-8} in an environment with a carrier-to-noise ratio of 21 dB has a nice margin operating over a carrier-to-interferer ratio of 21 dB and no Gaussian noise. The carrier-to-interferer ratio is dependent on the relative interferer-to-channel bandwidth, power ratios, and the frequency location of such interferers with respect to the center of the channel.

Most sources that are regulated by the FCC would fall in the narrowband category for most channel bandwidths. Table 7.2 shows the relative levels of ingress originating from the regulated sources and the coverage based on types of antenna use.

Nonlinear effects within the network also have an impact on ingress. When ingress traverses nonlinear components, the number of ingress peaks multiply. Special efforts have been made in the downstream path to provide a highly linear environment to transmit analog video. The most obvious nonlinear HFC component is the reverse amplifier. A reverse amplifier may not be driven hard enough to create nonlinearities if the upstream channel is relatively empty. A more subtle source of nonlinearities is metal oxidation at connectors and electrical contacts throughout the network. This metal-oxide interface may form junction diodes. In this case, it may not be necessary to reach high power levels to generate nonlinear components.

The cause of ingress noise pickup in the HFC can be divided into three sections:

1. The trunk and feeder network, which uses rigid coaxial cable

2. The drop wiring, usually an RG 6 coaxial cable run

3. The residential wiring, most likely with several RG 59 cable runs installed

A well-designed and well-maintained HFC network should, in principle, have a negligible amount of ingress picked up in the rigid-cable portion of the network. In the drop portion of the network with braided shielding and F-connectors that may become loose, and/or with the likelihood of no grounding at the home end of the drop, the possibility of ingress pickup increases substantially. In residential wiring, the RG 59 cable is installed by the builder or owner and is subject to damage and/or misuse. The residential wiring has limited shielding and, in addition, it may suffer from poor grounding, reflections, bad connector installation practices, and so on. These conditions are prone to have the residential wiring act as antennae. Any effort in isolating the coaxial home wiring subsplit bandwidth from the rest of the network will have a beneficial impact on the HFC network.

TABLE 7.2
Relative Effects of
RF Ingress Noise

Transmitter type	Transmitter distance from home in km	Transmitter power in W	Relative received power	% homes covered by transmitter radiation	% homes prone to ingress pickup	# homes affected in a FSA 500
Shortwave	100	5000	−26.8	100	20	100
	1000		−46.8			
	10,000		−66.8			
	100	250,000	−19.8	100	20	100
	1000		−39.8			
	10,000		−59.8			
	100	500,000	−16.8	100	20	100
	1000		−36.8			
	10,000		−56.8			
	100	1,000,000	−13.8	100	20	100
	1000		−33.8			
	10,000		−53.8			
Ham	0.5	100	−11.7	12.5	20	12
	1		−17.7			
	5		−31.7			
	10		−37.7			
	0.5	500	−04.7	12.5	20	12
	1		−10.7			
	5		−24.7			
	10		−30.7			
	0.5	500	00.0	12.5	20	12
	1		−06.0			
	5		−20.0			
	10		−26.0			
CB	0.5	5	−26.0	100	20	100
	1		−32.0			
	5		−46.0			
	10		−52.0			

CAUSES AND APPROACHES TO SUPPRESS NOISE. There are many approaches to suppress or avoid ingress in HFC networks. Since these approaches are not mutually exclusive, they could be combined to improve the performance of the network. The guidelines shown below will have to be done on a plant-by-plant basis. Network performance is effected for both the upstream and downstream channels, although the upstream channel is more pronounced in the overall performance.

- Aligning the amplifiers properly in the reverse direction.
- Combating ingress noise. This battle seems uncontrollable at times. Leaks in the cable system are potential ingress points. Everything that exists in the over-the-air RF spectrum between 5 to 40 MHz has a potential for trashing the upstream path. The obvious sources of this electromagnetic interference at this frequency range include most if not all electrical gadgets found inside a home. That includes all FCC-conforming RF power levels such as CB and ham radio transmission, 15.734-kHz horizontal sidebands from leaky TV sets, RF computers, hair dryers, power-line interference, electric neon sign interference, international shortwave broadcasts, electric motors, vehicle ignitions, garbage disposals, washers, nearby airplanes, high-voltage lines, power system atmospheric noise, bad electrical contact, electronic toys, and so on.
- Almost 70 percent of the source of this ingress noise is generated at the subscriber drops. Low-quality coax is in use for the subscriber drops. Radial cracks and cracks in the shield's foil are the main source of leaks and hence ingress noise. The do-it-yourselfers are also contributing their share of system leakage when installing their in-home wiring using older, bad, or loose connections. One effective approach to improving the network performance is through the upgrade of adequate coaxial residential wiring and adding good connector and good grounding practices. This, however, may prove costly.
- Path distortion. This distortion is due to unintentional diode in the signal transmission path. These connectors create virtual diodes. Diodes are created when two different metals bond and a thin oxide due to galvanic erosion eventually builds up between them. Harmonics of 6-MHz signals will result when passing through this nonlinear diode junction.
- Impulse noise. This noise is the least understood, but it is created by the high-voltage corona effect.
- The most basic and simple approach is to avoid the channels with ingress. This can result in a very limited upstream bandwidth, available especially if the channels are 6 MHz wide.

■ Reducing the channel bandwidth adds robustness to the system, since it reduces the group delay distortion and enables the use of higher-order modulation schemes. This approach may not be economically feasible in some regions.

■ Isolating the subsplit region from homes that do not require bidirectional services would improve the performance of the network. Another version of that filtering approach, but for the homes requiring bidirectional services, is the network interface unit location at the home-drop boundary. That unit, if properly grounded, could reduce ingress problems of the drop as well.

■ Error correction is another practical approach that will improve the transmission/performance and add the necessary CNR margin.

■ A frequency-agile system is one method used to reduce noise impairments. The approach is to select a carrier frequency in the return path where noise is not present. This means that noisy return-path spectra will be marked as not usable. The following are the conditions that must be satisfied:

The system must be able to determine that the RF channel is impaired.

The system must be able to determine the type of impairment (narrowband vs. broadband).

There must be sufficient spectra available to allocate alternate (standby) RF channels.

The system must be able to dynamically and automatically select alternate RF channels and notify associated equipment.

Frequency-agile systems function well by avoiding noise. This, however, created the debate in the industry, and some feel that if fully adopted and deployed it would become a liability in the future when interactive services demand more and more bandwidth from the upstream resources. The argument maintains that frequency agility is useful if the noise source were a narrowband and not a broadband noise component. Ingress noise is the only type of noise that meets this criteria. Frequency agility is not an effective strategy for dealing with impulse or amplifier noise, as the noise is broadband in nature. This implies the noise components affect all carrier frequencies.

PHASE DISTORTION AND GROUP DELAY VARIATION. Phase distortion is present in frequency-dependent components. In a low-pass filter, for example, deviation from linear phase precedes the amplitude variation at the band edge. This implies that, even when the amplitude measured shows no variation at or near the band edge, it will suffer the effects of phase distortion. In addition to the phase distortion at the band edges, reflections can also cause phase and amplitude distortion. In both cases, phase distortion will cause intersymbol interference. In the upstream bandwidth, the dominant cause for intersymbol interference is phase distortion at the band edges rather than reflections at the home.

7.3 Noise Types and Network Characteristics

Noise in the cable network is generated in both the downstream and upstream directions. Noise in the upstream direction is more difficult to eradicate and is most pronounced when evaluating the overall network performance. IEEE 802.14 developed channel models to identify the source of noises in the cable networks. Mathematical modelings were used to measure the effect of these noises and hence use these parameters to guide in the development of the physical layer interfaces.

7.3.1 Noise Characteristics in the Upstream Direction

Figure 7.1 best illustrates the type of noise and the characteristics of the upstream channel. These impairments are described in subsections as shown below.

HUM MODULATION. Hum modulation is amplitude modulation due to coupling of 60-Hz AC power through power-supply equipment onto the envelope of the signal. Figure 7.2 shows the equation representing hum modulation. In this equation, $x(t)$ is the input signal, $y(t)$ is the output signal, A is the hum amplitude (e.g., for 5 percent hum modulation, $A = 0.05$), and $m(t)$ is the hum waveform. A 60-Hz triangle with peak amplitude $= A$ is used for modeling purposes.

MICROREFLECTIONS. Microreflections occur at discontinuities in the transmission medium, which cause part of the signal energy to be reflected. The model is a finite impulse response (FIR) filter, consisting of a delay line with weighted taps representing the reflections. Figure 7.3*a* shows the filtering operation, and Fig. 7.3*b* shows the FIR filter structure.

Figure 7.1
Impairments for
upstream CATV
channel.

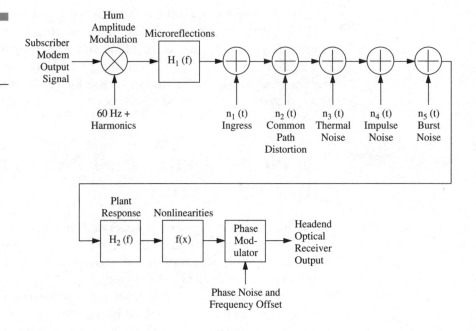

MICROREFLECTION TAP VALUES. The tap values $h_0, h_1, ..., h_{N-1}$ of Fig. 7.3b defines the delay and level of each microreflection. The tap values are obtained by analysis of measured delay spread data as follows. The mean reflection level at each delay is computed. A piece-wise linear approximation, using two segments, the first decreasing linearly (in dB) with delay and the second constant, is fitted to the mean data. This gives a simple delay profile model. Impulse response realizations are then created by generating random numbers with standard deviation proportional to the delay profile at each delay. The impulse responses are renormalized such that the total echo energy equals the energy in the original data, under the constraint that the first tap, which represents the desired signal, has a value of $h_0 = 1$.

SOURCE AND DESCRIPTION OF INGRESS NOISE. As described previously, ingress noise is the unwanted narrowband noise component

Figure 7.2
HUM modulation.

$$y(t) = x(t)[1 + Am(t)]$$

$x(t) \longrightarrow \bigotimes \longrightarrow y(t)$

$1 + Am(t)$

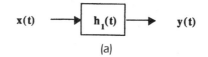

Figure 7.3
Microreflection: a, diagram; b, structure of microreflection FIR filter.

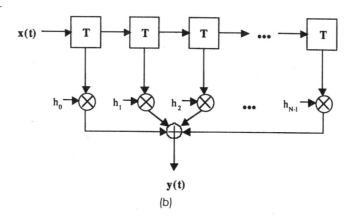

that is the result of external, narrowband RF signals entering or leaking into the cable distribution system. The weak point of entry is usually the drops and faulty connectors. Ingress noise creates apparently random transmission interruptions, usually on a single channel or closely associated channels. It also appears on a spectral display as an unwanted carrier, typically of a short time duration, and it is a narrowband signal, less than 100 kHz in bandwidth. You will find time duration and carrier bandwidth variations as a result of the various signal sources.

Common ingress sources, shown in Fig. 7.4, can be any open-air RF transmission such as CB, ham, civil defense, aircraft guidance broadcasts, international shortwave, AM broadcasters, and the like. Factors that contribute to ingress noise being introduced into the cable distribution system include loose connections, broken shielding, poor equipment grounding, noisy motors, poorly shielded RF oscillators in the subscriber household, and all electrical home appliances and toys. To shield against these narrowband interfering carriers is inversely proportional to the square root of the frequency. Since the upstream transmission is at the lowest frequency of the network's passband, the noise summates at the trunk. This causes the upstream transmissions to be most affected by the ingress noise.

The single most reliable way of reducing ingress noise is to ensure that the cable distribution system is mechanically sound when initially installed. This includes ensuring that all connectors are mechanically fit;

Figure 7.4
Typical ingress noise
sources.

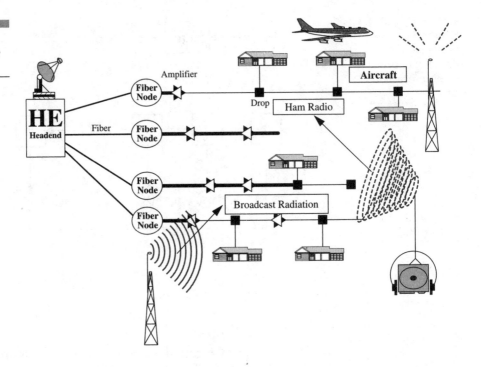

securing all cable system terminations and making sure the shield is properly grounded through the connectors; locking down all bolts and screws for the network devices; and ensuring that all unused taps, ports, and access test points are properly terminated.

INGRESS NOISE MODEL. The model for ingress consists of a FIR filter excited by white Gaussian noise, as shown in Fig. 7.5. The filter $h_{ing}(n)$ is designed to turn the white noise into peaks and a shaped noise floor corresponding to measured CableLabs data. The result is a sequence of narrowband sources, each of which is modulated with Gaussian modulation corresponding to random AM modulation or many narrowband users sharing a radio band.

TRANSMISSION IMPEDIMENTS. There are several types of RF transmission impediments (noise) found in the return path of a two-way cable distribution network. Each type provides a "unique" contribution to the problem. However, transmission impediments are not easily identified and often not properly differentiated. Thus, they are often bundled together under the generic category of ingress noise.

Ingress noise as a category is most often used as a broad description of the cable plant's noise problem, but ingress noise itself, which has greater

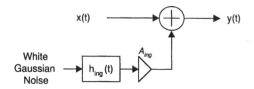

Figure 7.5
Ingress noise and common-path distortion model.

impact on upstream transmissions, is only one of the more common types of noise.

ADDRESSING INGRESS NOISE. Addressing ingress noise is an ongoing issue within the industry that has received much greater focus with the advent of two-way transmissions over the cable plant. Historically, there have been two approaches within the industry for dealing with ingress noise. These include making certain the cable connections are mechanically tight, as noted above, or using an RF spectrum segment that is unaffected by the ingress noise. Recent indications by the major MSOs indicates a preference for the second approach. The MSOs seems to be relying on the application's electronics and utilizing solutions like dynamic frequency agility and forward error correction. This may not be the most cost- and/or performance-effective approach.

First Pacific Networks conducted an experiment to identify major causes of a typical HFC distribution network. Over time, it was found that the major cause of ingress noise is corroded cable components and mechanically loose fittings in the drop portion of cable plants. It was also discovered that the upstream path of the cable distribution network had never been properly aligned. These combined factors allowed for the leakage of external RF signals into the distribution network, thereby creating a transmission impediment.

Though this may seem to be surprising, it is not unusual. Some of the factors on cable operators' to-do lists are:

- Improvement of cable-plant drop construction materials, specifically at the drop portion
- Alignment of the cable plant with amplifiers in particular
- Mechanical tightness of the cable plant—loose connectors
- Presence and strength of external RF carriers
- Relative strength of the signal carrier to the ingress signal (noise)

COMMON-PATH DISTORTION. Common-mode rejection is due to nonlinearities in the passive devices in the cable plant. The nonlinearity is created as a result of corroded connectors. As mentioned previously, it is common in distribution legs for oxides to form between two metal surfaces, creating a point of contact diode. This diode may appear in the ground portion of the connector, creating common-mode distortion and allowing penetration of ingress noise to the system. The effect of this diode creates distortion in both the upstream and downstream paths.

In the downstream direction, common-path distortion consists of signals that are cross-modulated and reflected into the upstream as the result of unintentional nonlinearities in the cable system. The source of such nonlinearities is corroded connectors. The model for common-path distortion is similar to the model for ingress: a FIR filter excited by white Gaussian noise, as shown in Fig. 7.5. The result is a sequence of signals separated by 6 MHz, each of which is modulated with Gaussian modulation, corresponding to difference frequencies between downstream TV channels and/or data signals. The gain constant A_{cp} determines the common path distortion level.

Amplifiers to a lesser extent also contribute noise in the return plant. Mistermination is usually the main cause of this phenomenon and may cause instability and oscillation.

THERMAL NOISE. White noise is generated by random thermal noise (electron motion in the cable and other network devices) of the 75-ohm impedance, terminating resistance, operating at 68°F with a 4-MHz channel bandwidth. Unlike the downstream path, all the white noise from all terminations in all the amplifiers in the cable system are funneled into the headend in the return channel. This funneling effect exasperates and increases the noise measurement as it propagates through more amplifiers and connections in the cable network toward the headend. For example, the 75-ohm terminator generates thermal noise. This noise is carried through each return amplifier, which adds its own noise contribution to the headend. All distribution points have their own 75-ohm terminator, and hence each leg adds its own noise contribution to the return system.

In a spectral display of a cable distribution network with no carriers or signals present, the white noise appears as a constant, but random and irregular, across the passband. This establishes the reference power level that must be exceeded for effective and reliable transmission. Thermal noise is modeled as additive white Gaussian noise and is shown in Fig. 7.6.

IMPULSIVE NOISE. Impulse noise is one major problem in two-way cable systems and the most dominant peak source of noise (a short burst duration of less than 3 seconds). White noise is generated in both the

Figure 7.6
Thermal noise model.

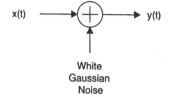

$x(t) \longrightarrow \bigoplus \longrightarrow y(t)$

White
Gaussian
Noise

downstream and upstream paths. Impulse noise is mainly caused by 60-Hz high-voltage lines and any electrical and large static discharges such as lightning strikes, AC motors starting up, car ignition systems, televisions, radios, home appliances, washers, and so on. Loose connectors also contribute to impulse noise. Impulse noise creates random transmission interruptions on multiple return channels on both the upstream and downstream. It appears on a spectral display as an increase in the noise floor across multiple channels and across the entire passband.

There are two kinds of impulse noise:

1. Corona noise (shown in Fig. 7.7)
2. Gap noise

CORONA NOISE. Corona noise is generated by the ionization of the air surrounding a high-voltage line. Temperature and humidity play a major role in contributing to this event. A corona is best described as at least a 300-kV energy discharge into the air from a high-voltage line, often located on the same poles or conduits as the CATV cable.

GAP NOISE. Gap noise is generated when the insulation breaks down or connector contacts corrode. Such failures pave the way to the entry of lines discharge of 100 kV. This discharge or arc has a very short duration (in μsec), with a sharp rise and fall. The sources are most likely to be automobile ignition and household appliances such as electric motors. It is modeled as a periodic train of filtered impulses as shown in Fig. 7.8. Parameter values are shown in Table 7.3 and defined as:

A_{imp} = the impulse amplitude
p_{imp} = the impulse period
d_{imp} = the delay of the first impulse

Furthermore, the filter $h_{imp}(t)$ represents the pulse shape of the impulse noise. The pulse is modeled as a narrow rectangular pulse with width T_{imp}.

Figure 7.7
Typical impulse noise.

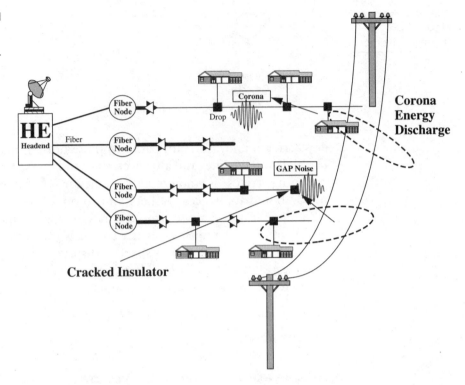

A more complex model of burst noise would randomize the amplitude and period.

BURST NOISE. Burst noise is similar to impulsive noise but with longer duration. It typically results from several sources: corona discharges from power lines, discharges across corroded connector contacts, automobile ignitions, and household appliances with electric motors. It is modeled as gated, filtered, additive white Gaussian noise, as shown in Fig. 7.9. The periodic time-gating function $g(t)$, shown in Fig. 7.10, provides a simple model of the amplitude of the noise envelope (representing its bursty nature) vs. time. These parameters (values shown in Table 7.4) are:

A_b = the burst amplitude
w_b = the burst width
p_b = the burst period
d_b = the delay of the first burst

Moreover, the filter $h_b(t)$ represents the frequency dependence of the burst noise, which generally is more pronounced at lower frequencies.

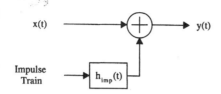

Figure 7.8
Impulsive noise
model.

A more complex model of impulsive noise would randomize the repetition period.

PHASE NOISE AND FREQUENCY OFFSET. Phase noise arises in frequency-stacking multiplexers that occur in some return path systems. It may also occur in laser cavities but is less pronounced. Phase noise is also produced in the subscriber and headend modem oscillators. Phase noise is modeled as filtered and integrated Gaussian noise, as shown in Fig. 7.11. Frequency offset is also shown. The following parameters apply:

$h_\phi(t)$ = the low-pass noise-shaping filter response, which can be tailored to approximate the desired phase noise rolloff vs. frequency, including DC gain and corner frequencies
Δf = the frequency offset
φ = a constant phase offset

PLANT RESPONSE. The cable plant contains linear filtering elements which are dominated by the diplex filters that separate upstream frequencies from downstream frequencies. The model of the plant response is a FIR filter designed to approximate the composite response, as shown in the Fig. 7.12.

NONLINEARITIES. Nonlinearities include limiting effects in amplifiers, the laser transmitters in the fiber node, and the laser receiver in the

TABLE 7.3

Impulsive Noise
Parameter Values

$h_{imp}(t)$: impulse response for impulsive noise	Rectangular pulse shape
T_{imp}: width of impulsive noise	100 ns
A_{imp}: burst amplitude	Depends on desired impulse noise level
p_{imp}: burst period	20 msec
d_{imp}: delay of first burst	5 msec

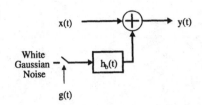

Figure 7.9
Burst noise model.

headend. The model used to represent these effects is a cascade of 2 zero-memory nonlinearities $f_1(x)$ and $f_2(x)$ as shown in Fig. 7.13. $f_1(x)$ represents the gradual limiting effect of RF amplifiers. It is modeled as a hyperbolic tangent function $y = \beta \tan h(\alpha x)$. $f_2(x)$ represents an AM laser nonlinearity. A simple three-segment piecewise linear model consisting of a linear segment and two constant segments is used.

Nonlinearities can limit the useful dynamic range of the channel. In the presence of strong ingress or impulse/burst noise, or if authorized users do not respect allocated power constraints, intermodulation products and harmonics are produced in the nonlinearities. These spurious products add to the noise floor and may fall at locations throughout the spectrum.

7.3.2 Noise Characteristics in the Downstream Direction

Noise contribution for the downstream channel was also modeled and used by IEEE 802.14 to evaluate the proposed physical layers described in Chap. 5. The following diagrams address issues beyond noise. They are intended to give an overview of the downstream channel in all its aspects, when used in the HFC system.

Figure 7.14 illustrates the type of noise and other factors defining the characteristics of the downstream CATV channel.

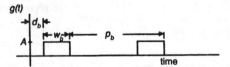

Figure 7.10
Burst noise gating function $g(t)$.

TABLE 7.4

Burst Noise

$h_b(t)$: impulse response for burst noise	None (delta function)
A_b: burst amplitude	Depends
w_b: burst width	5 msec
p_b: burst period	15 msec
d_b: delay of first burst	20 msec

FIBER CABLE. The fiber affects the digital signal in two ways:

1. There is significant group delay due to the high-modulation frequency of the signal in the fiber. The resulting group delay is specified by $g(f)$.
2. White Gaussian noise is added to the power P_{nf}.

PLANT RESPONSE. In any network topology, an impulse response is created between the headend and a subscriber modem. Impulse response is defined using words like *tilt* and *ripple.*

TILT. The tilt is a linear change in amplitude with frequency and an approximation to the frequency response of the components in the network. Tilt due to the cable and amplifiers is approximately canceled by equalizers in the amplifiers. For a well-designed system, the tilt should be similar for all modems. In the channel model, the tilt is set at 2 dB per 6 MHz.

RIPPLE. The ripple is a sum of a number of sinusoidal varying amplitude changes riding on top of the tilt and is a measure of the effect of micro-

Figure 7.11

Phase noise and frequency offset model.

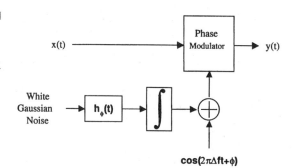

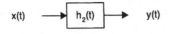

Figure 7.12
Plant response
model.

reflections in the network. The frequency of a ripple depends on the delay of a microreflection. The greater the delay, the faster the ripple. However, the greater the delay, the lower the power due to loss in the cable. Hence, there is an inverse relationship between the frequency and power of a ripple. To simulate a ripple, multiply the frequency response by

$$1 + \alpha e^{-j(2\pi f \tau + \theta)}$$

where α is the amplitude of the ripple, τ is the frequency of the ripple, and θ is a random phase shift in the microreflection. The amplitude is related to the ripple by the equation

$$\alpha_{TOT}(dB) = -11.926 - 12.69\tau(\mu s)$$

where α_{TOT} is the total power in a 200-ns range around τ. Therefore, given a set of values for τ, the tilt and ripple models can be generated. Up to 20τ-values will be randomly generated with uniform distribution over the range [0 μs, 5 μs] to obtain a random channel frequency response for use in simulation. In each 200-ns range, α_{TOT} will be shared among the ripples that have τ within that range. When specifying a channel response for a simulation, the center frequency of the modulation should be specified along with the reflection delays and amplitudes. Alternatively a random set of channels could be generated with the statistics given above, but tens of channel responses should be generated and used to estimate the performance of the modem. This model was used to assess the overall performance of a modem by averaging its performance over a large number of channels picked randomly from the ensemble. Generally, the user will demodulate the frequency band of interest and take its inverse discrete Fourier transform to obtain an impulse response.

Figure 7.13
Nonlinearities model.

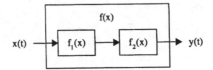

Figure 7.14

Impairments in the downstream CATV channel.

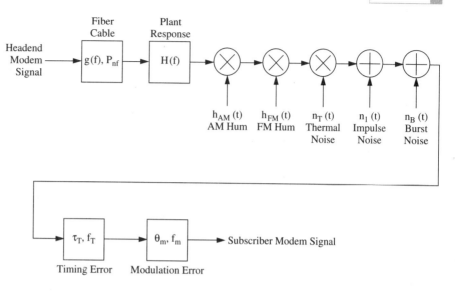

Figure 7.14

Impairments in the downstream CATV channel.

REFERENCE POWER LEVEL. For the purposes of simulation, the power of the headend signal is assumed to be 1. All other power levels are relative to this. Channel impairments are described below.

AM HUM MODULATION. AM hum modulation is amplitude modulation caused by coupling of 120-Hz AC power through power supply equipment onto the envelope of the signal. The equation, also represented in Fig. 7.15, is:

$$y(t) = x(t)[1 + Am(t)]$$

where: $x(t)$ = the input signal
$y(t)$ = the output signal
A = the hum amplitude; e.g., for 5 percent hum modulation, $A = 0.05$ (5 percent is default)
$m(t)$ = the hum waveform—a 120-Hz square wave with unit amplitude is used for modeling purposes

FM HUM MODULATION. FM hum modulation is frequency modulation caused by coupling of 120-Hz AC power through power supply equipment. The AC power signal modulates the signal on the line, causing it to shift both up and down in frequency. The equation, as represented in Fig. 7.16, is:

$$y(t) = x(t) \cos[\beta m(t)t]$$

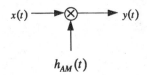

Figure 7.15
AM hum modulation
model.

where: $x(t)$ = the input signal
$y(t)$ = the output signal
β = the hum modulation factor
$m(t)$ = the hum waveform—a 120-Hz square wave of unit
amplitude is used for modeling purposes

THERMAL NOISE AND INTERMOD.   Thermal noise is modeled as white Gaussian noise, shown in Fig. 7.17, with power defined relative to the power at the output of the plant response. Intermod is caused by nonlinearities in the system, generating harmonics of other channels. It is also modeled as white Gaussian noise. The power at the output of the plant response is the product of the power at the input to the channel model and the plant gain. The total noise power due to the thermal and intermod noise relative to the plant response output is –40 dB.

BURST NOISE.   Burst noise wipes out the signal for a period of time. In the downstream direction, it is due to laser clipping, which occurs when the sum total of all the downstream channels exceeds the signal capacity of the laser. It is modeled as gated, filtered, additive white Gaussian noise, as shown in Fig. 7.18. The periodic time gating function $g(t)$, shown in Fig. 7.19, provides a simple model of the amplitude of the noise envelope (representing its bursty nature) vs. time. The parameters are defined by the following quantities:

A_b = the burst amplitude
w_b = the burst width chosen uniformly from a range [10 μs, 100 μs].
p_b = the burst period, set to 1 s.
d_b = the delay of the first burst set either to 1 μs or 0.25 s.

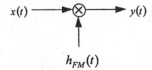

Figure 7.16
AM hum modulation
model.

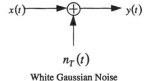

Figure 7.17
Thermal noise model.

$x(t) \longrightarrow \bigoplus \longrightarrow y(t)$

$n_T(t)$

White Gaussian Noise

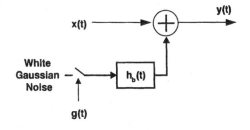

Figure 7.18
Burst noise model.

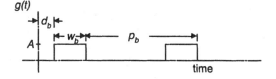

Figure 7.19
Burst noise gating
function $g(t)$.

In addition, the filter $h_b(t)$ represents the frequency dependence of the burst noise. At present, we assume the noise is white.

PHASE NOISE AND FREQUENCY OFFSET. Phase noise is created in the headend and subscriber modems. The characteristics of the phase noise will depend on the RF components in the modems. A typical phase noise characteristic was provided for simulation and comparison purposes. Phase noise is modeled as filtered and integrated Gaussian noise, shown in Fig. 7.20. Frequency offset is also shown in the figure. The following parameters apply:

$h_\phi(t) =$ the low-pass noise-shaping filter response, which can be tailored to approximate the desired phase noise rolloff vs. frequency—it includes DC gain and corner frequencies
$\Delta f =$ the frequency offset

GAIN. The gain of the network is kept relatively constant for all users. However, there will be a variation of ± 15 dB in the signal at the receiver. This is also chosen randomly with a uniform distribution.

Figure 7.20
Phase noise and frequency offset model.

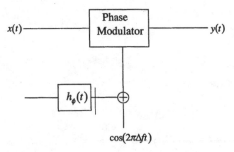

Figure 7.20
Phase noise and frequency offset model.

CHANNEL SURFING. Channel surfing causes microreflections to appear and disappear. Because the significant sources of channel surfing are close to the receiver, a large but slowly changing ripple in the frequency domain will appear or disappear. This means each 6-MHz channel observes a fairly constant change in gain. Experimental measurements have shown that the gain change is less than ±1 dB and takes about 100 to 200 ns. Hence the channel surfing simulation is a gain change of +1 dB or a gain change of –1 dB in the channel response with the change occurring linearly over 100 ns. The change occurs randomly with a +1 dB channel following the last 1-dB change and vice versa. The performance of the modem during a channel-surfing gain change and burst noise should be assessed.

Parameters used in these models contributed to the PHY development described in Chap. 5.

Chapter 8

Management of the High-Speed Cable Modem

8.1 Overview

In the previous chapters, the physical and MAC layers of the cable modem were described. In the next three chapters, the management and various other cable modem services that are unique to the HFC system will be explored. IEEE 802.14 is working closely with other bodies to resolve these remaining issues. It is unlikely that all of them will be completed in time for release 1 of the IEEE 802.14 standards.

This chapter describes the basic management operations of the cable modem. The focus will be on the needs for managing devices in a cable TV system. Some requirements are peculiar to a cable TV network, while others are generic to most networks. The management aspects that deal with HFC operation will be briefly identified. HFC management topics are beyond the scope of this book and may differ from one cable operator to the next.

This chapter is divided into two sections. Section 1 will describe cable modem management functions. Powering up the cable modem requires a host of tasks that need to be initialized before the cable modem can be declared functional.

Section 2 will describe traffic management as it relates to available bit rate (ABR) services. These two sections are not necessarily related, but future solutions may require management messages to fully address ABR implementation.

8.2 Cable TV Management: General Aspects

The OSI network management concept for managing nodes can be divided into the following functional groups:

1. Configuration
2. Performance
3. Accounting
4. Fault isolation
5. Security

Configuration management deals with addressing, network parameters, access control, and levels of service parameters. **Performance man-**

agement deals with traffic statistics and RF signal quality. **Accounting management** deals mainly with traffic statistics. **Fault management** deals with error statistics, RF signal quality, and network testing. **Security management** deals with encryption support at the application level.

There are certain characteristics that are unique to the CATV network. To manage a cable TV network effectively, the cable modem link manager must be able to monitor or have control over such entities as:

■ Access control

■ Levels of service (CBR/VBR/ABR)

■ Frequency agility

■ Receive/transmit power and distance from the headend

IEEE 802.14 completed some aspects of the management stated above, at least the topics that influence the cable modem. Description of all the HFC network management aspects is beyond the charter of IEEE 802.14. Therefore, only layer management associated with the cable modem will be described.

8.3 MAC Management Protocol

MAC management protocol deals mostly with cable modem initialization and basis error monitoring and encryption. Below is a summary of the most important procedures needed for the cable modem.

8.3.1 Overview

MAC management protocol includes a method of attaching stations to the headend, and it is independent of the transmission protocol. In an HFC topology, all stations transmit data to the headend (many to one), and the headend sends messages to all stations (one to many). MAC management protocol assumes that a single headend's signal will be present on any one of the *m* downstream channels (as shown in Fig. 8.1) and that multiple stations may be active at one time. There may be one or more *n* upstream channels.

IEEE 802.14 designed a message-based protocol based on IBM, General Instrument, Com21, and other proposals. These proposals are described

Figure 8.1
Downstream/
upstream functional
channel
configuration.

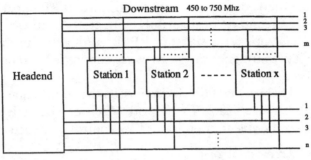

below. Some of the features are adopted in release 1 of the IEEE 802.14 standard; the rest will be adopted in later releases, including the associated MIBs. The IEEE 802.14—adopted management protocol was designed to have minimal overhead yet provide a rich set of options and parameters. The protocol developed was open-ended and designed to support future management messages and transport dependent functions such as proxy signaling or proxy ABR.

8.3.2 Access Control Parameters

Attaching a cable modem (referred to as a *station*) to an HFC system is not friendly, technically or otherwise. A host of initialization steps must be performed to make it "plug and play." The concept developed by IEEE 802.14 is functionally similar to connecting the phone to the wall jack, but, unlike the phone, a set of parameters must be initialized before the cable modem can be declared operational. Some of these parameters are:

- Synchronization
- Authorization
- Local address assignment
- Ranging and power calibration
- Assignment of default upstream and downstream channels
- Assignment of encryption information

These parameters are described below in a form of messaging protocol. Security and encryption is described in Chap. 10.

8.3.3 Station Initialization/Basic Operations

Initialization and a set of management messages including state machines are required in order for a station to become active in the HFC network. Below are descriptions of some of the critical procedures that are needed for that purpose.

JOIN PROCESS.   The join process is the procedure used by a station to attach to the network. The process includes several phases as described below.

INVITATION SEEKING.   When a station wishes to participate in the HFC network, it will scan all known downstream m channels as shown in Fig. 8.1. This is similar to pushing the radio seek button to search for a clear radio signal. Once a quality signal is established, the station will lock. It will then listen for an INVITATION message. Each headend that has sufficient available bandwidth to support an additional subscriber will transmit an INVITATION on a periodic basis. The attaching station will then successively tune to each of the m downstream channels until it receives an INVITATION message. Once received, the station will then compare its tuner frequency with the one sent by the headend in the INVITATION message. If it compares (functioning accurately), the station will send an INVITATION_RESPONSE response message to the headend using the suggested upstream channel recorded in the INVITATION message. Multiple upstream channels are likely to be provided in the INVITATION, since some topologies and frequency assignments prohibit a single upstream channel from being used by all stations that receive a given INVITATION.

If the INVITATION message contains more than one upstream channel on which to transmit the INVITATION_RESPONSE, the station should successively attempt to respond to each one of the upstream channels until it receives a positive response from the headend. Since it is possible for a station to see all downstream channels that are being transmitted to its fiber node, a station is likely to be listening to the wrong downstream channel, or a channel that does not have a direct return path to the headend. Proper assignment of upstream frequencies is very important. For this reason, the INVITATION_RESPONSE message contains the downstream channel frequency at which the subscriber is tuned.

When an INVITATION_RESPONSE is received by a headend, it compares the received downstream frequency value with its actual downstream frequency. If the two compare, the subscriber is known to be listening to the correct downstream channel. If the frequencies do not compare, the network management will be notified of this misalignment. This may be the result of frame(s) transmitted by a station before it is ranged.

RANGING. Ranging is a process by which the headend determines the round-trip delay of data destined to a specific station. Accurate ranging of stations permits a TDMA-like slotted channel mechanism on the upstream, and less guard time will be required between different stations. With precise ranging, all station transmitters along the entire length of the CATV system will be aligned in terms of timing such that, if every station on the channel began transmitting on the upstream channel, the first symbol of each would arrive at the headend receiver at exactly the same instant the first symbol of the downstream frame was leaving the headend transmitter.

During the ranging process, each station transmitter is downloaded with a transmit timing offset value. IEEE P802.14 specified a maximum CATV length of 50 miles (80 km). The ranging process has the effect of positioning each station in a virtual timing space such that all stations appear to be within zero propagation delay of the headend. For example, using the difference between its current time and a station's local time at the time a frame is transmitted, the headend can determine a time correction value to be assigned uniquely to each station. The INVITATION and INVITATION_RESPONSE messages exchange all the data necessary to perform ranging, but additional ranging exchanges may be performed on any station via the RANGE_REQUEST message.

The headend also determines a power level adjustment value to guarantee optimal performance.

CONFIGURATION. After ranging is completed, stations will be provided with configuration parameters via the ASSIGN_PARAMETERS message. This frame must be acknowledged by the station so that the headend is guaranteed that the station will not continue to attempt to join, which would result in excessive INVITATION_RESPONSEs being sent.

8.3.4 Cable Modem/HFC Dialogue Examples

Following is a dialogue between the headend and cable modem to perform some of the critical management tasks needed for proper cable modem operation. The procedures are described as examples to facilitate the message flow sequence.

CHANNEL ASSIGNMENT EXAMPLE. As described above, a station attempts to access the network by surfing the various downstream m channels shown in Fig. 8.1 until it finds a valid signal. Once locked, the station sets a timer and waits to receive an INVITATION message. The INVITATION message informs the station which of the upstream channels are available. The timer will expire if no INVITATION message is received. The channel is assumed to be unable to handle additional traffic, in which case the station scans the next downstream channel until it receives an INVITATION before the timer runs out. The station will then send an INVITATION_RESPONSE and indicate the downstream channel to which it is currently tuned. If the first INVITATION_RESPONSE goes unanswered, it could be for one of several reasons. If the downstream frequency in the INVITATION_RESPONSE does not match the frequency to which the receiving port is set, the port should report a frequency assignment error to network management. If the INVITATION_RESPONSE was corrupted or collided with another frame, the station may not know it, but the stations will attempt several retries. The INVITATION_RESPONSE may have been damaged during transmission due to a power level that was too high or too low for the headend to accurately perceive. Hence, the INVITATION_RESPONSE should be transmitted at varying power levels when it goes unanswered by the headend. After the station tries several times unsuccessfully using varying power levels, the station then tries another downstream channel seeking invitation, and the process repeats.

SCENARIO OF A CABLE MODEM JOINING THE SYSTEM. The headend periodically transmits an INVITATION that indicates to surfing stations that the current downstream channel has available bandwidth for new subscribers. The INVITATION is transmitted to the new station HFC group address (broadcasting), since newly attaching stations do not yet have an HFC address assigned. After waiting a random amount of time to avoid multiple responses, a station will respond with an INVITATION_RESPONSE that carries the station's IEEE address, product certification ID code, and the downstream channel from which the station received the INVITATION. Once the INVITATION_RESPONSE is received, the headend verifies that the response was transmitted by a station with the correct downstream frequency. If the downstream frequency is not correct, the headend will stop the process and assume that the station is not responding correctly. If the downstream frequency is correct, the headend assigns the attaching station a unique HFC address and other operational parameters by transmitting an ASSIGN_PARAME-TERS message. Since the ASSIGN_PARAMETERS message is also transmitted to the new station HFC group address, all receiving stations must inspect this message's IEEE address field to see if it is in fact intended for them. The ASSIGN_PARAMETERS message must be acknowledged by the station. The headend calculates the round-trip delay of a message by determining the difference between the station's current time and the headend's current time. Since the station's time is initially set by the headend's time via the INVITATION message, the difference between the two is the round-trip delay plus a very small delta (the time needed to set the station time after reception of an INVITATION). The station time is derived by the recovered receiver clock, which ensures that all stations increment the time count at the same rate. The protocol allows a station to be ranged at any time, as characteristics of the cable may change over time and temperature. The RANGE_REQUEST is used to request a range update.

EXAMPLE OF A STATION REQUESTING DIFFERENT CHANNEL.
A station may request a different downstream channel to meets its QOS needs. In such a case, the station will transmit a REQUEST_NEW_CHAN-NEL message to the headend. Once that message is received, the headend will determine if another channel is available, and, if so, it will transmit a JUMP message to the station. The JUMP will force the station to transmit a DISCONNECT, change its downstream frequency to that specified, and cause the station to enter the SEEKING_INVITATION state. It is also possible for a JUMP message to be issued by the headend without solicitation from the station. This may be necessary for the headend to perform load balancing and dynamic bandwidth allocation.

EXAMPLE OF ROUND-TRIP CORRECTION. The round-trip correction (RTC) value is used to delay a station's upstream timing relative to its downstream timing. At time $T = 150$, the headend transmits an INVI-TATION. This message contains the headend's current time. Thirty time units later, a newly attaching station receives the INVITATION and sets its local time to that specified in the INVITATION, namely 150. After waiting a variable and nonarchitected amount of time, in this case 20 time units, the station transmits an INVITATION_RESPONSE. This response indicates the station's current local time, namely 170. The station's timer is incremented based on a signal derived from the downstream receive signal; this ensures that the station and the headend keep time at exactly the same pace. When the headend receives the INVITATION_RESPONSE at time $T = 250$, the headend will immediately subtract its current time from the time found in the response. This value is the round-trip delay time to this particular station and is equal to $250 - 170 = 80$ time units. This single calculation reflects the time it took the INVITATION to reach the station $(180 - 150 = 30)$ plus the time it took the INVITATION_RESPONSE to reach the headend $(250 - 200 = 50)$. The period of 20 time units the station was idle between receiving the INVITATION and transmitting the INVITATION_RESPONSE is transparent. The station can wait any amount of time before transmitting its response. Assuming that the maximum theoretical round-trip time to the most distant station is 100 time units, then the round-trip correction value is $100 - 80 = 20$. An ASSIGN_PARAMETERS message is transmitted to the station, and the station uses the round-trip correction of 20 for future transmits. This process ensures that all stations have a delay such that they appear to be the theoretical maximum distance away from the headend.

EXAMPLE OF A STATION PUT ON HOLD. The headend is unable to respond immediately to an attaching station. A JOIN_HOLD message is sent to the requesting station, preventing it from moving too quickly to the next downstream channel seeking an INVITATION message. A station receiving this hold message should wait until a time-out occurs. The station responds with the INVITATION_RESPONSE as normal, but the headend is unable to determine immediately all parameters necessary for the ASSIGN_PARAMETERS message. Instead of transmitting the ASSIGN_PARAMETERS message, a JOIN_HOLD message is transmitted. Once the headend has determined, all the information necessary for the ASSIGN_PARAMETERS message, it will be transmitted to the station, which acknowledges it as normal. Following this, the station behaves as in any other case.

EXAMPLE OF A CHANNEL CHANGEOVER. The headend monitors the error performance on a per-channel and per-user basis. When channel performance degrades below a certain threshold, then all users on the channel will be transferred to an unused frequency channel, if available, and the same framing structure and slot assignments will be maintained. If an unused frequency channel is not available, then the headend will assign each station individually to other channels following the basic dynamic bandwidth assignment algorithm. The headend will also detect the case when a certain user is experiencing unacceptable error rates while other stations on the upstream channel are not. In this case, the headend can reassign that particular station to another channel to see if that clears the error.

STATION REMOVAL. The headend can at any time use the RE-MOVE_STATION message. Stations will go off-line once their IEEE address has been verified. The duration parameter in the REMOVE_STA-TION message allows the headend to turn off a station temporarily. This can be used by the headend as a maintenance tool to find the source of network errors. Stations that are removed must go off-line, wait the specified period of time, execute their diagnostics, and rejoin if the station diagnostics pass. The headend may permanently disable a station. This will cause a station to disable its transmitter and its local interface. The station may continue to receive messages from the headend, monitoring the UNREMOVE_STATION message. This function is especially useful for disabling stations that have unpaid bills. This function may also be used to disable stolen devices or faulty, noise-generating stations.

8.3.5 List of HFC Command Messages

The HFC command messages are as shown below. Some of these messages will be incorporated in the IEEE 802.14 release 1 standard. Others may be included in later releases.

SYNCHRONIZATION TIME-BASE MESSAGE. This message is broadcast periodically on all downstream channels. It includes Time-Stamp information used to synchronize stations for upstream TDMA transmission as well as transport delay.

DEFAULT CONNECTION MESSAGE. This message is broadcast on all downstream channels at a certain interval. It contains the power-up upstream channel, the upstream slot, and the minimum power level.

POLL REQUEST MESSAGE. This message is transmitted from the headend to a single station (addressed to its box identifier) on all downstream channels when there is a major reset such as a neighborhood power outage. This allocates every other slot of an upstream channel to a station and sends the poll request message to each of those stations. The poll request message contains the power-up upstream channel, the upstream slot, the old power, and range offset values for that station.

STATION IDENTIFICATION MESSAGE. This message from the station to the headend is transmitted on the power-up upstream channel if the station is powering up (either for the first time or on a major reset operation in response to a poll request). It is transmitted in the appropriate slot. It contains the station identification and its downstream channel. This message is also transmitted in the following cases:

- A station identification message had been transmitted earlier, and the station receives a power-on collision message within a certain time interval. In this case, the station performs a backoff and retransmits the message.

- A station identification message had been transmitted earlier, and the station has not received any response message from the headend within a certain time interval. In this case, the station increases its power and retransmits this message. This is repeated until the station receives the local address assign message or the max power level has been reached.

- The station received a logical address message from the headend asking for its box identification number.

POWER-ON COLLISION MESSAGE. This message from the headend is broadcast on all downstream channels in the case where individual stations are attempting to come online. (It identifies the upstream channel and slot in which a collision occurred.) Stations that receive this message and have just transmitted a station identification message perform a backoff and retransmit the station identification message.

LOCAL ADDRESS ASSIGN. This message, from the headend to station, is transmitted to a single station on the station's downstream channel. It contains the station box identification number and the unique HFC unicast VC that will be used in future unicast communication with the station.

RANGE AND POWER CALIBRATION MESSAGE. This message from headend to station is transmitted by the headend to a single station on the station's downstream channel after it has transmitted a local address assign message to the same station. It contains the range offset value and the power control value and is addressed to the station's local address. This message may be transmitted to the same station in response to any received upstream slot to cause it to adjust its power and range offset.

STATION ACTIVATE REQUEST.   This message from the station to the headend is sent on the station's default upstream channel in its upstream slot to request bandwidth.

STATION DEACTIVATE REQUEST.   This message from the station to the headend is sent when the station detects that it has no active connection and wishes to transition to the inactive state.

SWITCH DOWNSTREAM CHANNEL.   This message from the headend is sent to a station, asking it to change its downstream channel.

DOWNSTREAM SWITCH SUCCESS.   This message is sent by the station on its upstream channel in response to the switch downstream channel message after it has tuned to the new downstream channel and found the initialization connection message.

DOWNSTREAM SWITCH FAILED.   This message is sent by the station on its upstream channel in response to the switch downstream channel message after it tunes to the new downstream channel and is unable to hear the initialization connection message.

REPROVISION MESSAGE.   This message is sent from the headend to a particular station to reassign its channel and slot assignments. A station requiring more upstream bandwidth during active sessions shall receive this message to add additional slots to its upstream transmission pool.

UNICAST SILENCE MESSAGE.   This message is sent from the headend to the station's downstream channel, telling it to stop transmitting on its upstream channel. (This implies a deallocation of upstream resources.)

BROADCAST SILENCE MESSAGE.   This message is broadcast from the headend on all downstream channels and informs every station to disable its upstream transmitters and stop accepting data.

MULTICAST SILENCE MESSAGE.   This message is transmitted from the headend to all stations on a particular downstream channel and informs every station on that channel to disable its upstream transmitters and stop accepting data.

RETRANSMIT REQUEST. This message is sent from the headend to the station on the HFC command broadcast VC on all downstream channels to request a retransmission. This message contains the sequence number of the next expected cell.

LOGICAL ADDRESS REQUEST. This message is sent from the headend to the station's downstream channel to read the station's box identifier. The station will respond with a station identification message.

MODULATOR PARAMETERS MESSAGE. This message is sent from the headend on the station's downstream channel to read the modulator parameters used for upstream transmission.

MODULATOR PARAMETERS RESPONSE. This message is sent by the station in response to the modulator parameters message.

BUILT-IN TEST MESSAGE. This message is sent to the station's downstream channel, asking it to run its test.

BUILT-IN TEST RESPONSE MESSAGE. This message is sent by the station on its upstream channel after running its built-in test. The results of the test are returned in this message.

DOWNLOAD RAM MESSAGE. This message is transmitted by the headend to a single station on the station's downstream channel to cause this station to download the data contained in the message. This message may be used to download test procedures or any other data to the station.

GOTO MESSAGE. This message is transmitted by the headend to a single station on the station's downstream channel to cause this station to perform an unconditional jump to the station location specified in the message. This message may be used to reinitialize the station or to cause it to execute a downloaded test procedure.

TEST RESULTS MESSAGE. This message is transmitted by a station after it has executed the software test procedure that had been downloaded to the station by the headend. The results of the software test procedure are sent to the headend via this message.

SNMP STATISTICS REQUEST MESSAGE. This message is sent on the station's downstream channel to retrieve and/or reset SNMP-related statistics that are maintained by the station.

SNMP STATISTICS RESPONSE MESSAGE. This message is sent on the station's upstream channel in response to the SNMP request message.

CONFIRM DOWNSTREAM SWITCH MESSAGE. This message is sent by the headend to a station after the headend has told the station to tune to a new downstream channel. This message is sent to the station on the new downstream channel.

ENCRYPTION ASSIGNMENT MESSAGE. This message is sent by the headend to a station to provide it with a new traffic encrypting key to be used to encrypt/decrypt cells.

ENCRYPTION CONFIRMATION MESSAGE. This message is sent by the station to the headend in response to an encryption assignment message. It confirms receipt of a new traffic encryption key. The contents of this message are encrypted with the new encryption key.

"ARE YOU ALIVE" MESSAGE. This message is sent to a station on its downstream channel to verify that the station can listen to the headend on the station's current downstream channel.

STATION ALIVE MESSAGE. This message is sent by the station to the headend in response to the "are you alive" message.

NETWORK BUSY MESSAGE. This message is sent by the headend to the station during station activation when it cannot allocate the guaranteed minimum bandwidth to the station.

CANCEL NETWORK-BUSY MESSAGE. This message is sent by the headend to a station when bandwidth resources become available. This message is sent to a station that had earlier been denied its minimum bandwidth allocation.

DECOMMISSION MESSAGE. This message is sent by the headend to a station when the headend wants to decommission the station and put it into an unauthorized state.

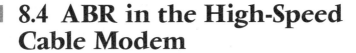

8.4 ABR in the High-Speed Cable Modem

8.4.1 Introduction

Supporting ABR traffic in the 802.14 HFC network is more complex than had been anticipated. There were several proposals made, and most suggested that the headend would play a role in managing ABR services. Management messages may play a role so that the headend can proxy ABR control of all the connected stations. Scheduling ABR traffic from the headend enables:

- Reconciliation of clustering and concatenation inside the HFC

- Efficiency with spacing according to ATM flow dictates at the exit toward the ATM switch

- The most efficient traffic profile for the ATM core, allowing minimum transit traffic

8.4.2 ABR Review

Before describing the mechanism of ABR over HFC, it would be useful to describe briefly the basic concepts adopted by the ATM Forum rate-based method for ABR control. Chapter 3 described ABR mechanism in more detail. For the purpose of this description, we assume a cable modem station is connected to a WWW server.

Referring to Fig. 8.2, the rate-based closed loop control dynamically dictates to the stations the rate to be used at any time according to congestion conditions along the route of each connection. It takes place end-to-end and works per connection (VP/VC). Switches indicate congestion by use of the EFCI (explicit forward congestion indication) bit of the cell header or the CI bit of the resource management (RM) cells.

The EFCI mechanism is used for backward compatibility with existing ATM switching nodes using this mechanism to control congestion. Table 3.2, regarding PTI coding, describes EFCI.

RM cells are issued every N number of cells (typical value 32) of the connection originating from the source-end system (SES) and are looped back from the destination-end system (DES) to the source. Switching nodes that can support this scheme can set an explicit rate based on the

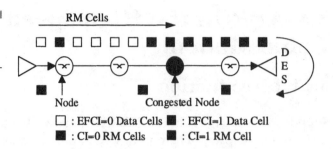

Figure 8.2
End-to-end ABR rate-based mechanism.

local conditions if it is lower than the one set by the previous switch along the route. This rate is looped back and reaches the source, which must obey it until a new update arrives. Switches that support only the EFCI bit can reach the same result with the help of a single-bit positive feedback scheme, guiding the source to reduce the rate successively after each round trip by a rate decrease factor (RDF). The end result is that any bottleneck occurring anywhere in the ATM WAN can be shared fairly by all connections going through it. Hence, per-connection control is of utmost importance. As a result, it is possible that a station with two connections open to different destinations may have to reduce its rate of one VC at the same time that it is permitted to raise the rate of the other.

ABR MAC-CONTROLLED QUANDARY. An important aspect of this framework is that flow control takes place at the ATM layer and not at the higher layers as in previous store-and-forward networks. This has a great impact on the HFC MAC, which, as the arbiter of medium access, it is also the function that controls the access rates. The problem is that the rationale underlying the rate control comes from expediencies originating far away in the network, which therefore are irrelevant or even in conflict with the expediencies governing the local access. Thus, the above framework poses the following problems to the 802.14 MAC:

■ The MAC targets terminations (stations) and not VCs. This choice aims at the efficiency arising from shorter permit and request fields and lower incidence of exhaustion of queues. When the queue is exhausted, a new contention is required. Contention is an inefficient access method.

■ The traffic of RM cells, particularly those looped back in asymmetric connections, takes up upstream bandwidth, which is at a premium. This can be prohibitive, since the load of RM cells is proportional to the rate. Downloading big WWW or ftp chunks of information at the available downstream rates will create very high levels of upstream RM-cell traffic controlling the downstream rate.

■ If the sources were to follow the rate dictated by the rate-based control, then the piggyback mechanism could not be efficiently used, since the cells for each connection would be separated by many slots, breaking more often than not the chain of piggyback requests. This is unwelcome, since efficiency would greatly suffer. Furthermore, cell concatenation, which further increases efficiency by not repeating the PHY overhead, would be out of the question, since it blatantly violates the traffic control requirements, representing prolonged peak rate emissions. If made permissible by means of a high GCRA tolerance, it would result in a much higher reservation of resources by CAC, meaning high inefficiency and tariffs.

■ Policing the dynamic rate per VC should not take place at the network entry, as happens with any ATM access network, to protect the highly sensitive shared medium from noncooperating users.

■ Even if we let the RM cells reach the users, and even if no clustering and concatenation were encouraged by the MAC policy, still the headend would most likely need to shape the traffic, since the multiplexing at the HFC is done without taking into account in the scheduling the real arrival times because of its distributed nature. Such a multiplexing is much worse than a FIFO scheduler, and the egress traffic is highly distorted as judged by the strict standards of the ATM CAC and UPC mechanisms. There are doubts even if the traffic from ordinary FIFO multiplexers would be acceptable to the ATM traffic controls at times.

■ Alternatively, the policing tolerance size that would make it acceptable would largely cancel the rate control effect and again cause inefficiencies in the core network. A spacer unit at the headend seems difficult to avoid if compliance to the traffic-control dictates of ATM is to be achieved.

It is important to note that the interpretation of the time scale at which the rate is to be measured has been determined indirectly and is reflected in the tolerance parameter of the GCRA algorithm that will be enforced by the network policing units. This is true for any ATM source. Therefore, regarding an ATM traffic contract, it is not enough to satisfy a compliance to a longer-term average, e.g., to send some clumps of cells separated by silent periods such that the average rate is the dictated one. Two such streams with the same long-term average but different burstiness are not equivalent in their traffic statistics and switch resource requirements. Therefore, they will not suffer the same loss level if sent via the same network. For this reason, at CAC (Connection Admission Control) time, the network will reserve more resources for the more bursty connection, or, for the same resources, less traffic of the more bursty type can be admitted. This can be expressed in a less rigorous but more picturesque way as the more bursty the traffic, the more bandwidth it consumes for the same average load. Only multiplexers applying a scheduling policy with respect to target connection profiles can decrease the burstiness of connections, while FIFO scheduling always increases the burstiness and HFC multiplexing even more. The issue is, if the increase stays within the limits initially negotiated by the CAC, the system should not be allowed to increase more than has been accounted for by the modification to the CAC algorithm to take the idiosyncrasies of HFC into account. But then we cannot commit a large number of resources to this, because, even if such increases are taken into account, there is still waste that goes on all the way through all the links and switches until the very destination where our jitters connection ends up.

Given the fact that ABR traffic is controlled per VC, one could be tempted to make each VC a MAC user (i.e., by issuing permits to target VCs and not just terminations). However, this method creates more problems than it solves. How is the station to notify the arrivals per VC in a piggyback? Permits also would require twice as much overhead.

The above considerations explain the motivation for proposing the following framework for the support of ABR traffic in HFC. The scheme is independent of any specific MAC protocol.

8.4.3 The VC-Based Backpressure Mechanism

IEEE 802.14 MAC makes use of the efficiency of the piggyback requests to reduce contention to just the first cell of a station going from "idle" to "active." Then piggybacked requests are used until queues are empty. This strategy is efficient inside the HFC, but, as explained, it is in direct conflict to the rate control philosophy of ATM networks because it produces streams that are much more resource-consuming than necessary. Instead of choosing between the inefficiency inside the core network (i.e., cell clustering) or the inefficiency inside the HFC access network (i.e., by spaced cell emissions), we can have the best of both. This can be done by controlling the rate from the headend where ABR cells destined for the ATM core network could be buffered and spaced in accordance with the rate flow control. This shaping of traffic would take place at the headend and avoid loading the HFC part with the load of RM cells, since the RM flow is thus terminated at the headend. The target of permits is assumed to be per station and service class. An example of the classes could be:

- A high-priority class for CBR services. It could also cover demanding real-time VBR, which is supported on a peak-rate basis.

- A class devoted to VBR with multiplexing gain.

- A class with lower priority that covers ABR traffic. (Signaling cells will probably require head-of-line priority.)

The number of classes is not important for the mechanism except that the VCs belonging to ABR must be recognizable by the headend. The permits must contain an identifier of the service class they address, as is the case with many of the MAC proposals. The requests could also be per service class. Buffering at the headend and rate controlled emission would be useless without a mechanism to control the station rate per connection (i.e., a backpressure mechanism operating on a VC level). This way, we can throttle a connection that experiences a bottleneck somewhere in the core ATM network, as it is indicated by the arriving RM cells, while still servicing another one from the same station. Without backpressure, the buffer at the headend would overflow, and, if the ABR flow specification is not per connection, we cannot support it.

The backpressure mechanism requires four extra bits in the permit field. So, it incurs an insignificant bandwidth penalty downstream, where it is not scarce, while none is taken from upstream, since the feedback is implicit. Although attached to the permit field, so as to identify the station to which the backpressure refers, the backpressure bits operate on their own. The buffer at the headend is logically allocated per VC. It has an upper and a lower limit. When the upper limit is reached for a certain connection, then it orders the station to stop sending cells from this VC. The opportunity to do this is the moment it receives a cell from this station carrying this VC. Then, with the next permit going to that station (irrespective of how many slots later), it sends the four bits with the format 0XXX. The zero is the stop code, and the other three comprise a code to be used to identify the VC when the command to start again will be issued. At that time, both the station and the MAC controller create an association between the VCI and the short code. If more permits are sent in the time of one round trip, then obviously the association can only refer to the last cell that was sent one round trip delay earlier and not the more recent cell.

Thus, no confusion can arise for any network length if the stations have in their parameter list the value of the round trip in slots.

When the lower limit of the buffer is reached, then, with the first permit for that station (which could be prompted by another request of the station or just by the headend for the stopped VC), the backpressure command 1XXX is sent, which indicates starting again the previously stopped VC, now identified by the code XXX. The lifetime of this three-bit code is the duration of the VC stoppage. It can cover up to 8 stopped VCs per station at any time, but in fact the same code XXX may be given to more than one VC, so a much larger number of VCs can be supported. In this case, upon receiving the command 1XXX, the station first sends one cell from a randomly chosen VC having the code XXX. If it is not the one that should be restarted, then the MAC responds with stop again (0XXX). Upon receiving this, the station will try another VC until it gets the right one. The penalty is losing a few cells upstream, but little harm is done. The upper limit of the buffer is chosen to allow a margin of two or three more cells since it is a common buffer space and the allocation to VCs is only logical. More important is that, even with earlier-than-necessary cells, the bandwidth is not wasted. This way, very few bits are consumed for this function. The ABR rate control dialogue should be seen as taking place between the MAC controller and each station independently, using as a vehicle the permits that are addressed to the relevant station. All these exchanges happen independently in between those for the other stations and whenever the permit target provides the opportunity.

Upon receiving a command to stop a VC, the station will send with the upstream slot an updated request containing the number of waiting ABR cells without the ones of the stopped VC, so as not to cause unnecessary permits to be issued by the permit allocation algorithm. There will be no contention again for any cell of the stopped VC. The station will wait until it receives a permit with restart. To avoid increasing the processing delay, the stop command could be sent just one cell before the upper limit of the logical buffer allocated to the VC, so upon reception of the command, the station can still send this last cell if it was already in the send register.

Regarding robustness, we rely on the protection existing for the whole permit field. But in the event of a corrupted command, there are two possible effects: restarting a VC that should not be restarted—which will be easily stopped again—or stopping a VC that should not be stopped. For the latter case, it is proposed that the stopping of a VC should have a timeout in its validity, after which the station could attempt to send again.

8.4.4 Other Implications

With this solution, since at the headend buffer easy access to VCs is possible, two more very important prospects arise:

1. First, more advanced scheduling algorithms can be used. The situation is analogous to the egress queues of ATM switches, and the same expediencies apply, since the jitter introduced by the switch will not be larger than that introduced by the HFC network.

2. By policing the connections and passing the action to the input, the HFC users are protected.

The MAC closed-loop mechanism can play this important role, giving the HFC network the chance to protect itself by denying further access to cells of this connection after the first violation (in addition to tagging or rejecting them). This can only be done while still within the HFC network. The policing that will take place at the switch entry will protect the switch but not the HFC network. The backpressure mechanism will give the required per-VC effect to the policing action so that no other connections will have to suffer. The alternative could be to install policing units at the entry points. This is a more costly proposition, and it will be difficult to update the units with the policing parameters.

CBR traffic will not be rebuffered at the headend. The same is true for VBR. At a 3-Mbps upstream link rate, the prospects for rt-VBR traffic do not seem significant. However, if the rate goes higher, it may be possible to have significant support for this kind of traffic, too. In this case, some study is required to determine the best support method between the following approaches:

- Use peak rate allocation (probably best as a starting point).

- Allow clustering for efficiency with some possible increase in the tolerance parameter of the GCRA at the PCR and SCR policing units. The issue of efficiency inside ATM should be raised here again (clustering could make the SCR practically equal to the PCR).

- Use shaping at the headend for efficient switch-resource allocation of ABR.

In any case, the VBR traffic PCR and SCR can be policed (without buffering) at the headend, and, as with ABR, the action can pass to the entry by denying one grant for each cell found noncomplying. Increasing requests will be met with increased denials as long as conformance is not satisfied. Thus, the overflow will occur at the entry queues of the offending user. It can be further restricted to the relevant VC by means of a similar mechanism to that proposed for ABR backpressure, but this may not be necessary. So with CBR permits inherently imposing policing and the proposed policing for VBR and ABR, complete protection at the entry is assured without extra cost of H/W-based policing units at the entry points, which would need to intercept signaling messages to obtain the required parameters and would increase the cost of the modems.

The category of UBR has no service guarantees. There is no CAC, and it is not policed. Requests for UBR cells should not be mixed with ABR, and the MAC should only grant permits if no other traffic is requesting the bandwidth. Therefore, UBR cells should not contend at will because the philosophy of their support would be violated if they could take access from other classes. Therefore, only certain minislots marked by the MAC could also be accessible to UBR contention at moments of low load. Then, only the slots carrying UBR cells could contain a piggyback request for other UBR cells. A support of UBR traffic along these lines could be an interpretation of the UBR class in the HFC environment.

9

POTS over HFC

9.1 Overview

POTS (plain old telephone service) is one of the services cable operators are offering today (or soon) over the HFC system. The Telecommunication Act of 1996 gives cable operators the go-ahead to compete in this regulated market. The challenge of providing voice services is not only a matter of added revenue, it will soon become a key service that is needed if the MSOs are to compete in this highly competitive new interactive market.

The business case of providing POTS over HFC is very attractive. The access war has just started to the displeasure of some of the RBOCs. The new regulation is forcing the traditional telco operators to unbundle services and offer them to the marketplace. Unbundling services is one sensitive subject some RBOCs hope will go away. What unbundling of services means, among other things, is that your friendly telco service provider must lease residential copper wires to others who wish to compete in this access market. What the RBOCs get in return is entry to the lucrative long distance business, which was denied them before the Telecommunication Reform Act. This has a profound impact on the way the RBOCs' strategic planners worked in the past, specifically when setting priorities in their network's migration and evolution. Service unbundling also shifted some of their valuable resources to allow them to refocus their energy from VOD (for example) to a more reliable source of revenue. This service unbundling will open the floodgates to all newcomers as well as the traditional interexchange carriers (IXCs) such as AT&T, MCI, Sprint, and others. Everyone will be able to compete directly with the traditional RBOCs as access providers. In fact, many organizations are already doing just that. The IXCs' infrastructure is service-ready for handling local calls in terms of POTS service offerings, billing, network management, IN, trunking, and a host of other network elements needed to service calls.

Cable operators will be in the middle of this access war. They have less interest in leasing the copper wires from the RBOCs but must face the challenge of providing reliable voice service over HFC, as will be described later in this chapter. Cable operators may have an advantage in the short term by offering consumers a package deal that includes basic cable TV service with combinations of premium TV channels, CNN, POTS services, and so on. Even if POTS reliability is not fully addressed, it can still be an attractive service for families with teenage children. This second voice line, referred to in the industry as the *teenline*, will be instrumental in keeping the peace in the family.

In this chapter we will describe the following:

- The strategies and challenges for a POTS service offering by cable companies
- Telco voice network summary description
- POTS over HFC solutions
- PCS over HFC

9.2 Key Service Offerings

Presently, it looks like all players, individually or in partnership, are differentiating their future interactive service offering by providing high-speed Internet access, video on demand, broadcast entertainment, and telephony services. U.S. West and Continental CableVision is one of the few examples witnessed by the industry in the broadband convergence revolution. The cable operators do have an attractive and enviable infrastructure that is conducive to deployment of broadband services. The cable modem can provide a high-speed interface to the cable network that has plenty of bandwidth to spare. ADSL and VDSL technology has evolved much faster than expected, and now it is the choice of most of the RBOCs for providing high-speed access service. Fiber-to-home is a technology that is still costly as far as its deployment as an access interface.

It is expected that service offerings will be bundled and a customer will be offered an attractive combination package, luring him or her to change service carriers. In fact it will be a reality—and in some places it is already—that, much like today, the customer will be interrupted during dinner hour to hear a sales pitch peppered with gifts to switch carriers. The gift may be in a form of so many minutes of free calls, frequent-flier mileage bonuses, and so on. This gives consumers options that were never available in the past. Service differentiation becomes a luxury or the criteria upon which the consumer can base his or her subscription decisions.

9.3 Service Differentiation

Operators, be they telcos or MSOs, must distinguish themselves in a variety of ways. The key differentiators are:

- Time to market
- Quality of service offering
- Customer services

The cable operators are challenged by all of the above. The one that is most challenging and pronounced is customer services. The first two can be handled if cable modems are deployed.

Telco operators with ADSL deployment can easily meet these standards. As mentioned earlier, cable operators, despite this disadvantage, do have an advantage of bundling services, including CATV distribution. However, they must provide at least as good or better POTS to compete effectively with the telco reputation. Cable customer service reputation is admittedly quite bad, but then none of the services they offered in the past were considered essential or life-threatening.

The good news/bad news survey for cable operators, recently published, revealed the following:

- Seventy-eight to eighty-four percent of customers who have never experienced a problem with a service will probably or definitely buy from that provider again.

- Eighty-two to ninety-two percent of customers with complaints that were handled satisfactorily will probably or definitely buy from the provider again.

- Nearly 40 percent said they would be likely to purchase local telephone service from a provider other than traditional local exchange carriers.

- Of those willing to switch, 67 percent said they would choose an interexchange carrier.

- Forty-nine percent said they definitely would *not* purchase local telephone service from cable companies.

These statistics are putting more pressure on cable operators, who must address the customer service dilemma and the menacing reliability problems facing them when they offer POTS.

The following section will briefly describe the telco STM network before proceeding with how to handle POTS over HFC. This should give us an idea on how reliability is addressed and should reveal the cultural differences that are responsible for shaping the communication networks.

9.4 The Traditional STM Network

The telecommunication network of today operates in a synchronous transfer mode (STM), sometimes referred to as *circuit switching.* From its inception, this STM network was developed to handle voice; hence, real-time was demanded for such synchronous operation. The description below is not intended to be comprehensive by any means. It is only intended to give an overview of the telco mind-set when providing services and illustrate the discipline associated with the reliability that must be built into a network.

9.4.1 Modern Voice Network

Figure 9.1 depicts a simplified but typical configuration of a voice network that was or is being deployed by the RBOCs. The network hierarchy, in simplified terms, can be classified into three network nodes:

1. Access network

2. Core switching

3. Service platform

The description below may be insulting to some, but for the purpose of illustrating basic switching philosophy it should be adequate.

ACCESS NETWORK. The access network is the portion of the network that is closest to the subscribers. A digital loop carrier is normally installed in neighborhood areas providing access to customers via copper loop (a pair of copper wires). The mode of operation is point-to-point; hence, no one else is sharing the wires (so no MAC is needed). This pair of wires, identified as *tip and ring,* is used to perform several functions. It carries:

- Analog voice signals (digital signal if it is an ISDN line)
- Power to the end-user device (20 milliamps DC) to detect off-hook condition
- Ring voltage
- Coin collect, when used in coin-operated telephones (by reverse tip and ring polarity)

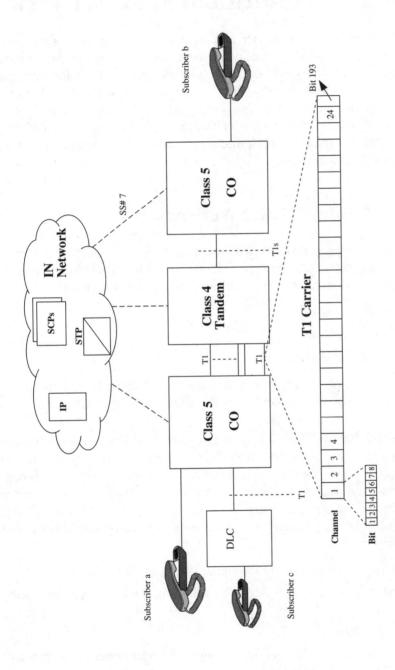

Figure 9.1
Modern STM network.

DLC = Digital Loop Carrier
CO = Central Office
STP = Signaling Transfer Point
SCP = Signaling Control Point
IP = Intelligent Peripheral

DIGITAL LOOP CARRIERS (DLCs).   The DLC is a 1:1 mux node that connects subscribers to a class-5 switch. Class-5 switches are where all switching functions are performed and where the services are provided. The rationale for deploying DLC is quite simple. It is much less expensive to connect, for example, 24 subscribers to a nearby DLC box and carry the information on a single T1 line to the central office (CO) than to have all 24 pairs of wires hooked up to the central office. A T1 line connecting the DLC to the central office can serve 24 subscribers, each occupying a single independent voice channel. The mapping is one-to-one; that is, 24 subscribers are each assigned a particular channel in the T1 line.

A more sophisticated T1 interface like the Bellcore recommendation of TR8 is also used today to assign dynamically one of the 24 channels in the T1 line when a subscriber goes off-hook. The signaling bit in the T1 line (bit 193) is used as a messaging system between the DLC and CO, among others, to synchronize the T1 frame. More than 24 subscribers can be served by the T1 (24 channels) and, with good traffic engineering, the system can perform with very low probability in blockage. A more sophisticated TR-303 interface has recently been developed by Bellcore to handle ISDN lines. In TR-303, one of the 24 channels in the T1 line is dedicated for signaling. More than one T1 line can be connected from the DLC to the office. T1 redundancy comes in all flavors, with N:1 being the norm.

It is important to note that the interfaces discussed are transparent, and CO has the sole responsibility of handling the call, at least as far as services and call features are concerned. For example, if all the T1 lines connecting to the central office were to be cut, then services to the connected subscribers would be disrupted. Networking the T1 lines to more than one CO would make a more reliable system if one set of the T1 lines is cut or if one of the COs goes out of service. Some operators install a remote switching unit instead of a DLC. If the T1 lines are cut or the CO dies, the small remote switch can still handle calls sent out to that community.

T1 CARRIER.   The T1 is a digital carrier that operates at 1.544 Mb/s. It is divided into a 193-bit stream. Bit 193 is used for synchronization, T1-line framing, and signaling (using the alternate frame approach). Each subscriber will be assigned one of the 24 channels full-time during the life of the call. A channel contains 8 bits and is used to sample the voice at a rate of 125 μsec (8 kHz). Transport of this sample at 125 μsec is maintained throughout the network. Bit 8 of each channel is sometimes referred to as *rob-bit signaling* and is used alternately as a signaling bit to send dialing information (calling party number, etc.). Digital repeaters are used to condition the digital T1 line, are equivalent to the amplifiers deployed in the cable network, and are powered by DC current carried by the T1 copper wires. Alarm logic and signals are monitored by the host CO, verifying the repeater's operating conditions. If a repeater fails, the CO will reroute the traffic using other T1 trunks that are available in the network.

CLASS-5 CO. Class-5 STM switches can handle up to 100,000 subscribers. The line cards in these switches are periodically scanned for the off-hook condition to handle a subscriber call. The off-hook state is detected by DC current flow due to the closure of the tip and ring loop. The class-5 switch, based on the dialing information received from the subscriber, performs number translations and routes the call to its destination. The processors in the class-5 switch perform the control functions, while the switch fabric is used to carry the 8-bit voice samples through the core.

If the called party is not in the class-5 local area (a long distance call), then the call is dispatched up in the hierarchy to a class-4 or tandem switch.

It is important to note that:

1. Every call-affecting element in the switch is duplicated to maintain reliability. Among the most critical are the following:
   - Power supplies powering the electronics
   - Line cards and associated controllers
   - T1 carriers

2. Control platform (duplicate call-handling controllers) usually operate in an online and standby mode. When new software is loaded into a switch, the class-5 switch is reconfigured to operate in a hot-standby mode; for instance, if the new software crashes, the old working software will take over in the duplicate processor (hot standby) to keep the switch operational. Other duplicated peripherals are:
   - Core fabric
   - Peripheral devices, including the billing magnetic tape drives
   - Translation databases

AC power is used indirectly to power any central office. DC batteries (–48 V) are used instead to power all aspects of the electronics and relays. Redundant power converters are used to power the traditional 5-, 3-, or 25-volt integrated circuits. The DC power is also used to ring a phone and/or extend 20 milliamps current to detect an off-hook condition. AC power is used only for charging the –48-volt batteries. When AC power fails, the batteries must have enough energy to operate for 8 hours. Diesel-powered AC generators are used when AC power fails.

Services and features associated with the call are embedded in the class-5 software platform. The class-5 switch software is the most complex to develop. Real-time call handling is only one aspect of the complexities of having up to 100,000 subscribers.

CLASS-4 CENTRAL OFFICE. Class-4 switches belong to the interexchange carriers (IXCs): AT&T, Sprint, MCI, and so on. It functions similar to the class-5 switch in terms of reliability. The class-5 switch routes the call to that particular IXC class-4 office based on the user's long distance subscription.

INTELLIGENT NETWORK (IN). The intelligent network is a newcomer to the networks, and the RBOCs began deployment in the 1980s. Class-4 and class-5 switches are connected to one or more IN networks using signaling system 7 (SS7) links. SS7 links operate at 56 or 64 Kb/s. SS7 is a well-defined messaging protocol specified in ITU, ANSI, and Bellcore.

The IN idea is quite simple and unique. Developments of features offered to subscribers in class-5 switches became very expensive and time-consuming. The RBOCs were looking for new revenues, and the idea is that independent software vendors can develop these features with a well-defined API and locate them in IN. Traditionally, revenue-added services were embedded in the class-5 switch. IN was conceived to reduce the cost of developing these services and features and accelerate time to market. The idea is that information on how to treat a call can and should be separated from the switch software and therefore can be located in an IN complex independent from the switch. IN has two basic functions:

1. Call management
2. Call control

One example of an IN call-management mode is number portability. A subscriber forwards his call to a Paris telephone number. That number is then stored in the SCP (signaling control point) in the IN. The class-5 switch will then automatically route all the subscriber's calls to the number stored in the SCP (which is the Paris number). Other features such as call forwarding, distinctive ringing, call trace, call waiting, and the like, can be migrated to and reside in the IN database. All 800-number translation databases reside in SCPs.

In a call control mode, the IN signaling transfer point (STP) is used as a packet switch and functions as a relay to perform out-of-band signaling. This signaling channel is used to set up a call between class-5 or class-4 switches. All aspects of the call treatment are performed using this SS7 messaging system. SS7 decreases call setup time, increases call-handling time, and reduces or makes obsolete the MF receivers hardware, enabling the network to use its resources more efficiently. All IN components are duplicated including SS7 links, redundant STPs and SCPs, and all associated control platforms.

9.4.2 HFC Solutions for Voice over HFC

Architecturally, in an HFC environment, the idea of handling voice is not all that complex. The voice is somehow transported bidirectionally between headend and the stations. The headend processes the voice call using various switching technologies.

There are three possible approaches to handle voice over HFC:

1. Using the traditional STM technology (POTS over HFC)

2. Native ATM voice switching

3. PCS (personal communication services)

All three solutions will most likely be deployed in an HFC network in combinations or individually. The most likely scenario is to provide POTS over HFC to meet the demand and time to market. Brief descriptions of the three solutions are shown in this and the following section.

POTS OVER HFC.   In this configuration, as shown in Fig. 9.2, a switch or digital loop carrier would be located at the headend. The headend controller receives the voice channel from the cable modems located in the home and simply passes it on to the DLC or the switch for further processing. If the headend is equipped with DLC, then T1 trunks carrying voice circuits will be forwarded to a nearby class-5 switch for call handling. Bellcore specification of TR-303 at the T1 trunk interface will be the best choice.

The headend could also be equipped with a fully featured class-5 switch; in this case, the capabilities would include:

- Local switching
- Billing
- Bypassing the local exchange carrier and interfacing directly to interexchange carriers (long distance carriers), thereby saving user access charges

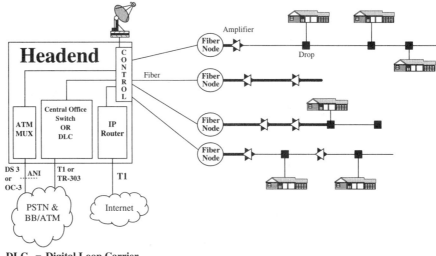

Figure 9.2
Featured headend
with POTS and IP
capabilities.

DLC = Digital Loop Carrier
PSTN = Public Switch Telephone Network
T1 = T1 Carrier

The headend can also be equipped with IP routers for handling IP ser-vices and connectivity to the Internet network. The headend controller will separate IP traffic and hand it to the routers for further processing.

The ATM mux (could also be an ATM switch) is an ATM access node carrying ATM connections to the ATM switches in the public network. The headend controller can segregate ATM traffic, which could be data, video, or CBR services. The ATM Forum defines this interface to be the ANI (Access Network Interface). The specification of this ANI interface will be available in late 1997 or early 1998. This architecture is the most likely scenario that will emerge in the near term if not the long term.

The question that remained is how best to transport voice traffic and, more specifically, voice samples from the station via cable modem to the headend. IEEE 802.14 made provisions to accommodate voice synchro-nization so the headend and the cable modem can lock to the 8-kHz clock. The simplest and most cost-effective way to transport voice samples is simply to transport this DS0 (125-μsec voice sample) using one byte of the ATM payload.

In 1990, Alcatel ABS proposed such an approach in IEEE 802.6. The voice samples were transported in one or more marked octet of DQDB cells. (A DQDB cell is the same as an ATM cell.) Lucent and others in IEEE 802.14 also proposed to carry STM voice samples in the ATM cell, and Lucent emphasized the need for dynamically allocating these periodic samples based on the traffic behavior. This voice sample is segregated at the headend and delivered to the STM node.

One possible potential problem with this approach is that voice is handled at the MAC layer. All the electronics will require power to keep the MAC operational and to maintain call-handling capabilities. If the AC fails due to power outage, voice handling would be affected. Batteries, if used, can only be effective for a short period. We all know how useless our modern AC telephone sets are during power outage—we all look for the old tip and ring telephone set in order to make a phone call.

Handling voice at the physical layer is one solution that can survive power outage. Only the physical-layer electronics need powering, and that limited amount of power can be delivered externally from the cable plant via the coax.

VOICE OVER ATM. As mentioned in previous chapters, one of the prize features of broadband is to integrate all services under a single ATM switching platform. Theoretically, voice can be carried in packets over ATM in an adaptation layer. It is then transported directly to an ATM switch in which individual calls (voice packets) are switched. The broadband switch treats and routes the packet in a manner similar to the ongoing UNI 3.1 or UNI 4.0 presently specified in the ATM Forum.

ATM end-to-end architecture is as shown in Fig. 9.3. In such a configuration, the ATM switch or the access node, as defined in the ATM Forum, will treat and transport all the integrated services to a broadband switch. The headend will take on the responsibility of an access node with an ANI interface extending to the broadband network. Connections can be voice, data, or video. To work on an ATM end-to-end basis (native ATM service), a voice-over-ATM adaptation layer needs to be defined.

ITU, the ATM Forum, and ANSI are working on a new AAL for landline voice application. This new AAL2 was a spinoff of the AAL-CU (ATM Adaptation Layer—Composite User), which was first conceived and developed in ANSI and ITU and is now being specified in the ATM Forum. The AAL2 is in the final stages of standardization in ITU and was described in Chap. 3.

BACKGROUND. Since its inception, ATM was plagued by how best to handle voice packet in an ATM cell. Europe was promoting the idea of full-service integration and defined the ATM cell with a 32-byte payload (more efficient in carrying voice and no echo canceller needed). The

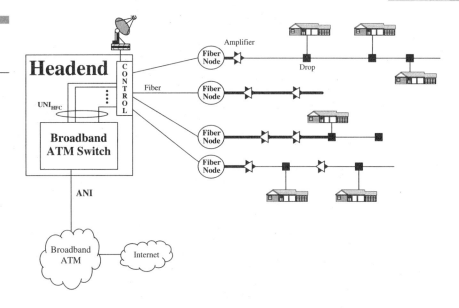

Figure 9.3
Native voice over
ATM services.

United States was more partial to the video application and defined an ATM cell with a 64-octet payload. The United States proposed a composite user approach, where several users shared a single ATM cell payload. The $(32 + 64)/2 = 48$-byte payload was a compromise agreed to in ITU (in the late 1980s) and made almost everyone happy. The ATM cell has a 48-byte payload plus the 5-byte header.

To maintain good voice quality, a voice sample must be transmitted every 125 μsec continuously (as presently provided in today's STM network). In an ATM environment, if a telephone unit waits to fill the entire ATM cell payload (48 bytes) with voice packets and transports them to the other end to the called party subscriber, then more than 3 msec would have elapsed. The 3-msec delay is the recommended value set by the standard body. If such a delay (over 3 msec or 6 msec round trip) is experienced, then an echo will be perceived by users, especially all who have the 2–4 wire bridge transformer, and most do. An echo canceler can be installed on the line cards or trunks to remedy the echo problem, but this will, no doubt, be an expensive proposition. If only a few samples (a few bytes) occupy the ATM cell payload (instead of the entire 48 bytes), the delay will be short; hence, no echo will be experienced. The downside of this solution is that the efficiency of transporting voice over ATM will become extremely poor. With AAL2, the ATM cell payload is shared among several users, each injecting short packets of voice samples.

9.4.3 A Solution for PCS over HFC

One attractive alternative of providing voice in HFC is to introduce PCS. PCS is the new technology in wireless communications services, and it has several attractive features that appeal to a good part of the population. Unlike cellular, PCS is digital and therefore accommodates excellent-quality voice as well as messaging services. PCS offers several advantages over cellular, but comparing both technologies is beyond the scope of the book.

PCS IN AN HFC INFRASTRUCTURE. It seems that PCS is very friendly to the HFC infrastructure and cable plant in general. CDMA (call division multiple access) technology at 1900 MHz can be installed cost effectively by deploying the miniature low-power antennae in the HFC plant. The remote antennae can be installed using a tap in various locations in the distribution network (onto the aerial cable). The subscriber signal (1900 MHz) is picked up by these antennae and converted into the upstream band (between 5–40 MHz), which sends it on its way to the headend. This PCS signal, collected from that HFC community, is filtered from the TV signal at the headend and then relayed to the switch and the rest of the landline network. This is as illustrated in Fig. 9.4. Terminating a call to an HFC subscriber, the reverse applies. The downstream signal is transported in one of the 6-MHz bands.

THE PCS/HFC ADVANTAGE. PCS over HFC has several advantages over cellular or even the telco-based network infrastructure. The antennae of the cable-based PCS system are hung on aerial cable and can be relocated or moved in the distribution plant to accommodate that geographical area. Additional antennae can be added just as easily to increase capacity and serve growth in that market. Since the HFC infrastructure is already in place, therefore, cable operators will have the advantages over other network infrastructures where tower installation cost, power, network maintenance, base stations, and so on, are already part of the HFC network.

The other noticeable advantage is the centralized modulation and the common electronics of the additional radio channels that are centralized at the base station. The new radio channels can be assigned or reassigned dynamically to meet the capacity and market growth.

Having noted all the advantages, that is not to say that deploying PCS in a cable network is not risky business. Building a PCS network is an expen-

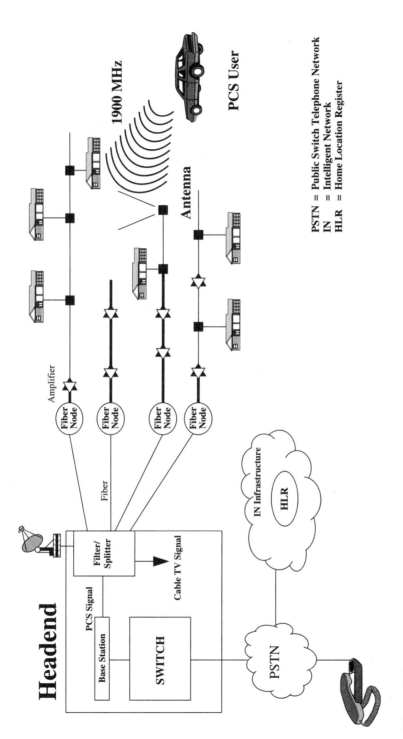

Figure 9.4
PCS in the HFC infrastructure.

421

sive proposition. Recently the FCC auctioned the licenses for the PCS spectrum band for a total of 120 MHz. A 30-MHz license runs around $15.00 per population. To serve a city of one million can cost as much as $15 million. This is only for the right to provide the service. This will add up to a lot of money if cable operators are to serve the U.S. population in general.

PCS service offering is based on the bids rewarded by the FCC. The market right now bears as many as eight major players, so it can be concluded that the market is going to be very competitive. Joint ventures are forming rapidly to corner the market. Cox Cable recently joined Sprint, Comcast Corporation, and Tele-Communications Inc. in a venture that offers service bundling of a variety of telecommunications and video services, including local and long distance phone, wired and wireless telephone, cable television, and high-speed Internet services. Cox is promoting a plan called Alternate Access to reduce billing to their customers. This allows business to connect directly to their preferred long distance carrier, bypassing the local phone company's charges.

9.4.4 MSO Challenges

Reliability is one topic the MSOs must address to compete effectively with the RBOCs when offering voice service over HFC. The HFC network has several cost advantages over other networks but lacks other important reliability features. Among the most critical are:

- Amplifiers at the HFC distribution network
- AC power outage
- Security
- Network management

The more challenging requirement that telephony in particular poses for cable systems is the need for reliable service. Bellcore and telephone standards specify that failures, on average, must not exceed 53 minutes of downtime per line per year. Cable operators are aware of this stringent reliability requirement and that their customers are sensitive to outages; given that consumers are accustomed to highly reliable telephone service, the stringent standards may reflect the expectations of real customers.

AMPLIFIER REDUNDANCIES. The RF amplifiers in the HFC distribution network are shared among several hundred subscribers. A single failure of an amplifier could affect the whole community. Cable operators should furnish more reliable amplifiers or deploy redundant ones. The RBOCs also deploy amplifierlike regenerators to condition the digital T1 trunk. The T1 trunk in RBOC networks is redundant because the central office can reroute the call to another network in the event of a failure, thus maintaining high reliability. The maintenance is also very well defined and the failed trunk or any active elements in its path can be located immediately.

POWER OUTAGE.   Amplifiers are run by AC power in the distribution plant. If the power fails, then amplifiers and also active components in the HFC network will no longer function. The whole region will lose telephone service. An alternate AC power scheme should be provided to the amplifiers. As more and more fiber penetrates the distribution plan, this problem becomes less critical. However, this is unlikely to happen any time soon, so the MSO must find an accommodating solution to the power outage problem. The RBOCs provide enough power (−48 V DC) via the CO to a telephone set (tip and ring) to keep it operational. Power at the CO uses −48 V DC for all the electronics. Even ring voltage is derived from the −48-V batteries. The batteries are continuously recharged using AC generators. A similar problem exists for powering the cable modem. If AC power fails, the cable modem becomes useless. Powering the set-top box with AC in the past was appropriate. If the AC power fails, then the TV set is disfunctional in any case. But, if the cable modem is providing voice service, then an alternate power source must be provided to keep it operational. Powering the cable modem via DC current over the coax cable is one possible solution.

SECURITY.   In a shared medium environment, voice, data, and video are carried to a subscriber via the downstream broadcast channels. Broadcast channels are subject to all sorts of security problems from unwanted monitoring of phone conversations and viewing and data access habits to theft of service where attached units pretend to register on the network. This is further described in detail in Chap. 10. In the FTTC environment (RBOC's ADSL approach), the access configuration is point-to-point; therefore, service violation is limited to the user's premises.

NETWORK MANAGEMENT.   One of the powerful tools the RBOCs have is network management. It saves money in the long term by continuously monitoring and dynamically testing lines and remote nodes. It is very much like taking a physical on a periodic basis. That way, a problem is marked for repair before it fails. This sort of maintenance philosophy should be adopted by all MSOs if they are to be taken seriously in this new competitive market.

Study after study indicates that, if the problem is not solved rapidly and satisfactorily, the customer will likely change his or her access carrier.

Chapter 10
Security in the HFC Environment

10.1 Overview

Security and privacy in an HFC network is potentially much more difficult to address than that of other point-to-point architectures. HFC is a shared medium, and, as such, customers can access their own cable line to monitor others who are on the shared line. While no communication system that sends data outside physically secure premises can be considered perfectly secure, the existence of shared media enables potentially misbehaving users to have easier physical access to the network than in other cases. Introducing integrated services and digital cable architecture, as we have described in previous chapters, requires addressing and resolving several technical problems at all layers of an application.

In this chapter, we will first explore the security and privacy aspect in general and then focus on the security and privacy issues that directly relate to the cable modem. The intent is to give an overall picture of security/privacy and not necessarily how it is achieved. Security and privacy is being specified in various organizations and will probably take longer to implement, deploy, and test than most would like. ITU, the ATM Forum, and DAVIC have been working on these applications in phases. The ATM Forum phase-1 release is expected in late 1997 or early 1998. IEEE 802.14 has taken on the development of security at the access layer (i.e., the MAC). MAC-layer security will be described later in the chapter.

10.2 Security Objectives

Security has become a very important topic since the growth of the Internet. Almost all interactive applications require some sort of security. So when we talk of security, it touches nearly all aspects of our daily activities. Applications that must be secure range from:

- Commerce, such as banking and stock trading
- Entertainment, such as program theft
- Business, such as work at home and business transactions

The objectives of security in general touch three main players:

1. Users
2. Operators
3. Government agencies

From the user perspective, security means privacy and confidentiality, data integrity, accurate billing, and service activation and deactivation.

From the operator perspective, security means data integrity and privacy, maintaining a good reputation and trust with customers, accurate billing without fraud, correct network functionality, and accountability of service outage. In brief, network operators/service providers themselves need security to safeguard their business interests and meet their obligations to the customers and the public.

The public communities/authorities demand security by directives and legislation in order to ensure availability of services, fair competition, and privacy protection. Security and privacy also play a major role in government intelligence agencies and global trade. This will be described briefly later in the chapter.

Given the above, the ATM Forum security working group defined a set of categories that applies to the user plan, control plan, and services plan. Each plan contains a set of functions that need to be performed to address the overall security question. These functions are:

- Access control
- Authentication
- Data confidentiality
- Data integrity

These functions, when implemented alone or in combination, should address security issues at all levels. We will focus on the access portion of security and how that relates to HFC in particular in the sections that follow.

10.3 Security Threats

A threat is a potential violation of security. It can be directed at four different kinds of objectives: confidentiality, data integrity, accountability, and availability. Three kinds of threats may be distinguished:

1. An accidental threat, or a threat whose origin does not involve any malicious intent

2. An administrative threat that arises from a lack of administration of security

3. An intentional threat involving a malicious entity attacking communication and/or network resources

Items one and two will be resolved through standardization procedures. The focus will then have to be on intentional threats, which include:

- *Masquerade.* The pretense by an entity to be a different entity.
- *Eavesdropping.* A breach of confidentiality by monitoring communication.
- *Unauthorized access.* An entity attempts to access data in violation of the security policy in force.
- *Loss or corruption of information.* The integrity of data transferred is compromised by unauthorized deletion, insertion, modification, reordering, replay, or delay.
- *Repudiation.* An entity involved in a communication exchange who subsequently denies the fact.
- *Forgery.* An entity fabricates information and claims that such information was received from another entity or sent to another entity.
- *Denial of service.* This occurs when an entity fails to perform its function or prevents other entities from performing their functions. This may include denial of access to services.

To deal with these threats, a set of principal functional security requirements are identified.

10.3.1 Threat Analysis

Priorities will be derived from the individual assessments of the security threats and will depend on the respective network scenarios. It will be necessary to identify a list of the perceived vulnerability of possible threats to the system. In all respects, it is unlikely or not economically feasible to protect against every possible attack. The problem of providing security becomes one of determining how best and what to protect against, and hence which security mechanisms are required. In many cases, the subset chosen depends on many factors, including what must be protected, the cost and complexity associated with providing the protection, the ease of use of the protection mechanisms, the maintenance/management of the mechanisms, and so on.

10.3.2 Verification of Identities

The network must support the capabilities to verify the claimed identity of any user in a network. To deal with identification and authentication, the following security services should be made available:

- User authentication delivers corroboration of the identity of the user.
- Authentication service delivers proof that the identity of an object or subject is indeed the identity it claims to have.

The service used to counter masquerade can be divided into two classes:

- Peer-entity authentication to establish proof of the identity of the peer entity at one particular moment in time during a communication relationship.
- Data-origin authentication to establish proof of identity of the peer entity responsible for a specific data unit.

10.4 Controlled Access and Authorization

A network must support capabilities to ensure that users are prevented from gaining access to information or resources that they are not authorized to access. The security service to meet this requirement is access control. Access control service provides means to ensure that objects are accessed by subjects only in an authorized manner. Objects concerned may be the physical system, the system software, applications, and data. The network must also support the capability to keep stored and communicated data confidential. Confidentiality is needed for two purposes:

- To protect user-related network information
- As a service used by other security services, e.g., in order to handle cryptographic keys

The confidentiality service provides protection against unauthorized disclosure of exchanged data. Two kinds of confidentiality services are distinguished: data confidentiality and connection confidentiality.

10.5 Protection of Data Integrity

The network must support granting the integrity of stored and communicated data. Protection of data integrity is needed for two purposes: to protect the network's user-related information and to safeguard the service used by other security services. Security services can be divided for services dealing with the integrity of stored data and services for the integrity of communicated data.

10.6 Accountability

The network must support the capability that an entity cannot deny the responsibility for any of its performed actions as well as their effects. Individuals in a network must hold full responsibility for their actions. The security service to support this requirement is nonrepudiation, which provides means to prove that exchange of data actually took place. It comes in two forms:

- nonrepudiation—proof of origin
- nonrepudiation—proof of delivery

The above sections give an overview of the security topics that the network must deal with. In the following sections, we begin to focus on HFC service-specific security needs and associated anomalies.

10.7 Shared Medium Security/Privacy

Security in general is a problem for all telecommunication systems when introducing interactive services to the user. The HFC network presents a unique problem because of its shared medium access. The network itself cannot distinguish where downstream transmissions go. The signal is broadcast to all stations in the downstream direction. In the upstream transmissions, all signals are carried to the headend using the same shared bus. As stated earlier, sharing the common link brings about two possibilities:

1. A user might be able to receive transmissions intended for another user.

2. A misbehaving user can make transmissions while pretending to be another user.

Without a way to control this, cable operators would face several problems in delivering communication services, mainly:

- Service theft, where malicious users masquerade to obtain illicit access to services
- Privacy, where malicious users eavesdrop on private communications

Legacy LAN and other communication systems like Cellular have similar privacy and security problems. In the case of LAN, devices are connected with a network using a shared bus so that each station has access to all of the traffic on the network. These networks, however, are used in a corporate setting, and, as such, eavesdropping and service theft is not usually a major concern. LAN in the public environment is quite another story. In this case, the traffic is neither trusted nor secure; hence, computers use security systems that are designed at all levels to provide both authentication and privacy through encryption techniques. The security and privacy problems are also evident in the wireless communication systems such as cellular telephone service, where service providers have addressed the issue of service theft effectively but have done little to prevent eavesdropping.

10.8 Security and Privacy in the HFC Network

The security and privacy problems for HFC are different from the traditional point-to-point wireline networks. In the case of copper wires, which are dedicated to the user and connected directly to a line card at the central office, eavesdropping is not as easily done as it is in a shared medium line. Private conversations certainly cannot be monitored by users in other homes. Registering a device illicitly (service theft) on a dedicated line is near impossible. The operators know the identity of that line because it terminates physically at the site. In a cable network, the security problem is more difficult because many stations have physical access to the same wire.

As stated in previous chapters, the HFC two-way is many-to-one and one-to-many transmissions over potentially the same trunk. This is illustrated in Fig. 10.1. For example, a cable modem user from any home can eavesdrop on all stations connected to the bus. This HFC-related security and privacy is similar to the LAN situation but differs slightly in some respect. Because of the use of different upstream and downstream frequencies and directional splitters, it is not possible for a station to receive transmissions from other stations directly. However, all receivers on a branching bus receive the downstream transmissions from the headend, with all but the intended recipient normally discarding the packet. This is a somewhat better situation than in the typical LAN, where all transmissions are heard by everyone.

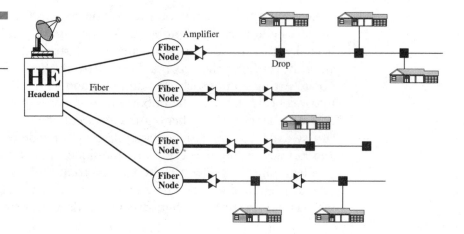

Some in the cable industry understandably question the need for privacy and the added complexities needed to provide it. Cable telephony could fall under the same category as cellular, cordless phone, or PCS. Eavesdropping on these systems can be just as easily done as on the cable modem, if not easier. It is then argued that, if a cable phone is to be used, especially as a second line, then privacy and its associated complexity may no longer be that attractive. Not long ago, even copper wires were shared among several telephone users. Distinctive ringing was used to identify a particular called party.

Modern cryptosystem techniques were designed to solve these access problems. IEEE Project 802.10 has developed a security architecture. The architecture provides for encryption at the link-layer control of the OSI reference model. It also provides for key distribution at the application layer. The security as expressed by IEEE 802.14 and the ATM Forum was required at the MAC layer. Hence, IEEE 802.14 created a requirement document that was used as a guideline for developing security at the MAC layer and hence HFC access.

10.8.1 Security Requirement at the MAC Layer

The intention is to specify sufficient access security mechanisms so as to make the security of shared media access networks comparable to that of non-shared media access networks. The following are stated as the minimum requirements.

AUTHENTICATION. The station must validate itself to the access network, which must determine whether the station device is entitled to service and bind it to the service profile and accounting record. The complexity of the authentication process must be minimized. This requirement suggests the use of a fast, symmetric protocol.

- Authentication shall be done with a minimum number of message exchanges.
- The amount of information and the number of bits in each message shall both be minimized.
- The station should be capable of being registered administratively with the access network.

ENCRYPTION.   User data between the station and the HFC interface must be made confidential.

REPLAY PROTECTION.   The access network must be protected against retransmission of a previous message by a source that is attempting to gain unauthorized use of network facilities.

KEY MANAGEMENT.   Maintenance of secret user information is critical to any secure network.

END USER CONVENIENCE.   The security functions must operate without requiring the end user to perform actions in support of the system security.

TRANSPARENCY.   The security functions must operate without degrading the level of service supplied to the user. Service degradation includes any factors that negatively impact the perceived quality of service, such as delays in service availability, additional noise, distortion, and so on.

EXPORTABILITY.   Shared media access networks will be used worldwide. Many countries have long regulated the use, import, and export of security functions. Access security functions should adhere to these export and import regulations.

DATA TRANSPORT RELIABILITY.   Shared media access networks must be tolerant of bit errors to minimize error extension. Single-transmission bit errors should not cause blocks of data to be corrupted by infinite error extension. The system must recover synchronization quickly after a data transmission interruption.

PHYSICAL SECURITY.　Shared media access security should not be dependent on elements of physical security beyond those inherent in HFC-type networks.

SECURITY RECOVERY.　Shared media access network security should be designed so that it is possible to reestablish secret information (e.g., keys).

UPGRADABILITY.　The security mechanisms selected for shared media access networks should not preclude the future use of new and/or improved security mechanisms, algorithms, and modes.

10.8.2 Encryption and Decryption

Encryption and decryption are performed using the DES algorithm as specified in ANSI X3.92-1981. The mode of encryption used is cypher block chaining.

There was also the general understanding that:

- Encryption should use proven cryptographic techniques.
- Encryption should be implemented in the hardware or software. It is desirable to be able to encrypt data carried over a variety of media. For low-speed links, software encryption is possible. For high-speed links, where a software implementation is not feasible, hardware should be used.
- Encryption should not be bandwidth-intensive. It is preferred that the encrypted data stream be the same size as the unencrypted data stream.
- Encryption should have low latency.
- Decryption should have limited error amplification. Multiple-bit error as a result of a transmission negates the higher-layer protocols and mechanisms to cope with the errors effectively.
- Encryption should be self-synchronous. It is possible, but rare, for data to be inserted unexpectedly, especially in the ATM network. Synchronous stream ciphers are unsuitable because such stream ciphers depend on maintaining position synchronization within the stream between the encryptor and the decryptor.

With these guidelines, IEEE 802.14 is incorporating cryptography into the MAC system so that the communication between a station and the headend could be private and authenticated. Done properly, this could address the eavesdropping and false-identity problems that propagate due to the nature of the cable shared medium network. The procedures are briefly described below.

KEY EXCHANGE. IEEE 802.14 is seriously considering the Diffie-Hellman key exchange protocol for HFC security. The Diffie-Hellman was chosen because it was designed to address the access needs as described below

PROVISIONING. In this function the headend allows the station to register. The station certification ID in the cable modem will be used to identify the station as legitimate user. Two steps are required to access the services: registration of the station hardware ID (certification number or serial number of he device) and the use of the Diffie-Hellman secret key exchange.

REGISTRATION. In this function, the station sends its unique ID requirements to the headend. The headend, aware of this new legitimate user (through provisioning), will proceed with secret key exchanges, attempting to register the user. The certification ID is simply used as authenticity (e.g., initial password) prior to the secret key exchange that will follow. If the headend is not provisioned to accept this ID, then registration fails. A hacker using a legitimate ID will be able to register provided that the legitimate user had not registered first. During this process, the ID information is transmitted in the clear; a hacker might listen to a successful registration transaction and record the ID information

SECRET KEY EXCHANGE. This operation uses the Diffie-Hellman exchange as described below. The possibility of hardware ID information obtained by a hacker, as described above, demands authentication. Authentication using a symmetric key technique requires that a common secret key be established prior to this communication procedure. Once this procedure is accomplished, then it can be used for a variety of security functions, including authentication and distribution of session keys for data encryption.

Assuming that a hacker obtains the legitimate ID device number during the registration procedure of a legitimate user, then he or she will be denied service later because the device ID in question is already in use or registered. If the hacker registers before the legitimate user (assuming both device ID and secret key were obtained illicitly), then service will be activated. This, however, is a futile exercise to the hacker, since the legitimate user will be denied access to the network when attempting to register. This denial of service to the legitimate user will prompt the operator to perform authentication by other means, thus revealing the attacker and unraveling this predicament. The Diffie-Hellman secret key exchange is a one-time procedure. Updating the secret key could be requested by an operator periodically to validate legitimate users who are on the network.

SECRET KEY AUTHENTICATION.   The authentication procedures incorporating the secret key can be used to verify the identity of the station to the headend. A hacker who obtains a legitimate ID number of the device must also obtain the correct secret key.

For legitimate users, keys are assumed to be present in the authenticating procedures prior to execution of the security message exchange protocols. In practice, both entities may have obtained the keys in a number of methods. Some suggested ways are:

- Manual entry by the service provider, e.g., obtaining the key via certified mail
- Coordinated entry between factory and service provider, e.g., the factory installs the key in the device, then provides a list of the keys installed in the devices delivered to the provider
- Entry through the use of a smartcard
- Entry by way of a legitimate entity (i.e., the post office) that provides the secret key

SECRET KEY EXCHANGE USING THE DIFFIE-HELLMAN PUBLIC KEY.   The Diffie-Hellman key exchange is used to establish a common secret key. IEEE 802.14 is in the process of adopting the Diffie-Hellman key exchange procedure. Take, for instance, a scenario of the key exchange communication between two parties, Alice and Bob. Alice could be an alias for the headend, and Bob is the alias for a cable modem. In principle, when Alice wishes to send a confidential message to Bob, she uses Bob's public key (usually located in a directory) to encrypt the message and send it. Bob uses his private/secret key to decrypt the message and read it. If a message is intercepted, only Bob's secret key can decrypt the message. In this procedure, anyone can send an encrypted message to Bob using the public key, but only Bob can decrypt it. Naturally, we assume that Bob will not share his secret key with anyone. Bob's private key cannot be deduced from the public key.

The above scenario can best be described in the following procedure:

1. Alice generates a random number x and calculates the quantity $X = g^x \bmod N$, which she transmits, along with g and N, to Bob. The message transmitted from Alice to Bob is called a secret key exchange request PDU message.

2. When Bob receives the request, he generates his own random number y and computes $Y = g^y \bmod N$. Bob transmits Y to Alice in a secret key exchange response PDU message.

3. Bob computes $S = X^y \bmod N$.

4. Alice computes $S' = Y^x \bmod N$.

The values of S computed by Bob, and S' computed by Alice, should be the same. Once established, the secret key can be used to support various security services for communications between Alice and Bob without the need for additional public-key computations.

The size of N is a key factor in determining how difficult it is for a cryptanalyst to determine S. N may be updated at the service provider's discretion to enhance the security of the system, in which case a similar procedure can be performed to update the secret key as required by the service provider. However, it is assumed that the key will seldom require an update. Fewer updates will reduce the amount of computation that is required of both Alice and Bob.

The value g is an integer less than N and is usually chosen such that it is a primitive mod N (i.e., it is capable of generating every number from 1 to $N - 1$ when multiplied by itself a certain number of times, modulo N).

To maintain a reasonable amount of security, IEEE 802.14 recommended values for N, x, and y to be at least 512 significant bits. The maximum allowable size is 1024 bits.

The above scenario is sometimes referred to as *electronic signature*. The output mathematical computational message is called the *digital signature*. This simple mathematical relation is verified by either party (Alice or Bob in the scenario above) as being genuine; otherwise, the signature may be fraudulent or the message might have been altered.

MAINTAINING STATION KEYS. IEEE 802.14 specifies two separate encryption/decryption keys that are to be maintained by the station. The *even/odd key* is identified in the message PDU byte and by the destination/source message. The setting of the bits determines the encryption or decryption to be performed. All keys are initialized to the "no encryption" state until the key exchange specifies otherwise. Once a message is received by an intended station, the MAC examines the even/odd encryption flag in the header and the proper decryption key chosen based on the setting of the destination address even/odd bit. Payload decryption is then performed or not performed accordingly.

Stations will always encrypt the information using one of the two keys. The key used will be the one set by the headend selected during the key exchange mechanism. The payload is the only field in the message that must be encrypted.

CRYPTOGRAPHY. Cryptography deals mostly with keeping communications private. Most transactions today require private communication, from purchasing goods on the WWW to simply performing banking or business transactions. The definitions of *encryption* and *decryption* are given below.

Encryption is the alteration of information into some garbled format. The purpose is to secure privacy by keeping the information secret to anyone else for whom it is not intended.

Decryption is the reverse of encryption; it is the reconstruction of the encrypted data back into some readable form.

Encryption and decryption require the use of a secret key to manipulate the data. Depending on the encryption mechanism used, a single key might be used for both encryption and decryption, while for other mechanisms the keys used for encryption and decryption are different.

Authentication is becoming a fundamental part of our lives. Our signature is given daily to communicate our agreement or seal contracts and documents and so on. As we move on into the information age, we need to replicate this practice electronically. Cryptography provides mechanisms for such applications. The mechanisms can be used to control access to an information-sensitive database, a pay-per-view TV channel, or a VOD/NVOD application. Modern cryptography is growing at a very fast pace. New tools will enable users to pay bills using electronic money or reveal information without revealing its content.

The modern cryptosystem provides mechanisms that are designed to solve exactly the sorts of privacy and security problems that are found in cable networks. A technique called *public key cryptography* (described above) is particularly useful for cable communication because it simplifies the problem of key exchange (providing a mechanism to communicate securely without an advance exchange of secret keys). Moreover, with authentication, it will guarantee that a transmission is in fact coming from the legitimate user of a secret key. Such systems are well-understood, secure, and software-ready. Providing a hybrid solution in the software is also implemented today, but users must be aware of the software vulnerabilities in the system.

SECRET KEY VS. PUBLIC KEY IN CRYPTOGRAPHY

SECRET KEY. Traditional cryptography is based on both parties sending and receiving a message and knowing and using the same secret key to encrypt or decrypt the message. This mechanism is known as *symmetric cryptography.* Security is compromised because the secret key must be communicated using some other medium such as a phone. Anyone who overhears or intercepts the secret key in transit can forge all the messages encrypted or authenticated using that key. Key management is the process by which the key is generated, transmitted, and stored. Therefore, all cryptosystems must deal with key management issues. Hence, all keys in a secret-key management must remain secret, and that presents yet another difficulty of how to provide secure key management in open systems with a large number of users.

PUBLIC KEY. The concept of public-key cryptography was first introduced in 1976 by Whitfield Diffie and Martin Hellman. The main objective was to solve the secret-key management problem as stated above. The concept is that each person gets a pair of keys, one called the public key and the other called the private key. Each person's public key is published, while the private key is kept secret; hence, the need for both users to share the same secret key is eliminated. Communications involve only public keys, and no private key is ever transmitted or needs to be shared by anyone. Bob need not call Alice on the phone to give her the secret key code while someone may be eavesdropping. Anyone can send a confidential message just by using public information; however, this crypted message can be decrypted only with a private key. Bob and Alice have the sole possession of that key. Public-key cryptography can be used for privacy (encryption) as well as for authentication and, logically, digital signatures.

ADVANTAGES/DISADVANTAGES. The following are some of the critical issues when comparing the mechanism for public and secret keys. It is not intended to be conclusive but meant to give an overview of both mechanisms:

- Public-key cryptography increases security and convenience. Private keys never need to be transmitted or revealed to anyone. In a secret-key system, the secret keys must be transmitted and are therefore subject to being discovered during their transmission.

- Unlike secret-key systems, the public-key systems provide a mechanism for digital signatures. Authentication using the secret-key mechanism requires the trust of a third party. Any one party can repudiate a transaction, claiming that security was compromised. Using a central database that keeps copies of the secret keys of all users is not fully secure. An attack on the database would trigger widespread forgery. Public-key authentication will not have this predicament—each user has sole responsibility for protecting his or her own private key.

■ Speed is one of the disadvantages of using public-key cryptography for encryption. Secret-key encryption methods are much faster than any currently available public-key encryption mechanism. A hybrid approach can be used to get the best results. For encryption, the best solution is to combine public- and secret-key systems in order to get the security advantages of public-key systems and the speed advantages of secret-key systems. In this case, the public-key system is used to encrypt a secret key that is used to encrypt the message. This protocol is referred to as a *digital envelope*.

■ Public-key cryptography may not be necessary in some cases. A secret-key cryptography alone could be adequate in some applications. This, of course, implies that a single authority knows and manages all the keys (for example, a closed banking system). Since the authority knows everyone's keys anyway, there is not much advantage for some to be public and others private.

■ The use of public-key cryptography was not intended to replace secret-key cryptography, but rather to enhance it. The first use of public-key techniques was for secure key exchange in an otherwise secret-key system.

10.8.3 The Case against Providing Security in HFC

It is apparent from the above that providing privacy and security for the cable modem is complex and potentially costly. There are some in the industry who believe that providing this sort of mechanism will unduly complicate the "plug and play" convenience the telephone user now expects. The question is, of course, how much security is warranted versus its threat. Following are some data that would be useful in making a case.

SOFTWARE BASE ENCRYPTION. Computing power is more than adequate today to support the encryption software. It is also possible to allow cryptosystems to be used only as needed, so that a specific application of a cable modem would not have to bear the added cost. For a cable system, the strongest arguments to use cryptosystem technology are to:

- Prevent theft of service
- Prevent unauthorized access
- Add privacy systems where support for security and authentication already exists—may add a little extra cost

An alternative to building cryptography into the MAC layer of the cable system is to implement it at higher levels in the network. PC applications that run on the Internet sometimes use cryptography to achieve security at the network level. The cost is born in the application and not at the MAC and therefore not at the cable modem, either. In the case where security is performed at higher layers, it will not be adequate to prevent eavesdropping and manipulation on the cable network. Without MAC-level protection, it is possible to eavesdrop on the time, amount, and destination of traffic as well as its content.

SECURITY AND "PLUG AND PLAY." It is anticipated that the cable modem will be considered as household equipment much like the model used today in the telephone companies. This allows consumers to purchase their own equipment and connect it to the network. The concern will then be the complexity a consumer will face when installing the cable modem to the network. A complex installation would be one burden to the cable operator, and a barrier to consumers is another. The ideal situation is to allow the cable modem to connect to the network and be provided with services immediately, much like it is done today when installing a telephone set. This is referred to as "plug and play." This may be an unfair comparison, because we have to compare installation complexity with an ADSL modem. With the ADSL modem, one expects less installation complexity simply because the connection is point-to-point; hence, authentication and device registration may not be needed. In any case, it will be much less complex than installing a cable modem device. Unlike point-to-point, it is not possible to associate a device with a subscriber. With all that complex automation, what happens if the device does not work? A subscriber cannot make a phone call to the cable operator to seek resolution because the device in question is delivering the telephone service to a home. Hence, the "plug and play" solution must also address when equipment does not work. Possible failures in these systems include the cable network itself, the drop and inside wiring, power failures, defective terminal equipment, and failures of equipment connected to the terminal device. The failure usually shows up as a service that does not work. The computer cannot make a connection, the phone does not work, or the TV has no signal. Occasional service calls by cable operators can easily eat their profits very fast.

With all the complexities described, we should not assume, for example, that the network will be fully secure and sufficient to keep users from building boxes that will violate security or privacy. Program theft and electronic money transfer is incentive enough for both hackers and criminals to violate and beat the security system. The history of violating security systems has taught us that preserving security and privacy will inevitably be an ongoing effort.

10.8.4 Reliability

Reliability of the cable network is one concern facing the operators. Future interactive applications require a high degree of reliability from the network. Matching the reliability of the telephone network not only is a matter of developing and deploying appropriate technologies but also requires a change in the way cable operators perform maintenance to the network. The psychology of the telco is that the problem is detected before a subscriber complains. In the cable network, equipment is rarely monitored, and equipment failure always results in a service call. Standards and nonpropriety solutions (open interfaces) are ways to combat security and reliability problems in the cable system. The large MSOs are beginning to recognize the power of standardization and are working in the direction of improving their network reliability. They still have a long way to go. The RBOCs, when procuring products, always insist on open interfaces and must conform to the toughest national and international standards. Did such policy increase the cost? On the contrary, it opened the competition to a multitude of competent switch vendors and reduced the cost of the equipment immensely. IEEE 802.14's developing the cable modem is one example of standardization in action. All major vendors are participating and positively contributing for developing the cable modem. All these vendors will be competing in the cable modem market, and cable operators and the consumer will no doubt reap the benefit of the reduced cost.

Cable networks need to meet tough standards at the network level. A particular cable modem must work to deliver the same services transparently when a subscriber moves to another home. This may sound strange when compared to the telephone network.

For voice service, reliability becomes a real challenge to cable networks. Bellcore specifications require that a POTS line has a downtime of less than 28 minutes per year. In a mux environment, the downtime must be less that 53 minutes per year. Power outage should not be a factor. For a cable network, this is quite a challenging goal. This would require cable operators to provide backup power for equipment within their networks as well as backup power for terminal devices that need to work when the power fails, as the average unavailability for electric power in the United States is 370 minutes per year.

Like telephone operators, cable operators must adopt a sort of online monitoring of the network nodes so they can detect poor network performance before it fails.

10.8.5 Cryptography and Government Control

Strong cryptography is one major concern to all government intelligence agencies. In the hands of a criminal, effective cryptography can be used as a weapon of war. During wartime, strong cryptography can impede the ability to intercept and decipher essential enemy communications. Consequently, strong cryptography is usually classified on the U.S. Munitions List as an export-controlled commodity, just like tanks and missiles. Government encryption policy is very much influenced by the agencies responsible for gathering domestic and international intelligence (the FBI and the NSA, respectively). The government is forced to balance the conflicting requirements of making strong cryptography available for commercial purposes—while still making it possible for domestic agencies to break those codes, when needed.

The Department of State or the Department of Commerce is responsible for evaluating products containing cryptography. Technical assessment is performed by the Office of Export Control at the Department of Defense's National Security Agency. NSA Export Control officers work with companies in order to determine exportability. Products are considered exportable if they do not require an individual validated license prior to shipment to the customer. The algorithm and keysize of the encryption play a role in determining product exportability.

U.S. security vendors are understandably upset about the export regulation imposed on their software products. One concern is the slow response of the export applications they submit to various government agencies. Another concern is that, in this global economy, the U.S. advantage may disappear, as these regulations do not apply to companies worldwide. This gives them time to catch up with U.S. cryptography technology. Recently, however, mass-market software applications utilizing RSA at 512 bits or less were considered exportable and are eligible for special State Department reviews to expedite the process.

11

HFC and HSCM Standards

11.1 Overview

Standardizing a cable modem for some cable modem companies is kind of unique. In the past, the MSOs have formed partnerships with vendors and developed product that was unique, proprietary, and service-specific. Some MSOs believe that standards impede ingenuity. Although this is true to a certain extent, the advantages of developing a standard-based product far exceed that of deploying a proprietary solution. Sometimes a proprietary solution may become the default standard. However, in an environment where networks must be interoperable in a competitive industry, a standard solution has been proven to be the best and most cost-effective alternative. With a standard product, the economy of scale will further reduce the cost of operation and open competition to the benefit of both operators and consumers.

Standards should be very important to the cable operators if they are to compete head-on with the well-organized, requirement/standards-conscious, and technically savvy telco operators. Reliability and network management are aspects that have been plaguing cable operators for a long time. Some of these problems can be traced directly to lack of a cohesive body of standards. The mere absence of MSOs from standards bodies is also contributing to the dilemma of interoperability and ease of networking.

Organizations and/or vendors who promise quick specifications and fast turnaround are doing a disservice to the MSOs. With few exceptions, MSOs are not in the habit of attending standards organizations, and, unfortunately, they show little desire to change. CableLabs, which represents 85 percent of cable operators in North America, is doing an excellent job in the standards bodies and is providing valuable and critical inputs. However, direct participation from all MSOs is still essential, not only to provide their local input but also to protect their own interests in the technical or business environment. The technical issues related to HFC and the cable modem are very complex. An honest technical debate is needed to flush out political decision or special agendas.

In this chapter, each of the organizations that deals with cable modem standardization or specifications will be discussed. Details of their affiliation and charters will also be disclosed.

11.2 Organizations Affiliated with the Cable Modem

There are several standards organizations, forums, and associations dealing directly or indirectly with standardizing or specifying cable modems over HFC. Among the most active are:

1. IEEE 802.14

2. CableLabs

3. ATM Forum

4. SCTE

5. MCNS

6. DAVIC

Direct and indirect liaison between these organizations, to a great extent, is well-established, and a good communication flow is maintained. Despite some overlap in the charters, cooperation has been very good.

11.2.1 IEEE 802.14 Role/Charter and General Information

In this section, the role of IEEE 802.14 in the development of a cable modem standard will be described. IEEE 802 embraces that role because of its expertise in shared medium networks.

IEEE 802.14 AFFILIATION.   IEEE 802.14 is a working group under the umbrella of IEEE 802 (International Electrical and Electronic Engineers). IEEE is an international organization affiliated to the U.S. ANSI standard body and affiliated internationally through ISO/IIEC JTO.

IEEE 802.14 HISTORIC PERSPECTIVE.   Dr. Graham Campbell of Illinois Institute of Technology (IIT) officially presented DQRAP (distributed queuing random access protocol) in the MAN WG of IEEE P802.6. This official DQRAP presentation was submitted in the July 1993 Denver meeting of the IEEE 802.6. DQRAP was inspired by IEEE 802.6 work on DQDB and dealt with developing a MAC for a cable modem to provide interactive high-speed services. In the November 1993 meeting, a study group was formed to begin work on cable-TV-oriented protocols in a more formal way. The first official interim meeting of the study group took place in Denver on January 20 and 21, 1994. In that meeting, TCI was instrumental in encouraging the group to proceed and provided valuable input. The IEEE 802.catv study group was then created and worked feverishly to develop the project authorization request (PAR). The study-group charter was to complete the PAR (described below) for eventual submission to the IEEE standards board. Once approved, a study group will be elevated to become a legitimate IEEE 802 working group

On the evening of Thursday, July 14, 1994, the IEEE Project 802 executive committee unanimously approved the formation of a new working group to produce standards for two-way communication over cable TV and similar systems. NESCOM/ANSI confirmed it and bestowed on it the name IEEE 802.14. The official name of the IEEE 802.14 project was given as "Protocol Standard for Cable-TV-Based Broadband Communication."

IEEE 802.14 ATTENDANCE AND VOTING RULES. Attendance to IEEE 802.14 is open to all. A meeting fee is imposed upon anyone who attends any of the IEEE 802 working group meetings. A participant, after attending two meetings at least 75 percent of the meeting time, becomes a voting member on the third meeting. One of the three meetings must be a plenary meeting. IEEE 802 plenary meetings are held three times per year. Before voting on any issue, a quorum of 75 percent of voting members must be present. On technical matters, a motion will pass with 75 percent of the votes. On procedural issues, 50 percent of the vote will be needed to pass a motion.

IEEE 802.14 PAR. A PAR defines the scope and the charter of any of the IEEE 802 working groups. The IEEE 802.14 PAR summary is:

1. To provide for digital communication services over a branching bus system constructed from fiber optics and/or coaxial cable as used in the cable-TV distribution network. The traffic types will be CBR, VBR, CO, and CL data.

2. Specifying a MAC/PHY will provide the capability of interconnecting 802.2-compatible LANs as well as emerging technologies such as video compression, and ATM will be maintained insofar as it is consistent with efficiency and good economics. Coexistence with existing analog video transmission and with forthcoming digitally encoded television signals will be maintained. It is expected that the standard will allow for flexibility in the assignment of frequencies

for transmission to and from the headend. Both symmetric and asymmetric data flow will be provided for, and data rates consistent with the capabilities of the media with round-trip distances up to 160 km will be accommodated.

CRITERIA/RATIONALE FOR IEEE 802.14 PAR. The rationale for developing the cable modem standard was one of the papers produced by the IEEE.catv study group. The PAR was used to develop the requirement document for the cable modem. Some relevant excerpts of the five tests needed for PAR approval are worth stating.

Broad Market Potential. The potential for cable-based interactive networks is as broad as cable television itself. The field is moving rapidly toward the incorporation of a very high degree of computing power in the set-top box, which was formerly a simple tuner and descrambler. From broadcast entertainment, the industry is seeking to move out toward a variety of new services, ranging from interactive games to video phone calls to information services to movies on demand.

In the Cable TV Protocol Study Group, we have seen broad participation by vendors. The major manufacturers of set-top boxes have also been participants and, on the computer side, most of the significant manufacturers of workstations, whose technology is expected to migrate into the set-top box of the future. Participation by the telephone industry has also been significant, with major vendors of telecommunications equipment, service providers, and research laboratories in attendance.

Compatibility with IEEE 802 Architecture. The keynote of the IEEE 802 standards family has been the provision of protocols that permit the use of a common medium shared by many users. The proposed cable-TV work adheres very well to this paradigm—in fact more closely than much other work now being undertaken within Project 802.

For reasons of reliability, easier maintenance, or use of media unsuitable to shared usage, there has been an evolution of the LAN field toward nonshared media radiating out from a hub. While most such hubs run protocols designed for shared media, some do not—802.9 and 802.12, for instance.

Cable television, however, is very definitely a shared medium. Efforts to make high-bandwidth connections available to each home through dedicated media, either twisted pair or fiber, have failed either because of insufficient bandwidth or high cost. Cable TV, however, has been an economical solution in the context of broadcast entertainment, and it promises to be economical in the provision of interactive services as well.

Distinct Identity.   Any work done to provide a protocol optimized for multiple services over the cable TV system would not lack a distinct identity. Several protocols have already been standardized for this type of physical medium, but they are not ideal for this application. In particular, the protocol from 802.3 lacks distance capability as well as the ability to provide for fixed-bandwidth or guaranteed-bandwidth services. With a maximum distance of 3.6 kilometers, its distance capability is too short for cable TV systems, which often range up to 80 kilometers in one-way distance. Likewise, the 802.4 standard does not work well in the cable TV environment. Designed to be used within factories and industrial complexes, its logical token-passing protocol expects all units in the network to be powered up at all times. If the token arrives at a unit that is powered off, the sequence is broken, and time must be taken to reestablish the polling through a relatively complex process. Consumer systems can be expected to be powered off more often than they are powered on.

Technical Feasibility.   Presentations in the Cable TV Protocol Study Group have indicated that technical feasibility should be no problem. Much of the technology needed has been developed under the aegis of high-definition TV (HDTV), which will involve digital transmission of compressed video. For example, data rates as high as 43 Mbps have been demonstrated within the 6-MHz television channel, and it is likely that this system will be used for HDTV and conventional-resolution TV, with many video channels multiplexed into the 6-MHz channel.

Some equipment is already on the market, meeting some or all of the service objectives laid out by 802.catv. Vendors such as Zenith and DEC are now marketing systems for sending data over cable TV systems. In addition, new protocols have been proposed that, while not yet implemented in silicon, perform very well in simulations that should be accurate indicators of how well they will run when actually constructed.

Economic Feasibility.   The convergence of the computer industry with the television and telephone industries is leading to set-top boxes with the computing power of the most advanced workstations. In addition, the need for video compression and decompression (or at least decompression) leads to extensive hardware logic with a capability in the hundreds of millions of operations per second. Predictions are that these boxes will have a retail price of only a few hundred dollars. In the midst of all this, the hardware and software needed to operate a MAC protocol is essentially lost in the noise. The physical layer is needed in any case, with the economics of scale likely to be tilted toward RF modem designs capable of running up to 43 megabits per second on the receiver side. Transmission on the reverse channel may not require high rates, but it is clear that the economies of mass production are likely to be available in this case also, and that one can safely undertake development of standards involving the transmission of 10 megabits per second and likely considerably more.

IEEE 802.14 SWGs. IEEE 802.14 members created three official subworking groups. They are:

1. Physical-Layer (PHY) Subworking Group (SWG)
2. MAC-Layer Subworking Group
3. Architecture Subworking Group

PHY-LAYER SWG. The PHY SWG was delegated to develop the standard for the cable modem physical layer. The charter of this subgroup is to develop the modulation coding scheme, which must also be optimized in terms of bit rate, efficiency, bit error rate, and availability to support the tree-branching multiple-access MAC over long distances for a large number of users. The PHY must be architected to support improvement in transmission technologies for the forward and reverse channels. An advanced PHY study group was also established to look into the long-term PHY technology advances, specifically in the DMT modulation technology where transmitters for the upstream scale to the needed bandwidth.

MAC-LAYER SWG. The MAC SWG was delegated to develop the standard for the cable modem MAC layer. The MAC layer is used to arbitrate transmissions on a shared medium accessed by multiple DTEs using a set of rules that are followed by each node. The cable-TV-network-based distributed environment, with forward and reverse channels, long propagation delays, and support of multiple types of services (data, voice, image, and video), makes the design of the MAC layer protocol a great challenge.

ARCHITECTURE SWG. The architecture SWG was tasked to oversee the linking, interworking, and functional interfaces between the MAC and PHY SWG. The SWG also developed the OSI reference model of the cable modem and its layering service architecture above the MAC and PHY. One of its main functions was also to ensure compliance with the requirement document created by IEEE 802.14 WG.

IEEE 802.14 WORKPLAN. As of November 1996, the WG agreed and/or committed to the following workplan to release the cable modem recommendations:

1. Initial draft	November '96
Fill holes, build toward 75 percent support	
2. Working group draft ready for WG ballot	July '97
Resolve negative votes	
Confirmation ballot	September '97
Resolve negative votes	
Further confirmation ballots, if needed	October/November '97
3. Sponsor ballot (formerly known as TCCC ballot)	January '98
Resolution of negative votes and confirmation	March through May '98
4. Forward to IEEE Standards Board	June '98

Suggested Comments

Manufacturers could begin ASIC design in ...	November '96
Model, refine, and begin software development and higher layer support in ...	July '97
Trials can begin about ...	January '98
First products appear for network operators in ...	March '98
Products available to consumers in ...	June '98

IEEE 802.14 WG STATISTICS AND GENERAL INFORMATION.
The following are statistics of the IEEE 802.14 WG meetings and contact for obtaining minutes or draft recommendations.

IEEE 802.14 meets three times a year with a possible additional four interim meetings per year. As of March 1997, about 500 technical contributions were submitted and discussed (from its inception).

- Average attendance per meeting during 1996 was about 110 participants, with a distribution list of over 400 representing 180 organizations.
- There are 154 voting members as of November 1996.
- All major vendors of cable modems and set-top boxes (STBs), including most major telco switch vendors, participate and contribute to the work.
- Recently, CableLabs and some MSOs regularly attended the meetings and gave technical and business direction to the WG.

IEEE 802.14 created a Web site (with its name) containing information relevant to the working group.

The official contact address to obtain standard documents from the IEEE office is:

445 Hoes Lane
P.O. Box 1331
Piscataway, NJ 08855-1331
908-562-33800
908-562-1571 (fax)

11.2.2 CableLabs

CableLabs was created in 1988. It is an R&D consortium sponsored by the North American CATV operators. The organization indirectly represents 85 percent of CATV subscribers in the United States, 70 percent of CATV subscribers in Canada, and 10 percent of CATV subscribers in Mexico. The staff and the facility are located in Boulder, Colorado.

CABLELABS' ORGANIZATION AND MISSIONS. CableLabs performs R&D projects for specific members or groups of members and also transfers relevant technology to its members. Presently, CableLabs is collaborating with the computer industry to develop HFC digital cable networks. Another function is to provide publications to MSO members only. However, some publications are accessible if CableLabs considers their distribution to the community of vendors beneficial to the cable business. Newsletters are published periodically. CableLabs organization is built around the following operational entities:

- *Advanced TV.* In charge of new TV techniques (video compression, HDTV)
- *Operations Technologies Projects.* In charge of all operations-related issues in existing CATV networks (improving connectors, etc.)
- *Network Architecture Design and Development.* In charge of designing new architectures (PCS, SONET hubs, etc.)
- *Clearing House.* In charge of conferences, publications, and external communication (Congressional and state government, public relations, etc.)

CABLELABS AFFILIATION. Like Bellcore, CableLabs is not affiliated with the American National Standards Institute (ANSI). Hence, it is not the mission of CableLabs to generate standards, but vendors might use CableLabs to check compliance of technical approaches or standards with the CATV market.

CABLELABS' NEAR-TERM DIRECTION. CableLabs' direction is set by the consortium members. The following priorities may not be in the same order for all members, but they do represent the consensus:

■ *Increased channel capacity.* The current average system supports 35 channels. Increased channel capacity is a defensive move to protect markets from DBS and telco erosion. Increased capacity will be achieved through video compression and digital modulation techniques as well as equipment upgrades that will increase bandwidth from the current state-of-the-art 550-MHz systems to future 1-GHz systems.

■ *Reliability of services.* Focus on solving customer complaints faster and upgrading the network to protect against power outages (backup AC power). Also looking into updating components (connectors, etc.) to provide more reliable service.

■ *HDTV.* CableLabs cooperates with the FCC-sponsored Advanced TV Test Center in Alexandria, VA, which is in charge of testing HDTV proponents systems for simulcast broadcasting. CableLabs has positively contributed to the technical community in terms of alignment of the HFC modulation technique with this project.

■ *New services*
High-speed Internet access
Video store to the home (VOD)
Personal communication services (PCS)
Computer program exchange
Interactive educational services

To achieve its goals, the Network Architecture and Design and Development Group is focusing on the regional hub concept. All headends would be fed from the hubs. The hubs would employ ring architectures with dual homing, and the hub nodes would be duplicated. This architecture (when deployed) will also serve as the platform for new service offerings.

CABLELABS CATV TEST MEASUREMENT CONTRIBUTION. CableLabs performed extensive measurements on several North American CATV networks, including Canada (Rogers). These measurements of the embedded cable plant (including house wiring) were used to determine the appropriate modulation schemes in that noise environment that will be needed for both the upstream and downstream physical layers.

CABLELABS HFC/TV-RELATED DOCUMENTATIONS. CableLabs recently released an RFP on HFC-related products. The specification was very instrumental in setting the direction of IEEE 802.14 WG when developing the cable modem standard. Other technical material that may be of interest or relevance to cable modem standards authored by CableLabs includes:

- *An Overview of Cable Television in the United States* (Walter S. Ciciora)
- *Cable's Role in the "Information Superhighway"—A Position Paper*
- *Digital Transmission Characterization of Cable Television Systems/SPECS Technology*
- *Regional Hub Field Test Proposal/SPECS Technology* (February–March 1993)
- *What Is a Cable Company/SPECS Technology* (July–August 1993)
- *News from CableLabs—SPECS—R&D Spotlight* (July–August 1993)

11.2.3 ATM Forum

The ATM Forum is an international nonprofit organization created with the objective of accelerating the use of ATM products and services through a rapid convergence of interoperability specifications. In addition, the ATM Forum promotes industry cooperation and awareness. Since its formation in 1991, the ATM Forum has created a very strong interest within the communications industry. Currently, the ATM Forum consists of over 700 member companies, and it remains open to any organization that is interested in accelerating the availability of ATM-based solutions.

The ATM Forum consists of a single worldwide Technical Committee, three marketing committees for North America, Europe, and Asia-Pacific, and the Enterprise Network Roundtable (ATM users group) through which ATM end users participate. The marketing group is very active in market research and recently commissioned a marketing study on ATM residential broadband market requirements.

ATM FORUM AFFILIATION. The ATM Forum is not affiliated with any standards organization. There are several MOUs established, such as ITU, IETF, and others. The forum works closely with the T1S1.5 (T1 committee/ANSI-affiliated) to establish the U.S. position and forward it to ITU. Liaisons are established with almost all other technical organizations.

RESIDENTIAL BROADBAND WORKING GROUP. The RBB working group was founded by the ATM Forum in February 1995. The RBB WG is chartered to bring ATM to the home and define a complete end-to-end ATM system specification, both to and from the home to a variety of devices like set-top boxes and PCs. Here, the emphasis is on defining and specifying the home UNI and access network interface (ANI).

The RBB WG is planning to adopt IEEE 802.14 PHY and MAC standards when available. The reference model of the access portion was generically developed to accommodate HFC, FTTC, FTTH, and so on. The RBB group is also working on specifying requirements of the core network and working with ETSI and ITU to coordinate requirements. Work has been started on security requirements on end-to-end HFC. This work is well-coordinated with IEEE 802.14 for wiring inside the home (home appliance interface). There are three proposals now under consideration:

- PCIA
- DAVIC interface
- IEEE 1394

ATM FORUM WG/STATISTICS/GENERAL INFORMATION. The ATM Forum is open to all. A yearly fee is imposed to become a principal voting member. An observer-status membership is also available with no voting right. Each company (principal member) or organization has one vote.

General statistics are as follows:

- As of April 1997, there was a total of 761 members: 206 principals, 398 observers, and 157 users.
- There is one worldwide technical committee.
- There are five full technical committee meetings per year.
- Typically 300 contributions are submitted per meeting.
- Approximately 650 participate per meeting.
- Seventy specs were approved, and thirty-four are in the works.

OTHER WORKING GROUPS OF THE ATM FORUM. The ATM Forum consists of 14 technical working groups, a marketing working group, and a user group. It is as shown in Fig. 11.1.

A summary of current activities is as described below:

Figure 11.1
ATM WG
organization.

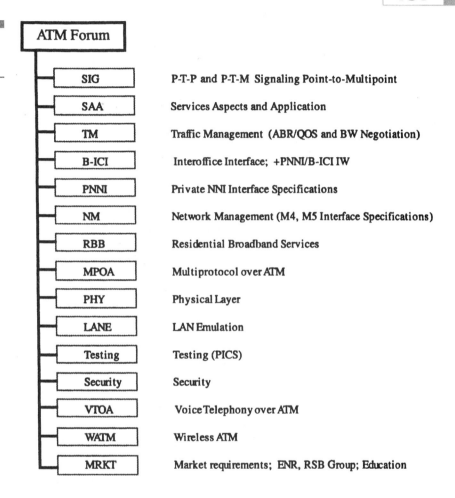

ATM Forum	
SIG	P-T-P and P-T-M Signaling Point-to-Multipoint
SAA	Services Aspects and Application
TM	Traffic Management (ABR/QOS and BW Negotiation)
B-ICI	Interoffice Interface; +PNNI/B-ICI IW
PNNI	Private NNI Interface Specifications
NM	Network Management (M4, M5 Interface Specifications)
RBB	Residential Broadband Services
MPOA	Multiprotocol over ATM
PHY	Physical Layer
LANE	LAN Emulation
Testing	Testing (PICS)
Security	Security
VTOA	Voice Telephony over ATM
WATM	Wireless ATM
MRKT	Market requirements; ENR, RSB Group; Education

Signaling. Defines procedures for call setup and negotiation of quality of service (QOS) for an end user connection. Specification will benefit users by enabling switched virtual circuits (SVCs), which are more flexible than permanent virtual circuits (PVCs).

Service aspects and applications (SAA). Specifies services such as application programming interfaces (APIs) and interworking with frame relay, SMDS, and circuit emulation services (T1/E1, Fractional T1/E1).

Traffic management. Provides specifications for traffic control to limit congestion and maximize bandwidth use. Starts work on UBR with MCR.

Broadband ISDN intercarrier interface (B-ICI). Defines interswitch communications in public networks. Presently, the group is working on interworking between B-ICI and PNNI.

Private network-to-node interface (P-NNI). Specifies the protocol by which ATM switches communicate within a private ATM network. The specification will enable multiprotocol-switched private networks.

Network management. Specifies the protocol MIB for each NM interface in the ATM network. Previously, NM WG identified and developed all NM aspects of the public and private ATM switches.

Multiprotocol over ATM (MPOA). Enables layer-3 protocols such as IP, IPX, and CLNP to operate directly over ATM, both between ATM-attached hosts and with hosts attached to other network technologies.

Physical layer. Defines the physical and electro-optical characteristics for interfaces and signals. Defines ways to place ATM cells into transmission frame structures of specific physical-layer data streams.

LAN emulation. Enables existing LANs to communicate with similar LANs and with ATM-attached stations over legacy protocols.

Testing. Establishes test suites to evaluate conformance, interoperability, and performance of implementations of the ATM Forum specifications.

Security. Develops requirements and information-flow specifications for ATM security for all aspects of security of ATM applied to user information, signaling information, and management information.

VTOA. Voice telephony over ATM; this group defines AAL-CU (AAL-2), a new AAL where voice samples are multiplexed inside the ATM payload.

WATM. Specifies requirements for wireless over ATM; expected to use available standards but will specify the air interfaces or the components not available elsewhere.

ATM FORUM MARKETING WG. The ATM Forum established three marketing working groups in North America, Europe, and Asia/Pacific. Joint meetings between them are regular. Charters of the North American marketing working group are:

■ Provide marketing and educational services designed to speed the understanding and acceptance of ATM technology.

■ Coordinate development of educational presentation modules and technology papers and facilitate exchange of information and requirements between the Enterprise Network. The Enterprise Network Roundtable, formed in 1993, consists of ATM end users. This group interacts regularly with the marketing group and provides valuable feedback.

11.2.4 Society of Cable Telecommunications Engineers (SCTE)

In August 1995, the Society of Cable Telecommunications Engineers was created to tackle interoperability and develop the appropriate interfaces. SCTE hopes to use IEEE 802.14 MAC and PHY when available, but they are going beyond the cable modem standards and dealing with interoperability specifications of HFC systems in particular. SCTE is open to the public. Both MSOs and vendors are members of that organization.

The SCTE charter reads: "Explore the need for SCTE involvement in the development of standards for digital video signal delivery through coordination of efforts with NCTA, the FCC, and other related organizations."

SCTE SUBWORKING GROUPS. There are seven SWGs, each dealing with specific topics. The SWG charters are as shown below:

Digital video subcommittee. The charter of this SWG is to develop standards for digital television services delivered by cable, including video and audio services, data and transport applications, network architecture and management, transmission and distribution, and encryption and access control.

Data standards subcommittee. The charter of this SWG is to advance the cable industry's interest in high-speed data delivery and develop standards for hardware interoperability.

Emergency alert systems subcommittee. The charter of this SWG is to develop filings to the Federal Communications Commission Office of the Emergency Alert Systems. The subcommittee will inform the FCC about cable's role in this country's Emergency Alert System, focusing on the industry's capabilities, limitations, and cost-effectiveness.

Maintenance practices and procedures subcommittee. This SWG is chartered to develop recommended practices for the proper maintenance and operation of cable television systems.

Interface practices and in-home cabling subcommittee. This SWG approved 69 standards for specifications and test procedures by midyear, and it has 12 standards in the proposed stage.

Design and construction subcommittee. This SWG is chartered to develop the coaxial basic construction and the fiber optics construction manuals. The SWG is also responsible for cost analysis of upgrade and rebuild.

Material management/inventory subcommittee. This SWG is working on automated material management and inventory management with real-time computerized control. This SWG is also responsible for the bar code system for cable television product packaging and shipping as recommended in the MMI IP 001 standard.

11.2.5 Multimedia Cable Network System (MCNS)

MCNS is an organization preparing a series of interface specifications for early design, development, and deployment of data-over-cable systems. The specifications are open, nonproprietary, and multivendor-interoperable. MCNS contracted with Arthur D. Little, Inc., to develop these interface specifications.

MCNS comprises four leading cable television operators:

1. Comcast Cable Communications, Inc.

2. Cox Communications

3. Tele-Communications, Inc.

4. Time Warner Cable

Partners of MCNS are:

- Rogers Cablesystems Limited
- Continental Cablevision
- Cable Television Laboratories

PARTICIPATION. Interested vendors are free to participate and contribute without fee or any other restrictions. Participants, however, must execute the information access agreement (essentially a nondisclosure and hold-harmless agreement).

MCNS GOAL. The objective of MCNS is to have accelerated specifications for developing a cable modem. The goal is to eliminate intellectual property, if possible, and simplify and minimize the costs of developing a cable modem. This facilitates manufacture of compatible equipment by multiple vendors. The protocols should be developed to support future upgrades and changes by negotiation at the session establishment of physical and higher-layer protocols. The specification must specify the protocols and algorithms to achieve seamless and reliable operation of the cable modems under all possible conditions, including robust fault recovery mechanisms and scalability of the algorithms. MCNS will consider the work done by IEEE 802.14 and others, and use the work and/or contributions to the extent that they are useful.

Furthermore, MCNS will attempt to structure the specifications it develops so that attractive work products emerging from other ongoing efforts can be easily incorporated into future revisions.

DATA-OVER-CABLE INTERFACE SPECIFICATION (DOCIS) PROJECT. The DOCIS project was prepared by Arthur D. Little, Inc., on behalf of cable industry system operators that provide service to a majority of cable subscribers in North America. The goal of the DOCIS project is to develop rapidly on behalf of the North American cable industry a set of communications and operations support interface specifications for cable modems and associated equipment over HFC. A reference model for a data-over-cable system has been established that includes data communications elements as well as the needed operations and business support elements such as security, configuration, performance, network fault, and accounting management.

PROJECT SCOPE AND TIMELINE. The scope of this project is described in the MCNS request for proposals. In summary, the project entails the development of interfaces between system elements organized into three phases:

PHASE 1. Phase 1 deals with the data communications aspects of the data-over-cable system at the subscriber location and the headend. Timeline for phase-1 specification release was July 1, 1996. Phase-1 interfaces include:

- The cable modem to CPE interface (typically a personal computer).
- The headend-based cable modem termination system interface. It would typically be connected to a backbone network of a server complex (e.g., local content servers; a gateway server to the Internet; and servers for security, spectrum management, network management, accounting, configuration, etc.).

PHASE 2. Phase-2 interfaces deal with the operations support system and telephone return interfaces. The timeline for phase-2 specification release was September 1, 1996. The interfaces are:

- The data-over-cable system operations support-system interfaces are used to support the management of the network, performance, security, accounting, provisioning, spectrum management, customer service, and network faults.

- Cable modem telco return interface is an interface between the cable modem and the public-switched telephone network.

PHASE 3. Phase-3 interfaces deal with the RF communications path over the HFC cable system between the cable modem and the cable modem termination system. The timeline for Phase-3 specification release was December 1, 1996. The interfaces are:

- The cable modem to RF interface is a single-port RF interface between the cable modem and the cable system coaxial drop. The specification includes all the physical, link (MAC and LLC), and network-level aspects of the communications interface. This includes RF levels and frequency, modulation, coding, multiplex, contention control, frequency agility, and so on.

- The cable modem termination-system downstream RF side interface and the cable modem termination-system upstream RF side interface are two separate ports on the cable modem termination-system equipment at the headend.

- Cable modem termination-system security management interface.

MCNS CONTACTS

Arthur D. Little, Inc.
Acorn Park, Bldg. 15-245
Cambridge, MA 02140
Tel: 617-498-5000
Fax: 617-498-7244

OTHER DATA-OVER-CABLE-RELATED DOCUMENTS/CONTACTS

Data Over Cable Service Interface Specification Project

Multimedia Cable Network System Partners (MCNS Holdings L.P.)

Comcast Cable Communications, 1500 Market Street, Suite 3400, Philadelphia, PA 19102

Cox Communications, 1400 Lake Hern Drive, NE Atlanta, GA 30319

TCI Technology Ventures, Inc., Terrace Tower II, 5619 DTC Parkway, Englewood, CO 80111

Time Warner Cable, 160 Inverness Drive, West Englewood, CO 80112

Cable Television Laboratories, Inc., 400 Centennial Parkway, Louisville, CO 80027

Continental Cablevision, Inc., The Pilot House, Lewis Wharf, Boston, MA 02110

Rogers Cablesystems Limited, 853 York Mills Road, Don Mills, Ontario M3B 1Y2 Canada

Data-over-Cable Service Interface Specification Project Manager and Consultants

Arthur D. Little, Inc., Acorn Park, Cambridge, MA 02140

11.2.6 DAVIC

DAVIC (Digital Audio Visual Council) is a nonprofit association established in Switzerland. DAVIC's goal is to be the home for all those who see the potential of digital technologies applied to audio and video. As such, there is no constituency because the range of industries represented in DAVIC covers content, service, and manufacturing companies. Total membership is 218 companies/organizations worldwide.

Membership in DAVIC is open to any corporation and individual firm, partnership, governmental body, or international organization. Associate member status is available to those who do want to take an active role in the precise technical content of the specifications.

DAVIC GOAL.   The purpose of DAVIC is to favor the success of emerging digital audiovisual applications and services by the timely availability of internationally ratified specifications of open interfaces and protocols that maximize interoperability across countries and applications/services.

The primary instrument is the production of technical specifications that maximize interoperability across countries and applications/services. DAVIC specifications are issued in the following versions: DAVIC 1.0, DAVIC 1.1, DAVIC 1.2, and so on. The DAVIC 1.0 version of the specifications allows the deployment of systems that support initial applications such as:

- TV distribution
- Near video on demand
- Video on demand
- Some basic forms of teleshopping

Future versions will expand on previous versions to provide more functionalities while keeping, as far as possible, backwards compatibility with previous versions.

DAVIC TECHNICAL COMMITTEE ORGANIZATION.   DAVIC established six technical committees (TC). They are:

1. Applications

2. Information representation

3. Subsystems

4. Physical layer

5. Systems integration

6. Security

A TC produces successive revisions of the DAVIC specification parts assigned to it. A TC chair will usually assign sufficient time to clarify an issue and will resort to membership voting when consensus cannot be reached. Specifications are drawn from existing standards, if available; however, each TC makes its own decision to balance other criteria such as technical merit, cost, and wide usage.

DAVIC WORKPLAN/TIMELINE.   The DAVIC workplan of relevance in the HFC environment is:

- Modulation and associated technology to transmit digital audiovisual information over terrestrial UHF and VHF TV channels
- Modulation and associated technology to transmit digital audiovisual information over 28–40 GHz (LMDS)
- High-bandwidth symmetrical digital connectivity
- Telephony
- Audioconferencing
- Video telephony
- Integrated videoconferencing

DAVIC systems offer integrated digital networking for telephony, video telephony, videoconferencing, and high-speed data interchange. This will allow the DAVIC infrastructure to be used for new residential, home office, and small business services. Access networks could be designed to take advantage of the synergy between entertainment TV and emerging telecom-based services.

TIMELINE OF DAVIC PUBLICATIONS

Publication of DAVIC 1.0	December, 1995
Call for proposal 4 issued	December, 1995
Deadlines for responses to CFP-4	February, 1996
Call for proposal 5 issued	March, 1996
Publication of DAVIC 1.1	September, 1996
Publication of DAVIC 1.2	December, 1996
Publication of DAVIC 1.3	December, 1997

DAVIC 1.0, 1.1, 1.2, AND 1.3. DAVIC 1.0 is organized into 11 parts.

1. Description of DAVIC system functions
2. System reference models and scenarios
3. Service provider system architecture and interfaces
4. Delivery system architecture and interfaces (HFC)
5. Service consumer system architecture and interfaces
6. High-layer and midlayer protocols
7. Lower-layer protocols and physical interfaces
8. Information representation
9. Security
10. Usage information protocols
11. Dynamics, reference points, and interfaces

DAVIC 1.1 will add more functionality to DAVIC 1.0. Some of the most important subwork items are:

- Switched video broadcasting
- MMDS
- Cable modem
- PSTN/ISDN enhanced broadcast
- Internet access
- Software download protocols

DAVIC 1.2 will add more functionality to DAVIC 1.1. Some of the most important subwork items are:

- Communications API for Internet access
- ADSL ATM mapping

- Part 10: basic security for DAVIC 1.0 systems
- Guidelines for Internet access
- Multiple STUs in the home

DAVIC 1.3 will add more functionality to DAVIC 1.2. Some of the most important subwork items are:

- Communicative services (telephony, conferencing, and multiplayer games)
- Home network
- Network-related control
- Multiple server and services
- Multicast technologies
- Extended JAVA functionality

Chapter

12

The Cable Modem
vs. ADSL

12.1 Overview

Today's network and infrastructure is unable to cope with the demand put upon it in terms of the new broadband services and bandwidth-intensive applications. The Telecommunication Acts of 1996 coupled with the explosion of interactive services is all the incentive required of the any operator to want to compete in the upcoming multibillion-dollar market. The two competing giants are the RBOCs and MSOs. The RBOCs (the $100 billion access market) are almost five times the size of its $22 billion cable competitors. They are determined to spend what it takes to deploy these new broadband services based upon infrastructures such as HFC or fiber-to-the-curb (FTTC) architectures.

Both the RBOCs and the MSOs have a unique infrastructure in place in both landline base and switching equipment to manage their traditional services. Leveraging on the landline infrastructure to carry broadband and interactive services is, presently, the most technically effective and rational solution and meets the time-to-market criteria. For the RBOCs, copper wire to the home is ubiquitous, so an ADSL (asymmetric digital subscriber line) modem makes business sense. For the MSOs, coax cable is the physical medium RF landline infrastructure. Therefore, the most optimum solution is to deploy cable modems.

There is every reason to believe that the consumer will enjoy this upcoming battle and reap the benefit of a true spirited competition in action. We can expect a cable or a telco operator in the near future to lure us in by offering attractive packages and throwing in POTS access charges for a mere few dollars. This may be wishful thinking, but we at least hope it will be similar to the present situation that makes customers feel wanted when negotiating with Sprint, AT&T, MCI, and now the RBOCs for long distance service.

This chapter will explore the following:

- Services under consideration by the RBOCs and MSOs
- ADSL technology overview
- Cable modem brief overview
- ADSL cable modem comparison (analyze the strengths/weaknesses of each solution)
- Who are the likely winners
- Alternate network architecture

12.2 Likely Applications

It is no secret that a battle between the RBOCs and MSOs is brewing. The two different cultures are bidding for the same business. The RBOCs are conservative, long-term planners that insist on open interfaces and product reliability and fully support and shape the standards. The MSOs are more focused on quick returns to meet their short-term business goals. To the MSOs, standards, open interfaces, and reliability are nice, but they are usually in the way. The MSO attitude on standards may be changing because MSOs are now cognizant that the era of multiservice multiprovider is a reality, and they must become more receptive to customer service and to the new competitive marketplace.

The key to survival in this multiprovider market is differentiation. The services that will be offered by both MSOs and RBOCs are likely to be:

- High-speed data services (Internet access)
- Broadcast, one-way entertainment
- Telephony (wireline and wireless)
- Digital NVOD or VOD
- Work at home
- Home shopping

Convergence of these services permeating across the millions of U.S. households will change the way people live, work, and play. By-products of this convergence will be a major source of growth and opportunity for the U.S. economy.

The major differentiators, as mentioned earlier, are the following:

- Time to market
- Quality of service

Although time to market is important in the short term, it will not be the ultimate deciding factor. It is only important if two other factors are satisfactorily met: customer service and cost. If the RBOCs were able to lure a premium customer from the cable company, it becomes extremely difficult for him or her to switch back if the price and service is acceptable. The RBOCs enjoy very good customer service reputation. That in itself will be a marketing tool they will use to go fearlessly after the cable customers who are primed to make the switch. When it comes to protecting their legacy market or in offering the new multimedia services, the RBOCs are more apprehensive of the IXCs like AT&T, MCI, and Sprint. The IXCs have an excellent record in customer service, billing, and, most of all, because of the stiff competition in long distance, they are well experienced in marketing their product and services.

Having noted all of the above, the battle is by no means over for the MSOs. In fact, it is just starting. Cable operators are, of course, aware of this wake-up call, and they are working feverishly to regain customer loyalty. The road may be rough, but they have been known to survive.

12.3 ADSL Technology Overview

In this section, we will describe ADSL technology with enough details to make the appropriate technical comparison with its cable modem rival.

12.3.1 Characteristics of ADSL

ADSL was originally conceived at Bellcore. AMATI and others cultivated the idea further and developed the DMT (discrete multitone) concept. Digital-subscriber-loop techniques allow, through the use of digital signal processing, to transport high-rate services over the existing copper-based infrastructure. The state of the art allows the transport of signals of a few Mbit/s over some kfeet of twisted pair cables while assuring a BER on the order of 10^{-7} or better.

The most important feature of ADSL is that it can provide high-speed digital services on the existing twisted-pair copper network in overlay without interfering with the traditional analog telephone service (plain old telephone service: POTS). ADSL thus allows subscribers to retain the (analog) services to which they have already subscribed. Moreover, due to its highly efficient line-coding technique, ADSL supports new broadband services on a single twisted pair. As a result, new services like high-speed Internet and online access, telecommuting (working at home), VOD, and others, can be offered to every residential telephone subscriber. The technology is also largely independent of the characteristics of the twisted pair on which it is used, thereby avoiding cumbersome pair selection and enabling it to be applied universally, virtually regardless of the actual parameters of the local loop.

The asymmetric bandwidth characteristics offered by the ADSL technology (64–640 kbit/s upstream, 500 kbit/s–8 Mbit/s downstream) fit in with the requirements of client-server applications such as WWW access, remote LAN access, VOD, and so on, where the client typically receives much more information from the server than he or she is able to generate. A minimum bandwidth of 64–200 kbit/s upstream guarantees excellent end-to-end performance, also for TCP-IP applications. These basic characteristics are reflected in two important advantages of the ADSL technology:

- No trench diggers are required for laying new cables, making it an optimal solution in advance of fiber deployment in the local loop.
- ADSL can be introduced on a per-user basis; this is important to the network operators, for it means that their investment in ADSL is proportional to the user acceptance of high-speed multimedia services.

The mature ADSL product combines the benefits of the DMT and ATM technologies, resulting in:

- Full bandwidth flexibility: upstream and downstream bit rates can be chosen freely and continuously up to the maximum physical limits. At initialization, the system automatically calculates the maximum possible bit rate, with a predetermined margin. The service management system can then set the bit rate to the level determined by the customer service profile, thus maximizing noise margin and/or minimizing transmit power.
- Full service flexibility: a random mix of services with various bit rates and various traffic requirements (guaranteed bandwidth, bursty services) can be supported within the available bit rate limits.

12.3.2 ADSL: A Technology on Unconditional Twisted Pair

A copper twisted-pair cable is the basic access connection of a telephone user to the telephone network. The frequency band used for the transport of the analog signal ranges from 300 Hz to 3.4 kHz. Thanks to the use of digital signal-processing techniques, digital subscriber loop (DSL) techniques are able to transport high bit rates over the existing copper twisted line simultaneously with the analog POTS. The basics of ADSL are also valid for other DSL systems that are currently available or planned, such as HDSL (high-speed DSL), SDSL (single-pair HDSL), and VDSL (very-high-speed DSL). HDSL offers 1.5 or 2 Mbit/s bidirectional data transport over two or three twisted pairs and is intended as a mere replacement for T1/E1 repeater lines in the distribution plant. Therefore, it is often referred to as repeaterless T1/E1. SDSL is a single-pair version of HDSL that targets to transport 1.5 Mbit/s bidirectional data over the full CSA (carrier serving area) range. SDSL can also be considered as symmetrical ADSL. The latter system (VDSL) is intended to provide very-high-bit-rate services, >10 Mbit/s over distances not to exceed 3 kfeet.

The ADSL transmission system offers an asymmetric capacity to the residential subscriber. In the direction of the subscriber, the ADSL system provides a capacity up to 6.1 Mbit/s (and above up to 9 Mbit/s). The bandwidth in the opposite direction (upstream) is in the range from 16 to 640 kbit/s. In general, the ADSL data rate is a function of reach and wire gauge as shown in Table 12.1.

The transmission channel capacity depends on the twisted pair characteristics and suffers from a number of impairments. The frequency-dependent attenuation and dispersion leads to intersymbol interference. Moreover, as coupling exists between wire pairs in the same binder or adjacent binder groups, crosstalk limits the transmission capacity of the copper loop. Furthermore, some subscriber loops have open-circuited wire pairs tapped onto the main wire pair, called bridged taps. The existence of bridged taps in the loop plant differs from country to country and depends upon the cabling rules used in the past. Their presence causes reflections and affects the frequency response of the cable, leading to pulse distortion and intersymbol interference. A loop can also be built up of wires with different diameters leading to reflections and distortion. Further, copper transmission suffers from impulse noise that is characterized by high amplitude bursts of noise with a duration of a few microseconds to hundreds of microseconds. It can be caused by a variety of sources such as central office switching transients, dial pulses, and lightning. Also, ingress noise from radio transmitters (e.g., AM radio stations) affects the transmission. Lastly, the impedance mismatch between the hybrid transformer, responsible for the split between transmitter and receiver, and the line impedance causes unwanted reflections. This problem can be resolved either by echo cancellation or by separation of

TABLE 12.1

ADSL Line Capacity

Reach	Downstream data rate	Wire size (AWG)
18,000 ft	1.5 Mbit/s	24
13,500 ft	1.5 Mbit/s	26
12,000 ft	6.1 Mbit/s	24
9,000 ft	6.1 Mbit/s	26

upstream and downstream transmission by means of frequency division multiplexing (FDM). To encompass the cited imperfections of the copper twisted pair, a highly adaptive transmission system is needed.

The ANSI T1E1.4 standard committee has selected discrete multitone (DMT) as the line code to be used in the ADSL transmission system. There are those in T1E1.4 who are trying to standardize yet other line codes. This is unfortunate because it would confuse the market and the consumer in particular. This attempt is contrary to the spirit of standardization, especially when the economy of scale is expected to play a role in reducing its cost. Moreover, multiple line codes will limit the choice a consumer would have if he or she opted to switch service providers. This assumes, of course, that they own a specific coding of the ADSL modem.

DMT modulation consists of a number of subchannels referred to as *tones*, each of which is QAM-modulated on a separate carrier. The carrier frequencies are multiples of some basic frequency (4.3125 kHz). The available spectrum ranges from about 20 kHz to 1.104 MHz. The lowest carriers are not modulated to avoid interference with POTS. The bandwidth used for upstream transmission is considerably lower than that used in the opposite direction. The upstream and downstream spectra can overlap if echo canceling is used. The alternative is the use of frequency division multiplexing (FDM), in which case no tones are shared by upstream and downstream transmission. Both alternatives are accepted by the ANSI T1E1.4 committee. The latter case is depicted in Fig. 12.1.

The transmit-power spectrum is flat over all used tones. The number of bits assigned to a tone is determined during an initialization phase as a function of the transmission characteristics as well as the desired bit rate. During operation, adaptation of this bit assignment is possible to compensate for variations in line conditions. In order to improve the bit error rate in the ADSL system, forward error correction (FEC) is applied. ANSI specifies the use of Reed-Solomon coding combined with interleaving. The additional use of Trellis coding is optional but may further reduce the BER or increase the SNR (signal to noise ratio) margin of the system for a given BER.

In the ANSI standard, special attention has been given to a service-specific interface, referred to as the channelized interface. This interface is based on a specific framing structure that can provide a combination of ADSL bearer channels envisioned up to now. In the current version, seven bearer channels can be transported simultaneously over ADSL: up to four downstream simplex bearers and up to three duplex bearers. The three duplex bearers could alternatively be configured as independent unidirectional simplex bearers, and the rates of the bearers in the two direc-

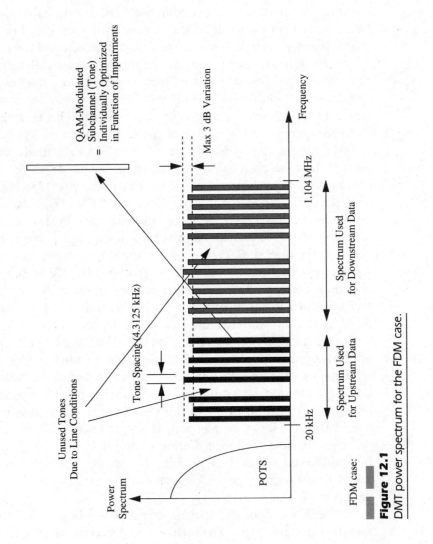

Figure 12.1
DMT power spectrum for the FDM case.

474

tions (network towards customer and vice versa) do not need to match. The bearer channel data rates can be programmed with a granularity of 32 kbit/s. Other data rates (noninteger multiples of 32 kbit/s) can also be supported but will be limited by the ADSL system's available capacity for synchronization.

12.3.3 ATM over ADSL

The variable bit rate inherent to ADSL ensures, in principle, the flexibility to support new services with other bit rates than those envisaged presently and takes advantage of the decrease of the required bit rate. This holds for video services, especially with MPEG2; images can be offered with reasonable quality at low bit rates from 1.5 to 6 Mbit/s. However, this inherent flexibility of the ADSL modem is limited by the definition of the framing structure combined with the definition of consistent physical interfaces. Therefore, it is advisable to offer a flexible transport mechanism that is future-safe, adaptable to the varying demands in bandwidth that can rely on a flexible switching technique. The ATM offers this flexibility, and when the ADSL system provides an ATM interface, a gradual evolution can be planned for the provision of B-ISDN services towards the residential subscriber. B-ISDN envisages a full-fiber-based access network, which involves large investments from the telecom operator companies, as the present access network is merely based on twisted-pair copper. By providing ATM over ADSL, ATM-based services can be offered to the residential subscriber right now while the access network is evolving towards a full fiber-based network (FTTC, FTTB, FTTH: fiber to the curb, the building, or the home). Meanwhile, for the subscribers and service providers, ATM cell transport over twisted pairs by ADSL technology remains transparent, and emerging new services can be introduced rapidly and independently of the state of evolution towards a fiber-based access network. As the ADSL transmission system offers a large bandwidth to the residential subscriber while still offering a reasonable upstream bandwidth capacity, it is specifically tailored to services that are asymmetric in terms of the bandwidth such as Internet services and VOD.

INTERNET APPLICATION. Some identify the Internet to be the "killer application" that eluded the industry in the last few years. In any case, ADSL is well-suited for Internet traffic, specifically for multimedia services like voice, data, and video, all of which are now ubiquitous on the World Wide Web. Downloading these services is tailor-made for ADSL.

The ABR service, specified in the ATM forum, will no doubt improve the traffic flow immensely.

VOD. As video compression techniques improve, several VOD services could be offered simultaneously to the subscriber by means of the ADSL system. With an ATM-based ADSL system, the VOD service is independent of the state of video compression, and full transparency is guaranteed between the VOD server and VOD set-top unit at the home of the residential subscriber. For the deployment of a VOD service, a mix of access network techniques (FITL, ADSL, coax) will be applied. When full ATM transparency is assured, there will be no need for interworking functions in the access network.

The envisaged network architecture for the ADSL system based on ATM is designed with the evolution of the access network in mind. The ADSL units at the central office side and subscriber side are referred to as ATU-C (ADSL transceiver unit, central office) or LT (line termination) and ATU-R (ADSL transceiver unit, remote terminal) or NT (network termination) respectively. Means are provided to operate the ADSL segment as an independent part in the access network. In this way, the ADSL system can interface on the LT side directly to an ATM switch in the central office or to a fiber-based access network with ATM transport. At the subscriber, the ATU-R can provide access to various types of service modules (SMs) or be integrated with a set-top unit for VOD service provision, as depicted in Fig. 12.2.

12.3.4 Description of the ADSL Termination Units

In an ADSL system, the DMT modem part can be isolated from the service interface. Between both, a specific interface can be defined. The modem part is completely independent of how different services are provided to the user and how these services interface with the network. Figure 12.3 shows a functional block of the ADSL system concept that has been developed by Alcatel Telecom. The system provides an ATM interface and basic ATM functions. Input from a channelized bitstream is also possible. The following paragraphs focus on the ATM functional blocks.

Mapping an ATM cell stream into a specific channel of the channelized interface could be done, but this is far from optimal. It introduces extra overhead and limits the full ATM flexibility. This is the case in the current version of the ANSI standard, where a downstream channel in

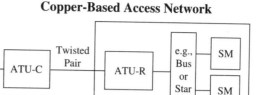

Figure 12.2
Access network
configurations.

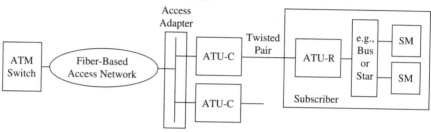

the framing structure can be reserved for ATM transport. For the upstream transport, a duplex channel should be reserved. Further, the ANSI standard puts unnecessary restrictions on the upstream and downstream ATM rates. To overcome these limitations, it is desirable to offer the ATM functional block a direct interface to the digital signal processing block. The ADSL modem interface ensures a transparent transmission of data over one unconditioned twisted pair. Besides, it provides two distinct transmission channels referred to as the slow and the fast channel as

Figure 12.3
Interfaces provided in
the ADSL system.

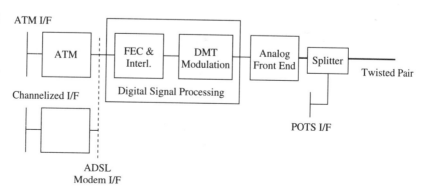

Figure 12.4
Fast and slow chan-
nels provided by the
DMT modem.

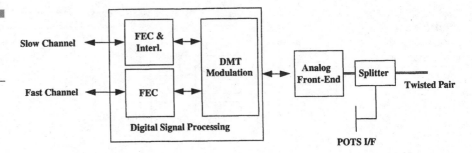

shown in Fig. 12.4. The data stream on the slow channel is interleaved after FEC, before it is passed to the DMT modulator. This improves the error correction capability at the expense of extra delay. The fast data is not interleaved and has a lower delay, while the error correction capability is smaller.

The maximum useful downstream and upstream bit rates depend upon the twisted-pair characteristics. The ATM functional block demultiplexes the received ATM cell stream from the slow and fast transmission channels. Both cell streams are treated in an independent way. The demultiplexing function is based on the ATM header content, for which a specific context table is provided that can be updated dynamically. When the ADSL system is integrated in a larger access network (e.g., in a hybrid fiber and copper access network), two quality of service (QOS) classes can be defined on the ADSL transmission system: the fast channel with a low delay but larger BER and the slow channel with better BER performance but larger delay. At connection setup, a choice will be made between fast and slow channel depending on the QOS requirements of the invoked service. The above mentioned multiplex/demultiplex function of ATM cells is based on the full header (including the VP). Therefore, the ADSL system can be incorporated in an access network where the QOS management is VP-based. An overview of the ATM functional block is provided in Figs. 12.5 and 12.6.

The ATM transport on the ADSL transmission system is cell-based. In the direction toward the DMT modem, the bit rate adaptation to the DMT modem characteristics is provided by means of idle cell insertion. The chip set is able to operate on both terminations: at the residential subscriber (ATU-R) and at the central office (ATU-C). The functions to be performed by the ATM interface are essentially the same for both ADSL terminations, but the transmit/receive bit rates are different. Therefore, the fast and slow buffers are able to serve the downstream as well as upstream bit rates. Cell delineation on the received ATM stream is based

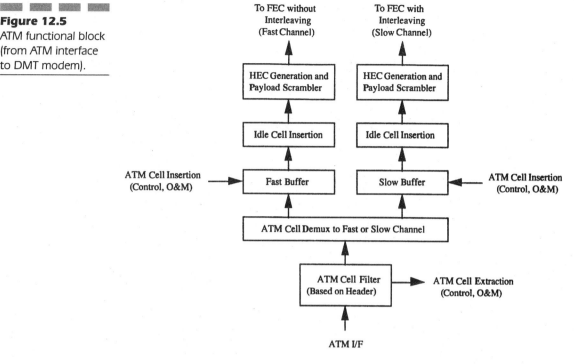

Figure 12.5
ATM functional block
(from ATM interface
to DMT modem).

on the HEC field in the ATM header and conforms to ITU recommendation I.432. Therefore, the correct HEC field is inserted in the ATM header at the transmit side (direction to the DMT modem).

In order to ease the delineation process, a payload scrambling is performed at the transmit side. Specific performance monitoring is done on both the fast and slow transmission channel by the HEC error detection and correction block. After idle cell filtering and cell extraction, both ATM cell streams from the fast and slow channels are multiplexed on a single ATM stream. An ATM cell buffer serves the output ATM interface.

The ATM interface can be used in a bus configuration: either as a slave component in the ATU-C, or as a master component in the ATU-R (see Fig. 12.2). The former allows connection of several ADSL systems on a single network termination (e.g., an SDH STM-1 or SONET OC-3c interface), while the latter permits several service modules to be connected to the ADSL system.

For operation and maintenance, the ATM interface provides insertion and extraction of ATM cells in both directions (from and to the DMT modem), on the fast as well as the slow channel. Cell extraction is based on the ATM cell header. The extracted cells are processed in an on-board controller (OBC). In this way, operation and maintenance of the ATM

Figure 12.6
ATM functional block
(from DMT modem to
ATM interface).

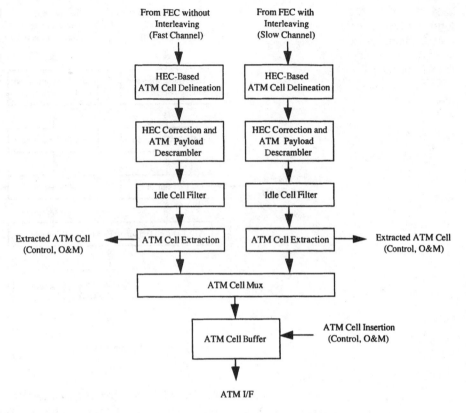

transport in the ADSL system can be performed autonomously. The cell insertion to and extraction from the ATM interface is also used for signaling purposes (e.g., for allocation of VP/VC resources in the ADSL system). Specific VP/VCs can be defined dynamically. The ADSL system provides means to perform tests in various loop-back configurations on the system in operation.

12.4 Cable Modem Technology Overview

The cable modem was described extensively in previous chapters. The overview below is intended to touch only briefly on topics that would be instrumental when comparing it with ADSL.

12.4.1 Summary of IEEE 802.14 Cable Modem

A conforming IEEE 802.14 cable modem has downstream data contained in one of the 6-MHz TV channels that occupy spectra above 550 MHz. The upstream channel assigned band is between 5 and 45 MHz. The downstream uses 64-QAM modulation technique and can deliver over 30-Mbps data rate. The upstream channel uses QPSK modulation technique and can deliver up to 2 Mb/s. Unlike ADSL, the cable modem must operate in a shared medium environment, and hence a MAC is needed to mediate the upstream traffic. The MAC protocol is ATM friendly because data is received and delivered using the ATM cell transport concept. The ATM cell is further segmented to improve system performance. Status and control information is looped back from the upstream to the downstream, so each station determines its pecking order.

A variety of cable modems are now available in the market, and most use QAM for downstream and QPSK for upstream (see Chap. 14). Some cable modems divide the upstream into frequency channels and allocate a channel to each user. Others combine the two multiplexing methods. Unlike ADSL, cable modem rates do not depend upon coaxial cable distance, as amplifiers in the cable network boost signal power sufficiently to give every user enough signal power. These factors, however, are built in the MAC, and, during initialization, time synchronization and power are aligned to the cable modem so that it can function properly. In the upstream, cable modem capacity will depend on the ingress noise and the number of simultaneous active users who are accessing the bus.

12.5 ADSL/Cable Modem Comparison

Comparing ADSL with cable modems is not an easy task. There are strengths and weaknesses in each of the technologies. The weaknesses of both will have more to do with the assumptions made, specifically in service offering. Without a criteria yardstick, the comparison becomes subjective and therefore unproductive. End-user expectation is the ultimate test. The consumer cares less about the underlying technology, be it ATM or Morse code. To the end users, the service and its ability to deliver the promised application is the only thing that matters. With that in mind, most agree that the services mentioned in Sec. 12.2 are applicable and hence will be used as a guide to compare ADSL and the cable modem.

12.5.1 Capacity

CABLE MODEM.   A cable modem user, in general, has a little over 30 Mb/s for the downstream. This bandwidth is shared among 500–2,000 users. With proper traffic engineering, applications can be performed without service degradation. For the upstream path, the shared bandwidth is about 2 Mb/s. This again should not present noticeable service degradation for all asymmetric applications in a properly engineered network.

ADSL MODEM.   The ADSL modem operates in a point-to-point application. The downstream bandwidth is not shared, but the bit rate varies depending on the quality of the copper line and terminating distance. The rate can be 1.5 Mb/s up to 6 or 9 Mb/s. Unless an ADSL subscriber resides in a rural area (farm), most users fall into the 6-Mb/s category. The ADSL modem can dynamically adapt its bandwidth to suit its environment even during the life of the call (being standardized). For the upstream, the unshared bandwidth also ranges from 16 to 640 Kb/s. For all asymmetric applications, ADSL can perform without any service degradation.

For IP-related traffic, most Internet/WWW servers operate at 56 kbps, with a few at 1.5 Mb/s T1 speed. This is likely to stay with us for the short term. Therefore, the inherent speed for the cable modem or ADSL cannot take full advantage of its available rate. Users, however, will experience a noticeable performance in speed compared with the 28.8-kbps analog modems. For the short term, both the cable or ADSL modem will operate adequately.

12.5.2 Throughput

CABLE MODEM.   In the long term, the cable modem will exhibit service deterioration if a large number of users attempt simultaneous transmission. This may require reengineering of the traffic more often than the norm. It will be based on viewing habit, time of day, and new bandwidth-hungry applications yet to be defined but surely forthcoming. A conforming IEEE 802.14 cable modem can hop onto other less-congested channels, but, at the end, the laws of physics prevail. Adding more upstream and downstream channels in the spectrum beyond what is specified today to meet the extra demand is technically feasible but costly. The amplifiers in the HFC network need to be replaced or at least modified. The cable modem may also need replacement if its frequency agility feature is not robust enough.

ADSL. The ADSL modem, assuming a 6-Mb/s rate, is fully dedicated to a single user. The limited upstream bandwidth will become a problem if video telephony applications become popular. Variable bit-rate MPEG2 and advances in video compression technologies could save ADSL if the user is not demanding high-quality video. The ADSL access node is also a mux. As more and more traffic is added, reengineering of the trunk capacity will be needed. A bigger capacity trunk, extending from the access node to the network, needs to be replaced to account for the traffic increase. Changing from DS1 to DS3 or to OC-3 is a way of life in the telco business environment. The procedure is routine.

Both cable modems and ADSL can survive the unpredictable future asymmetric hungry applications with acceptable performance. For the cable modem, it would be a more engineering-intensive upgrade.

12.5.3 Scalability

CABLE MODEM. The ADSL has an advantage over the cable modem. Subscription can only be made after modernization of the entire cable network into HFC. The $225.00 upgrade cost applies to homes passed.

ADSL. The ADSL is more scalable. An operator can provide service to any copper-based customer regardless of his geographic location. The copper customer will simply move to an ADSL access box located at the central office. When enough people subscribe, then an access node would be economically justified and deployed in that area. In this scenario, the ADSL has the advantage over cable modems.

12.5.4 Performance/Service Categories

CABLE MODEM. A conforming IEEE 802.14 should be able to handle CBR, VBR, and ABR services with proper traffic engineering. It is not yet clear how ABR services will be performed in the cable modem. If ABR services are to be performed by the cable modem itself (as it should), then the resource management (RM) cells from all the active cable modems will unduly tax the precious upstream bandwidth resources. The upstream bandwidth is shared among all active users. It is expected that the headend will act as proxy to all the active ABR connections (see Chap. 8).

ADSL. The ADSL can also handle CBR, VBR, and ABR services. For ABR, the connection is point-to-point, and hence it is not expected to have the cable modem limitation.

12.5.5 Security

CABLE MODEM. In a shared-medium environment, the cable modem is more vulnerable to misuse, eavesdropping, and service theft. All signals go to all cable modems on a single coaxial line, creating serious prospects of intended or inadvertent wiretapping. Encryption and authentication will be important parts of both systems but vital for cable modems.

ADSL. In an ADSL point-to-point architecture, eavesdropping will not be possible from the home. Copper lines are buried underground and are therefore inherently secure. Tapping the wire requires invading the line itself and knowing the modem settings established during initialization.

12.5.6 Cost

CABLE MODEM. The cable modem is expected to cost less than ADSL. Only one modem is needed to connect to a subscriber. The complexity of developing security modules on the MAC and the extra hardware of developing the MAC and RF tuners in the cable modem should be noted. Today's cable modem cost is around $300 to $500. Cable modems conforming to IEEE 802.14 are expected to be in that range.

ADSL. Two ADSL modems are required to connect a subscriber. It is expected that, with the economy of scale (more than cable modem) and competition, ADSL costs will drop. The inherently lower network costs of cable modems compared to ADSL access systems may be neutralized by higher infrastructure costs incurred when upgrading to HFC. The telco will not have to bear this modernization cost and can afford to offer an attractive package.

12.5.7 Voice Adaptation

CABLE MODEM.   The cable modem can provide POTS. If POTS functionality is embedded into the MAC logic, then it becomes vulnerable to power outage. Cable phone with voice modulated over its physical layer is a more attractive approach.

ADSL.   POTS over ADSL is provided over the physical layer. During power outage, the central office provides enough current to power the physical layer electronics and keep POTS operational.

12.5.8 Reliability

CABLE MODEM.   Cutting a CATV line in the street or losing above-ground cable in a windstorm will bring down all users on that line. A single noisy transmitter or defective cable modem on a shared bus will also bring down all users on that line. Amplifier failure can cause outage to the entire neighborhood. Each additional user creates noise in the upstream channel and hence reduces reliability. A well-maintained HFC network would obviate some of these problems.

ADSL.   A DSL modem operates on a point-to-point basis. Failure affects a single user only. ADSL itself suffers no degradation based on traffic or number of users accessing the network.

12.5.9 Internet Application Comparison Scenario

The present driver for both ADSL and the cable modem is high-speed Internet access. It is viewed as the market entry to jump-start all other broadband services and applications.

Politically, the White House is challenging the business community to structure the Internet so it can provide video multimedia access to the nation. The FCC vowed to support regulations in order to keep the Internet momentum in high gear. Internet is educational, entertaining, and instrumental in maintaining the U.S. competitiveness in this global economy.

Beyond high-speed Internet access, both technologies address remote access for work at home and telecommuting, distance learning, and access for the millions of personal computers in place today and to be sold over the next ten years. Below is a typical Internet access configuration for both the cable modem and ADSL.

TYPICAL INTERNET ACCESS FOR THE CABLE MODEM. Figure 12.7 illustrates a typical Internet access configuration for the cable modem in an HFC network. The IP router is located at the headend. In the shared medium arrangement, the headend concentrates all IP traffic and sends it to the IP router. Proxy servers or cache memory are optional and can be located at the headend. They contain copies of popular WWW pages and texts. These copies, like the Netscape home page and CNN news scripts, can be sent on demand to the requesting cable modem users with noticeable speed. This approach also mitigates congestion in the Internet network.

TYPICAL INTERNET ACCESS FOR ADSL MODEM. Figure 12.8 illustrates the alternative architectures for the ADSL modem. The access to the Internet network is using existing twisted-pair copper telephone lines. The ADSL modems connect at both ends of a subscriber's telephone line that was originally used for POTS. The POTS splitter, not show in the diagram, forwards the analog voice transparently to the POTS CO in a frequency below that of the ADSL domain. The receiver (home) ADSL modem connects directly to the Ethernet port of a personal computer or a local Ethernet hub. The access switch serves to concentrate access lines into router ports. A DS1 (1.544-Mbps) line from the router to the Internet can support five or more continuous users or as many as 55 subscribers with 10 percent usage. A DS3 (45-Mbps) pipe could support up to 1500 subscribers.

It is most likely that the access adapter includes ATM switch fabric and from there connects to the various Internet Service Providers located in the Internet network. A proxy server or cache memory can reside in the access housing.

12.5.10 Cable Modem Market Size

There are about 180 million basic CATV lines in the world now. Table 12.2 shows the U.S. statistics.

Most of the lines are old, one-way, and coaxial only. It was estimated that 6 million subscribers are upgrading from CATV to HFC. All major U.S. CATV companies have upgrade programs under way. With the exception of the U.K. and Belgium, the picture looks similar worldwide.

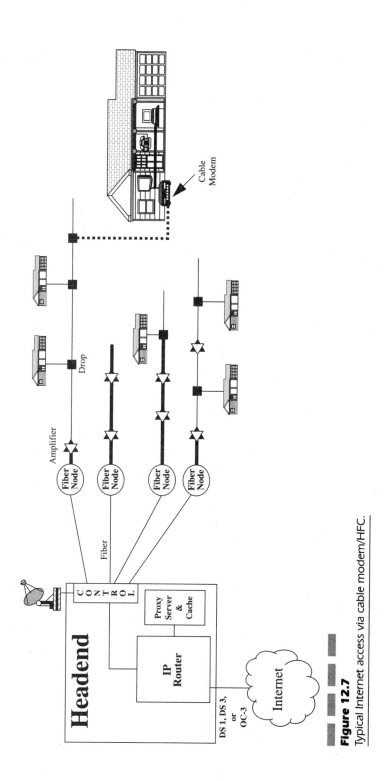

Figure 12.7

Typical Internet access via cable modem/HFC.

487

Figure 12.8
Typical Internet
access for the ADSL
modem.

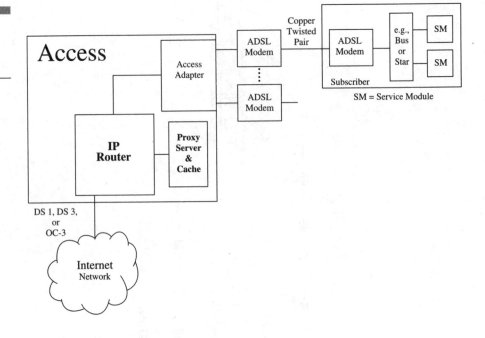

12.5.11 ADSL Market Size

ITU estimated that there are about 700 million copper telephone lines in the world today, 70 percent of which connect to residences. The rest are business and private lines. By the year 2001, the projection will exceed 900 million. In the U.S., as shown in Fig. 12.9, approximately 80 percent of

TABLE 12.2

U.S. Cable Statistics

Home passed	91 million homes	98% of TV households
Home served	56 million homes	61% of TV households
Growth of penetration	5% per year	
Percent of subscribers connected to 30–54 channels	60% of subscribers	
Rate of fiber deployment	100 miles/hour	
Average bill per month	$30	
Revenue of CATV business	$23 billion per year	
Revenue of advertising	$3.5 billion per year	

SOURCE: ADSL Forum

Figure 12.9
Copper line
distribution.

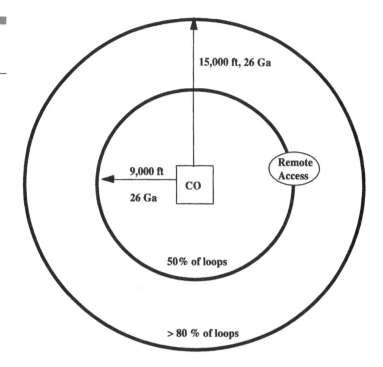

these lines can operate with ADSL at 1.5 Mbps, and 50 percent can support rates of 6 Mbps or more (see Table 12.1).

Variable-rate ADSL modems, with minimum speeds at 1.5 megabits, will enable connection to all users. When enough users subscribe to the service and a remote access node is economically justified, then the access could be relocated to that region. This will reduce copper distances and hence improve performance.

12.6 The Likely Winners

Based on the technical analysis above, cable modems and ADSL have comparable capabilities and can easily adapt into broadband/IP-based infrastructures. Serious deployment of both technologies, having similar capabilities, would not start until late 1998 or early 1999. Numbers of ADSL and cable modem deployments can rapidly grow from the low millions to tens of millions.

By the sheer numbers of copper access lines, it is apparent that ADSL or xDSL will dominate the future market. Moreover, telephone companies are already connected to the entire customer base, while CATV passes a smaller fraction. The economies of scale will play a major role in the ultimate outcome of who controls what share of the market. Based on ADSL forum estimates, telephone companies will achieve 70 to 80 percent market share in the United States.

When the untamed battle first started between the MSOs and telco, it was postulated that each will grab a 30 percent share of the other's market. The MSOs will provide cable phone and access charges to long distance carriers, while the telco takes 30 percent of the video market share. This theory has a lot of flaws. Based on the analysis above, that unlikely scenario implies that most users who reside in a modernized HFC network will switch to cable operators to provide them with POTS services. The cable operators can compete for the POTS services, particularly for a second line. The size of that market depends on the long-term strategies of the cable companies to improve their customer service image, network management, and reliability. Another problem the MSOs face is disfranchising. Eleven cable companies are serving the city of Philadelphia alone, so networking and billing becomes a real problem if all or some cable operators are to provide cable phone in that area.

The likely MSO strategy is to go after prime telco customers and offer attractive packages. This strategy gives the cable operators a jump start, but the competition will be fierce. Cable operators must also compete with the reputation of AT&T, MCI, and Sprint, who have already started competing for the same share of this market. The RBOCs are more apprehensive of them than the cable operators. The most likely and profitable scenario for the cable operators is to make alliances with the newcomers or traditional telco operators and leverage both of their strengths to compete effectively. TCI and two other cable companies joined with Sprint to bid on the wireless personal communication services spectrum. The friendliness of cable infrastructure to PCs and Sprint's reputation in customer service and billing make a good business match.

Telco executives feel their service reputation gives them a competitive edge over cable operators and will likely get 15 percent of the video market share. Bell Atlantic's recent poll of 500 of their customers found that 46 percent of cable subscribers said that they would switch their cable company if a similar service were offered by the phone company. Fifty-six percent said they will switch if the service includes VOD.

12.7 Other Competing Networks

It would be misleading to assume that the telco will compete with ADSL products exclusively. Other alternative networks are being built by telco operators or are investing heavily in that technology. HFC is also a solution being deployed by the RBOCs. The choice of building a particular network depends on the demography, economic model, and a host of other factors. Each network is going to be built differently in every town or city.

The three possible technologies, not including cable modem or ADSL, that are being considered by the RBOCs are:

1. FTTC
2. Wireless cable
3. Direct broadcast satellite (DBS)

The FTTC approach is to get as close as possible to the residential area and provide higher-bit-rate lines with limited reach. A subset of this approach is to provide two lines. One line will be coax and carry fully interactive, switched digital video. The other will be twisted-pair copper for voice.

Wireless cable is a digital radio transmission with 28-GHz capacity. Antennae are small enough to be pasted on a window. Once the small antenna is placed, then with a telephone jack and TV, a consumer will have video. It is obviously not practical in every terrain. It allows newcomers to enter this market who do not have an existing network of wires to build upon. The TV picture quality is excellent, but, so far, wireless cable is not adequate for fast access to the Internet.

The direct broadcast satellite (DBS) market is picking up faster than anyone expected. The appeal is program variety, selection, and video quality. Again, this medium is not well suited for two-way communication, at least presently. For slow-speed application such as Email, it can be attractive to the consumer who likes Email access but not necessarily Web surfing.

Chapter **13**

HFC in Europe

13.1 Overview

There has been a lot of interest in the cable data modem throughout Europe recently. The obvious appeal is the promise of delivering high bandwidths much beyond what is currently provided by the classical modems for telephone lines. European CATV operators are interested in the new business opportunities on their existing or upgraded cable plant.

New technology trials in Europe are usually funded by the European ministries. The European Union ACTS program (Advanced Communications and Services) is a leading initiative of the European Council of Ministers for European research and technology development in the framework of an objective known as Integrated Broadband Communications (IBC). Previous phases known as RACE 1 and 2 concentrated on key technologies, and today, the ACTS program, concentrating on services, comprises 120 projects for the testing of advanced telecommunications applications, especially interactive multimedia services, photonic technologies, and high-speed networks to be tested over a real broadband information infrastructure.

The European ATM pilot network is the pan-European test platform in preparation of the European Information Infrastructure interconnecting national hosts, known since April 1996 as JAMES (Joint ATM Experiment on European Services). This pilot network was initiated in 1992 by six public network operators and within 3 years had grown to cover the majority of countries in western Europe.

Since this is a nonprofit MOU, the majority of end users have been government agencies and research centers making experiments connected with national research networks, but more and more precompetitive ACTS projects financed by the EU are now ongoing.

ACTS conducted several field trials throughout Europe, and, like the cable operators in the United States, they are experimenting with various technologies and testing various offerings. We will only describe in detail one field trial sponsored by the Deutsche Telekom and two trials conducted by the ACTS program. The intention is to introduce the European flavor to the technological direction and planned services of HFC. The trials that will be described are:

1. Pilot projects in Germany (interactive services in Berlin)

2. ATHOC (ATM applications over hybrid optical fiber coax)

3. BISIA trial in Århus, Denmark

13.2 Pilot Projects in Germany (Interactive Services in Berlin)

The object of the pilot projects in Germany is to demonstrate multimedia capability over HFC. These projects, with 50 to 4000 subscribers, are tested to evaluate the various technical solutions.

The Berlin pilot system has been in operation since September 1996. Alcatel Telecom supplied the system, including servers and set-top boxes. The field trial "Interactive Video Services Berlin" was initiated by Deutsche Telekom as a showcase for all kinds of up-and-coming interactive services running over the existing fiber and CATV network. The project started in April 1994, and the first customer was connected at the end of August 1994. Since that time, the system has been running 24 hours per day. After years of operation, Deutsche Telekom has decided not to turn the system off, as originally planned.

13.2.1 Objectives of the Showcase

The objectives are both service- and technology-related. Various services are being demonstrated and tested on the users. The major challenge was, and still is, a lack of clear definition of services.

The main technical objectives and conditions are:

- The system should run on the existing Berlin fiber and coaxial cable network.
- No additional installations in the house or residential area would be needed.
- All services would be provided over the CATV outlet at the subscriber's premises.

Bidirectional communication carried existing analog TV and radio channels. New services are fully digital, using video MPEG1+ (MPEG1 at 2.5 Mbit/s, but with full vertical resolution as for MPEG2) format.

13.2.2 Services on Berlin Pilot

The services provided over the system include distributive as well as interactive services. Service highlights include:

1. Digital video broadcast.

2. Pay-per-channel (monthly charge of encrypted digital TV channels).

3. Pay-per-view (event-based-charge).

4. Top movies as near video on demand (15-minute start-time increments)

5. Interactive video on demand (IVOD) with the following subcategories:
   - Movies
   - Children's programs
   - Magazine/talk shows
   - Travel info/nature
   - Quiz/entertainment

   This classic VOD service allows fully interactive operation (start, stop, pause, forward, rewind) using an infrared (IR) control. Content is partially updated every two weeks.

6. Home shopping (video clips, still pictures, text overlays, catalogs with as many as 150 products, and alphanumeric input via IR control). Products are selected out of a catalog that is composed of full-motion video sequences, still pictures, and text. After the selection of an item, real ordering is possible with the IR control and a simple on-screen text editor for address, customer identification, and card numbers.

7. Health information channel. This channel resembles a medical dictionary based on feature films and medical movie sequences. At the end of the film, it is possible to retrieve additional information on a PC, which can be connected to the set-top box.

8. Telelearning with the subcategories:
   - Scientific movies
   - Interactive "Lernwelt Television 2000"

   In this service, two classes of complexity are offered: simple interactive VOD of scientific feature films including animations, and the service "Learning World TV2000." The second class is a mix of scientific films and educational games that have been tailored to display on the TV screen. As an example, 140 video clips (5 to 15 minutes in duration each) have been composed on the theme of ecology.

9. Information services with the subcategories:
   - Berlin city information
   - Info-Surf Berlin
   - Real-estate marketing

   These services are under the control of the Berlin city information and the Infosurf Berlin categories, where highlights of Berlin can be interactively retrieved as a mix of video, animation, and text.

10. Echo-TV is real-time digital encoding. A user connects via a broadband link to TV studios in Berlin. News, magazines, and sporting events are MPEG-encoded in real time and loaded on the special echo-TV server module in real time. Immediately after the end of an event, the stored content can be retrieved as interactive information on demand.

11. Pay radio.   This music service is operated on a pay-per-channel basis. It is possible to select different genres of music out of 92 types like classical, folk, rock, and cultural music, via the set-top box, TV set, and stereo equipment. When a piece is played, the title, the interpreter, and other essential information are displayed on the TV screen.

13.2.3 Architecture of Berlin Pilot Network

The architecture of the Berlin pilot network is as shown in Fig. 13.1. The modules are:

■ Real-time audio and video MPEG encoders, for broadcast, pay-per-channel, pay-per-view, and pay radio.

■ At the headend, a video server is installed that currently has a capacity of about 120 Gbytes. This is equivalent to 60 movies of 100 minutes each (2 Gbytes per 100-minute movie). An upgrade of the Berlin server to more than a 300-Gbyte capacity is feasible. Currently the server handles several hundred video files, with a size between 1-minute teleshopping clips and 2.5-hour overlength movies. The echo-TV service is implemented with a modification of this server technology that allows MPEG streams to be written on the disks in real time, whereas the main server is loaded with a special tape-loading station.

13.2.4 The Global Architecture of the Network

Fiber links up to 30 km in length feed CATV trunks of the existing Berlin coaxial cable network (illustrated in Fig. 13.2). Forty-eight users are connected to the central server. However, 240,000 CATV customers out of the 1.2 million CATV subscribers in Berlin are located in the service area of the pilot. No mutual interferences were created when the digital signals were added to the running CATV service.

The digital channels are transmitted within the hyperband (300 to 450 MHz) via fiber-optic feeders with a length of up to 47 km from the cen-

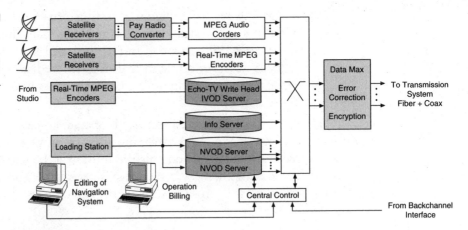

Figure 13.1
Block diagram of headend/server in the Berlin pilot system.

tral station to ten secondary CATV headends. There the digital channels are combined in the frequency multiplex domain with the CATV signals. Using auxiliary ports of the CATV equipment, care has been taken to insert new digital channels without any interruption or impairment of the existing analog CATV service. These precautions are necessary because currently about 240,000 CATV subscribers (out of 1.2 million Berlin CATV users) are connected to the network within the serving area of the trial system.

The back-channel is transmitted in the standard 5- to 30-MHz band from the set-top box via the coaxial in-house cabling to the conventional CATV service access point, normally located in the basement of the building. There the back-channel is extracted from the coaxial cable and sent by modem over twisted pair to the central control and server. This solution appears to be very attractive for trials where a few interactive service users

Figure 13.2
Architecture of Berlin pilot network.

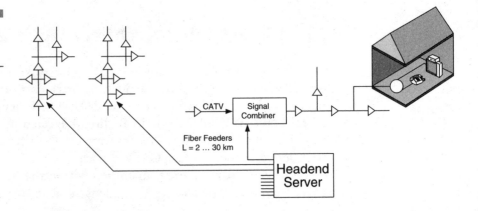

are widely dispersed in a large network, which cannot be upgraded with back-channel facilities for hundreds of CATV line amplifiers.

Due to the low number of customers who are dispersed over the urban and suburban Berlin city area, fiber links were used for the long distance, feeding the coaxial cells where the users are located. This situation represents in an ideal way the startup situation of new services when the subscription rate is very low and one multimedia headend has to serve a complete city.

13.2.5 Interfaces

The following interfaces are provided by the set-top box:

- 75-ohm coax for connection to CATV outlet
- 75-ohm coax for analog CATV feedthrough to the TV set
- 3× SCART for connection of TV set and VCR
- Stereo audio output for pay radio
- RS 232 for connection of PC or game console
- Infrared for all user interactions and control
- Smartcard for user identification and restricted access
- Digital interactive video server with loading station
- Real-time MPEG encoders (with C-Cube and Tadiran)
- MPEG video switch with 240×48 input-output ports
- Digital subcarrier transmission system for hybrid fiber coax network
- Set-top boxes with IR control and smartcard
- Satellite interfaces for pay radio (with DMX, Digital Music eXpress)
- Editing station for navigation system

MPEG encoding of all content and, on special request, the authoring of movie trailers for video on demand is done in Alcatel labs.

13.2.6 Test Insights

The Berlin trial verified that hybrid fiber coax technology is very flexible, in the sense that very small cells with only 20 homes passed (fiber-to-the-last-amplifier concept) and very large cells with 77,000 homes passed (fiber-to-the-serving-area concept) can be combined in a single system. Furthermore, it appears to be fully feasible to upgrade networks with today's analog CATV toward digital services by simply adding these signals somewhere in the running network.

On the service side, the trial discovered that hybrid fiber coax allows services to be deployed, even when initial penetration is very low. Digital transport over modulated subcarriers in the radio frequency domain allows a very flexible mix of services: simple analog TV and analog radio, digital broadcast, and digital interactive. In the near future, a migration from cable telephony over high-speed data broadcast toward full packet-switched connectivity will take place.

13.3 ATHOC

ATHOC stands for "ATM applications over hybrid optical fiber coax." ATHOC is a research project granted by ACTS (AC037). The prototypes will be tested in the field on CATV networks in:

- Stuttgart (Germany)
- Zurich (Switzerland)
- Aveiro (Portugal)
- Antwerp (Belgium)

These field trials are scheduled to begin in the second quarter of 1997. The objective of the trial will also demonstrate and test some services including the Internet.

In the headend, the ATHOC HFC equipment will be connected to an ATM core network via SDH links at 155.52 Mb/s, which carry ATM cells. These high-bit-rate pipes are ATM-demultiplexed to several lower-bit-rate pipes. Each one is broadcasted on the HFC access network in one 64-QAM channel with a total bit rate (including FEC) of 41.4 Mb/s. This bandwidth should fit in the 8-MHz channel bandwidth of European CATV networks. In the upstream direction, the total bit rate of the differentially encoded QPSK channels is 3.04 Mb/s, and the useful data of several of these channels will be ATM-multiplexed in the headend and then delivered to the SDH links.

To the customer's premises network, the ATHOC network termination (ANT) provides an ATM Forum interface at 25.6 Mb/s, which allows a high user bit rate connection to PCs.

The ATHOC prototype will have an in-band channel structure in line with decisions within the IEEE 802.14 working group. DAVIC 1.0 started with an out-of-band channel structure specification because it targeted mainly VOD applications. In the in-band case, the user terminal is tuned to only one 64-QAM downstream carrier, while for the out-of-band case, the terminal has a second tuner for receiving an additional low-bit-rate channel (e.g., QPSK format) on which control information (e.g., access control, video signaling) is sent. In the in-band case, the latter information is time-multiplexed with the normal data in the same channel.

The main advantages of the second control channel in the out-of-band case are that control information can still be exchanged between the user terminal and the headend when the user terminal is tuned on an application channel with another signal format (e.g., analog or MPEG-TS packets instead of ATM cells), and secondly, because the common control channel provides an easier reference for upstream slot synchronization to user terminals transmitting on the same upstream channel but tuned on different downstream application channels. The main disadvantages of the out-of-band channel structure are that this second control channel is not really needed and that it adds additional cost to the user terminal.

13.3.1 ATHOC Downstream Framing Alternatives

The ATHOC downstream frame on the HFC access network is almost the same as DAVIC. As in DAVIC, seven ATM cells are mapped in two DVB frames, according to the specifications of the digital video broadcast (DVB) workgroup. ITU adopted this frame structure in the J.83 annex A standard specification. A DVB frame (or alternatively annex A frame) is defined here as a synch byte followed by 187 bytes of contents (i.e., ATM cells), followed by 16 bytes of FEC (Reed-Solomon).

There was a lot of debate on ITU J.83 annex A versus annex B as basis for the framing of the future IEEE 802.14 standard. IEEE 802.14 later adopted both framing structures. These ITU annexes were originally specified for digital video broadcast, which transports MPEG packets. Aligning the cable modem frame format to some extent with the framing of current digital video broadcast applications has the advantage that silicon for these functions will be available before the cable modem standards are finalized. The market for this silicon will then also be bigger, leading finally to lower costs of user terminals.

The debate was that a system with downstream framing according to annex B requires 1 to 2 dB less CNR than one with an annex A frame, due to the additional Trellis Coded Modulation (TCM). This TCM is part of the annex B specification, which at this moment has no equivalent in the annex A frame. The performance is also confirmed by measurements from CableLabs. However, it is argued that current CNR specifications nationally imposed on CATV networks are so high (because of the quality required for analog TV distribution) that annex A also meets required performance. In that case, annex A would give about 1 Mb/s more of useful capacity for data (i.e., after subtracting the bits needed for FEC) in one 64-QAM channel. It is responded that other applications like analog TV broadcast have already taken all the specified-frequency bands of good quality such that the cable modem applications may have to search for remaining free frequency bands beyond the specified region that are of worse quality. However, in the case that the 1- to 2-dB improvement is needed, better TCMs are known that can be added to the annex A specification, leading to what could be called an "extended" annex A. The latter's performance would then even be better than annex B.

A disadvantage of annex B for data applications is its larger interleaving depth leading to more latency, which is 3.75 ms and 0.6 ms at 30 Mb/s for annexes B and A, respectively. The larger interleaving depth has the advantage that it can correct errors due to longer burst interferences. However, if this type of error cannot be corrected any more because the burst interference is too long, longer interleaving will create more errors in some cases. To solve these disadvantages, an "extended" annex B specification is proposed that has flexible interleaving including shorter interleaving depth options.

13.3.2 TDMA Mechanism in ATHOC

For the downstream frame of ATHOC, the function of the grant cell is to control the upstream TDMA method. A grant cell contains 12 grants, and each grant consists of a grant ID that refers to the identity of one user and thereby allows the use of the corresponding upstream slot. The user terminal (ANT) has to wait a time given in a time marker before it may transmit its upstream burst, and this time starts counting after the grant cell is received. Each ANT obtains its equalization delay after a ranging process during initialization. The ranging process is a known mechanism in which the headend equipment determines the distance from an ANT to that headend and from which the required equalization delay is calculated.

13.3.3 ATHOC Concluding Remarks

The first generation of interoperable cable modems will have a 64-QAM modulation for the downstream channel and a QPSK-modulated TDMA access for the upstream channel. However, there are still a lot of parameters that have to be specified in detail in the IEEE 802.14 work group before full interoperability can be achieved. Compromises will have to be made between different performance parameters and other criteria. Due to their high complexity today, some access technologies are excluded. However, with evolution of time and technological capabilities, their maturity will increase such that it may be expected that next generations of standardized cable modems with higher performance may be based on some of these alternative access technologies.

13.4 BISIA Trial in Århus, Denmark

The Århus trial is performed within the framework of the RACE (Research in Advanced Communication Europe) project R2121 BISIA (Broadband Interactive Services on Integrated Access).

13.4.1 Description of the BISIA Trial

BISIA is a field trial of an interactive broadband ATM-based access system (see Fig. 13.3) that will run on the CATV network of Tele Denmark Jydsk Telefon in Århus, Denmark. It started in October 1995.

The tree-and-branch CATV network is located in Århus, Denmark, in an area with three-story apartment buildings and a total of 872 subscribers. The capacity of the network is 60 analog TV channels. At present, 25 PAL TV programs are distributed, arranged as 7 VHF channels in the 47–230-MHz band and 18 UHF channels in the 470–862-MHz band with channel spacings of 7 MHz and 8 MHz, respectively. The CATV network was prepared for upstream transmission in the 5–25-MHz band and return amplifiers are installed for the trial.

The video server is part of the Danish BATMAN (Broadband ATM Access Network). The available services are home cinema and interactive news and include such features as still pictures, fast forward, and so on. A new CATV access node is installed in the CATV headend. On one side, this access node provides the modem functions for transmission on the

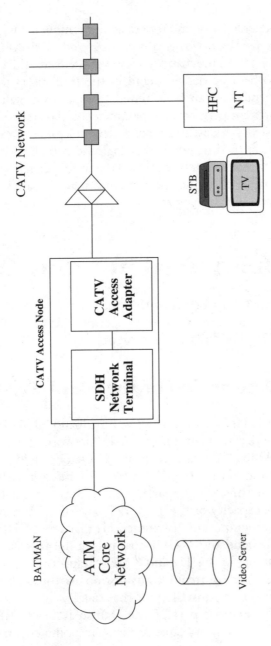

CATV Network

CATV Access Node

BATMAN

STB

TV

HFC NT

CATV Access Adapter

SDH Network Terminal

ATM Core Network

Video Server

BATMAN Broadband ATM Access Network

Figure 13.3
BISIA field trial configuration.

504

coaxial cable of the CATV plant, and to the other side it provides the interconnection to an ATM switch of the BATMAN via an ATM-based optical SDH link.

The customer premises equipment (CPE) consists of an HFC network termination (HFC-NT) and a TV set with a set-top extension unit (STEU). The functions are 64-QAM demodulation, ATM transmission upstream and downstream, MPEG decoding of a 2-Mb/s video service, and so on. The upstream access of an ATM-cell-based TDMA is implemented with FSK modulation at 2 Mb/s in the 5–25-MHz band.

The downstream transmission on the CATV network is at 34 Mb/s in an 8-MHz channel in the 230–470-MHz band. The modulation format is 64-QAM. This 34-Mb/s pipe consists of different ATM paths that can be assigned to different subscribers (ASUs) tuned on the same channel. The TDM frame reuses the 204-byte structure with synch (S) and Reed-Solomon (FEC) bytes from the digital video broadcast (DVB) specification for digital modulation on cable systems and from ETSI specification ETS 300 429.

ATM cells are packed in a 34-Mb/s stream and sent on the coax network in 64-QAM format in an 8-MHz channel. This 34-Mb/s data stream is composed of ATM cells, including the FEC bits per the European DVB standard.

13.4.2 ATM to DVB Alignment Scheme

DVB frame pairs are filled with seven ATM cells according to the custom alignment scheme for BISIA. The first DVB frame in the frame pair will be denoted as an A frame, the second as a B frame. During normal downstream communication, a continuous sequence of alternating A and B frames is transmitted over the CATV network. An important requirement is that transmission must be byte-aligned; that is, bytes that are sent to the DVB modulator and mapped to symbols inside the DVB modulator at the transmitter side must be mapped back to bytes by the DVB demodulator at the receiver side in exactly the same way. Maintaining byte alignment over the DVB modulation/demodulation stage is possible using the byte-aligned synch byte as a reference. Each DVB frame starts with a synch byte, which is equal to 0100 0111 = 47 hex. The three overhead bytes are located on the first payload byte of the first DVB frame (i.e., the second DVB frame byte, marked A) and on the first and last payload byte of the second DVB frame (i.e., the second and one-hundred-eighty-eighth DVB frame byte, marked B and D, respectively). In the first DVB frame, the first ATM cell starts immediately after the A byte, on the third DVB frame byte. The first DVB frame contains three full ATM cells and 27 bytes of the fourth ATM cell following one another. This makes the payload byte count $1 + 3 \times 53 + 27 = 187$ bytes, which is equal to the DVB frame payload length. In the second DVB frame, the fourth ATM cell continues with the remaining 26 bytes right after the B byte, and is followed by the full ATM cells 5, 6, and 7. At the end, a dummy byte is inserted in order to make the payload byte count $1 + 26 + 3 \times 53 + 1$, again equal to 187 bytes.

13.4.3 BATMAN

The left side of Fig. 13.3 shows the Danish BATMAN (Broadband ATM Access Network) to which the access node interfaces. It consists of the ATM core network and a video server with control stations. The ATM core network is based on 155 Mbit/s SDH connections and ATM cross-connects. Semipermanent ATM paths are set up from/to the video server, the control stations, and the Access Node.

The video server contains the video sequences. The services are test- and demo-types used to show how the system is functioning. Two such services are programmed:

1. Home cinema service, where films are selected individually

2. Interactive news service, where news items can be selected interactively

ATM PATH CONNECTIONS. To make a connection between the video server, the video server control station, and the user, ATM paths are established. In the BISIA system, three ATM paths are established for each subscriber terminal (CPE):

1. One VP/VC for the upstream signaling terminated at the video control station

2. One VP/VC for the downstream signaling originating at the video control station

3. One VP/VC for the digital video channel originating at the video server

The two signaling paths are set up when the CPE equipment is powered on. Via the upstream signaling path, a request is made for a downstream ATM path for the digital video channel. The VP/VC value for this ATM path is communicated to the CPE via the downstream signaling path. Via the upstream signaling path, requests can be made for other movies, other news fragments, still pictures, the program book, and so on. If there is a transmission error, the link will recover itself because the CPE will start sending the setup message again.

Chapter **14**

Cable Modem
Field Trials

This chapter shows all aspects that directly or indirectly relate to cable modems. They are:

1. Cable modem field trials summary
2. Cable modem commercial deployment
3. List of the MSOs
4. List of active vendors developing cable modems
5. List of available cable modems

Key players, vendors, operators, and service providers were contacted, and they provided the information directly or indirectly for the sections below.

14.1 Cable Modem Field Trial Summary

Cable company	Venue	Services
Adelphia	Coudersport, PA	High-speed data over cable
AdLink	Los Angeles	Regional ad insertion
Americable	San Diego, CA	Telephony
AOL	Virginia	Client-server solutions for Internet access
Cablevision Systems	Long Island, NY	Asymmetric broadband comms
		Cable telephony
		Online service over cable
City of Palo Alto	Palo Alto, CA	Telephony
COGECO Cable	(En Français) Canada	Montreal cable services
Comcast	Philadelphia, PA	Asymmetric broadband comms
		Online service over cable
Continental Cablevision	Boston, MA	Education
		Online service over cable
Cox Cable	Phoenix, AZ	Commerce
		High-speed data over cable
EPB Network Sys	Glasgow, KY	Telephony, data

Energy Network Sys	Little Rock, AK	Energy management, telephony
Helicon Cablevision	Uniontown, PA	Telephony
Hunter's Creek	Hunter's Creek, FL	Near video on demand
Jones Intercable	Alexandria, VA	Internet
MultiVision	Anaheim, CA	Cable telephony
Pacific Gas & Electric	Diablo Canyon, CA	Data
Palo Alto Cable	Palo Alto, CA	Internet access
Rogers' Cablesys	Toronto, Canada	Online service over cable
Service Electric	Pennsylvania	Cable modems for Internet
Shaw Cable	Edmonton, Canada	Business connectivity
Southern Co. Net	Atlanta, GA	Energy management, telephony
Sprint Telemedia	Kansas City, MO	Level 1 gateway
SW Network	Laredo, TX	Energy management
TCI	Arlington Heights, IL	Cable telephony High-speed data over cable
	East Lansing, MI	Distance learning/education High-speed data over cable
	Evansville, IN	Cable modems for education
	Glenview, IL	Cable modems for education
CI/J.C.	Sparkman Ctr. Englewood, CO	Online service over cable
Time Warner Cable	Elmira, NY	Cable modems—work-at-home, government Cable modem server complex Online service over cable
	Hawaii	Local government/towns
	Orlando, FL	In-home printer Network management hardware/software
TKR Cable	Warren, NJ	Telephony
TMN Networks	Toronto, Canada	NVOD network origination
Viacom	Castro Valley, CA	Asymmetric broadband comms Online services over cable

14.2 Cable Modem Field Trials in Detail

ARIZONA (PHOENIX). Ongoing commercial service for 100 subscribers.

- Sponsor: Cox Communications
- Cable modem: LANcity Personal
- Downstream/upstream: 10 Mbps/10 Mbps
- Services: @Home, a residential online service that offers Internet access, Email, world news, shopping, local content, and more

BOSTON SUBURBS (NEEDHAM, NEWTON, WELLESLEY). Commercial service for 31,200 subscribers, launched in 1996.

- Sponsor: Continental Cablevision
- Cable modem: LANcity Personal
- Downstream/upstream: 10 Mbps/10 Mbps
- Services: Internet access

BUFFALO. Commercial service, launching 1997.

- Sponsor: Adelphia Communications
- Cable modem: LANcity Personal
- Downstream/upstream: 10 Mbps/10 Mbps
- Services: Internet access

CASTRO VALLEY. Ongoing commercial trial for 250 subscribers.

- Sponsor: TCI
- Cable modem: Hybrid Networks Cable Client Modem 211
- Downstream/upstream: 10 Mbps/2 Mbps
- Services: Internet access

CHAMBLEE. Trial service for 100 subscribers, launched summer 1996.

- Sponsor: BellSouth
- Cable modem: LANcity Personal Peeks
- Downstream/upstream: 10 Mbps/10 Mbps
- Services: Internet access

CHARLOTTESVILLE. Commercial services, launching 1997.

- Sponsor: Adelphia Communications
- Cable modem: LANcity Personal
- Downstream/upstream: 10 Mbps/10 Mbps
- Services: Internet access

CHICAGO. Commercial service, launching in 1997.

- Sponsor: Cox Communications
- Cable modem: LANcity Personal
- Downstream/upstream: 10 Mbps/10 Mbps
- Services: @Home, a residential online service that offers Internet access

CONNECTICUT (HARTFORD). Commercial service, launching in 1977.

- Sponsor: Cox Communications
- Cable modem: LANcity Personal
- Downstream/upstream: 10 Mbps/10 Mbps
- Services: @Home, a residential online service that offers Internet access, Email, world news, shopping, local content, and more

ELMIRA. Beta test trial for 500 subscribers, launched in 1996.

- Sponsor: The Excalibur Group, a division of Time Warner
- Cable modem: Zenith HomeWorks Elite
- Downstream/upstream: 10 Mbps/2 Mbps
- Services: local content covering city government, schools, libraries, and the community

FAIRFAX. Ongoing trial service in 12 schools.

- Sponsor: Media General Cable
- Cable modem: Zenith HomeWorks Elite
- Downstream/upstream: 10 Mbps/2 Mbps
- Services: Internet access

FREMONT. Commercial service, launched in 1996.

- Sponsor: TCI
- Cable modem: LANcity Personal
- Downstream/upstream: 10 Mbps/10 Mbps

GEORGIA (ATLANTA).   Commercial service, launching in 1997.

- Sponsor: Cox Communications
- Cable modem: LANcity Personal
- Downstream/upstream: 10 Mbps/10 Mbps
- Services: @Home, a residential online service that offers Internet access, and Email

HAWAII (OAHU, MAUI, AND HAWAII).   Ongoing trial service for 130 public schools and 50 homes.

- Sponsor: Oceanic Cable, a division of Time Warner
- Cable modem: LANcity Personal
- Downstream/upstream: 10 Mbps/10 Mbps
- Services: Internet access

ILLINOIS (ARLINGTON HEIGHTS)

- Sponsor: TCI
- Cable modem: Motorola CyberSurfr
- Downstream/upstream: 10 Mbps/768 Kbps
- Services: Internet access

JACKSONVILLE.   Commercial service, launched September 1996.

- Sponsor: Continental Cablevision
- Cable modem: General Instrument SURFboard SB1000
- Downstream/upstream: 27 Mbps/1.7 Mbps telephone return
- Services: Internet access, multimedia services

KENTUCKY (BOWLING GREEN).   Commercial service, launched at Western Kentucky University in 1996.

- Sponsor: Glasgow Electric Plant Board
- Cable modem: Zenith HomeWorks Elite
- Downstream/upstream: 10 Mbps/2 Mbps
- Services: Internet access for residents

MARYLAND (BALTIMORE).   Ongoing trial service.

- Sponsor: Comcast
- Cable modem: Hewlett-Packard QuickBurst
- Downstream/upstream: 30 Mbps/3 Mbps
- Services: @Home, a residential online service that offers Internet access

MASSACHUSETTS (BOSTON).   Ongoing educational service at Boston College for 6000 dorm rooms and 100 offices.

- Sponsor: Boston College and Continental Cablevision
- Cable modem: LANcity Personal
- Downstream/upstream: 10 Mbps/10 Mbps
- Services: campus networking and Internet access

MICHIGAN (EAST LANSING).   Ongoing commercial service for 350 subscribers: $45 per month for 4-Mbps connection, $69.95 per month for 10-Mbps connection.

- Sponsor: TCI
- Cable modem: Zenith HomeWorks Elite, LANcity
- Downstream/upstream: 10 Mbps/2 Mbps
- Services: educational Ethernet, telemedicine

NEW YORK (BINGHAMTON/ELMIRA).   Commercial service, launched in 1996.

- Sponsor: The Excalibur Group, a division of Time Warner
- Cable modem: Zenith HomeWorks Elite
- Downstream/upstream: 10 Mbps/2 Mbps
- Services: Roadrunner, which gives users Email, Internet access

OHIO (AKRON).   Commercial service, launched September 1996; $39.95 per month includes the modem.

- Sponsor: The Excalibur Group, a division of Time Warner
- Cable modem: Hewlett-Packard QuickBurst, Motorola CyberSurfr

- Downstream/upstream: 8 Mbps to 25 Mbps/1 Mbps to 3 Mbps
- Services: Roadrunner, which gives users Email and Internet access

ORANGE COUNTY.   Ongoing commercial service for 200 subscribers.

- Sponsor: Cox Communications
- Cable modem: Motorola CyberSurfr
- Downstream/upstream: 27 Mbps/800 Kbps
- Services: @Home, a residential online service that offers Internet access

PALO ALTO.   Ongoing technical trial for 20 sites.

- Cable modem: Com21, ComPort
- Downstream/upstream: 30 Mbps/1.9 Mbps
- Services: Internet access for schools

PENNSYLVANIA (COUDERSPORT).   Commercial services, launching 1997.

- Sponsor: Adelphia Communications
- Cable modem: LANcity Personal
- Downstream/upstream: 10 Mbps/10 Mbps
- Services: Internet access

SAN DIEGO.   Ongoing trial service for 125 subscribers.

- Sponsor: Cox Communications
- Cable modem: Zenith HomeWorks Elite
- Downstream/upstream: 10 Mbps/2 Mbps
- Services: Prodigy
  Another trial service was launched in the fall of 1996:
- Sponsor: The Excalibur Group, a division of Time Warner
- Cable modem: Toshiba
- Downstream/upstream: not available
- Services: Roadrunner, which gives users Email and Internet access

SARASOTA.   Trial service, launching in 1997.

- Sponsor: Comcast
- Cable modem: Hewlett-Packard QuickBurst
- Downstream/upstream: 30 Mbps/3 Mbps
- Services: @Home, a residential online service with Internet access

SUNNYVALE.   Commercial service, launched in 1996.

- Sponsor: TCI and @Home
- Cable modem: LANcity Personal
- Downstream/upstream: 10 Mbps/10 Mbps
- Services: @Home, a residential online service that offers Internet access

SUNNYVALE.   Ongoing beta-test trial.

- Sponsor: TCI
- Cable modem: LANcity Personal
- Downstream/upstream: 10 Mbps/10 Mbps
- Services: @Home, a residential online service that offers Internet access

SYRACUSE.   Commercial service, launching 1997.

- Sponsor: Adelphia Communications
- Cable modem: LANcity Personal
- Downstream/upstream: 10 Mbps/10 Mbps
- Services: Internet access

VIRGINIA (ALEXANDRIA).   Ongoing commercial service includes modem, Ethernet card, and software for $39.95 per month.

- Sponsor: Jones Communications
- Cable modem: LANcity Personal
- Downstream/upstream: 10 Mbps/10 Mbps
- Services: Internet access, Jones Internet Channel

WASHINGTON (SPOKANE). Telecommuting trial service for 30 subscribers, launched in 1994.

- Sponsor: Cox Communications
- Cable modem: Zenith HomeWorks
- Downstream/upstream: 4 Mbps/4 Mbps
- Services: work-at-home program

YONKERS. Trial service for more than 500 subscribers, launched early 1996.

- Sponsor: Cablevision Services (now TCI)
- Cable modem: LANcity Personal
- Downstream/upstream: 10 Mbps/10 Mbps
- Services: Internet access

PALO ALTO. Commercial service for 150,000 subscribers, launched December 1995.

- Sponsor: TCI
- Cable modem: Hewlett-Packard QuickBurst
- Downstream/upstream: 30 Mbps/3 Mbps
- Services: Internet access

14.3 List of Areas of Commercial Deployment

AKRON AND CANTON, OHIO

- Sponsor: Time Warner Cable. Commercial deployment of Roadrunner service to 300,000 homes passed by two-way HFC plant
- Cable modem: Motorola and HP servers
- Services: $39.95 per month for unlimited Internet access, modem rental, and national Time Warner content

ALEXANDRIA, VIRGINIA

- Sponsor: Jones Intercable. First commercial deployment of Jones Internet Channel Service

- Cable modem: LANcity
- Services: unlimited Internet access and modem rental for $39.95 per month

CHAMBLEE, GEORGIA

- Sponsor: BellSouth
- Cable modem: LANcity
- Services: cable television and high-speed data services to 700 homes passed by its two-way HFC network. High-speed Internet access, including cable modem rental, is priced at $39.95 per month.

CHESTNUT-HILL, MASSACHUSETTS

- Sponsor: Continental Cablevision. Commercial deployment of HFC video and data network at Boston College serving 6000 dorm rooms, 2500 classrooms, and 400 administrative offices
- Cable modem: LANcity
- Services: provides connection to Internet access

COUDERSPORT, PENNSYLVANIA

- Sponsor: Adelphia
- Cable modem: LANcity
- Services: Power Link high-speed Internet service commercially available to 4000 homes on a two-way-active portion of the cable system. Unlimited access and cable modem rental for $39.95 per month.

EAST LANSING, MICHIGAN

- Sponsor: TCI
- Cable modems: LANcity, Zenith
- Services: 350 subscribers connected to TCI-MET service for Internet access, including schools, homes, and businesses. Unlimited Internet access is $45 per month.

EASTERN PENNSYLVANIA

- Sponsor: Service Electric/Blue Ridge Cable
- Cable modem: Zenith

- Services: trial Internet access and education services. Eight local cable systems and telcos participating.

ELMIRA, NEW YORK

- Sponsor: Time Warner Cable, with plans to launch a commercial service.
- Cable modem: Motorola and H-P server
- Services: 200 homes in Roadrunner service test of national and local content and Internet access.

FAIRFAX, VIRGINIA

- Sponsor: Media General Cable
- Cable modem: Zenith
- Services: trial Internet access for schools and employees. Twelve trial sites.

FREMONT, CALIFORNIA

- Sponsor: TCI/@Home. Twenty-five percent of the 69,000 TCI-cable homes in Fremont are passed by the two-way plant and are eligible for service.
- Cable modem: LANcity
- Services: first commercial deployment of @Home service is priced at $34.95 per month for unlimited Internet access, cable modem rental, and local content.

JACKSONVILLE, FLORIDA

- Sponsor: Continental Cablevision. Commercial deployment of one-way (telco-return) and two-way cable modems to 425,000 homes passed.
- Cable modems: LANcity, General Instrument. 10,000 GI SURFboard modems.
- Services: modem rental and unlimited Internet access priced at $34.95/month.

LONG ISLAND, NEW YORK

- Sponsor: Cablevision Systems
- Cable modem: Zenith, LANcity, Hybrid

- Services: trial of Internet access. Five hundred homes in Yonkers and Long Island are participating.

MERION, PENNSYLVANIA

- Sponsor: Comcast
- Cable modem: Hybrid Networks
- Services: trial of Internet access, local content, and AOL with 50 participants. Commercial service was available in 1996.

MONTREAL, CANADA

- Sponsor: Cogeco Cable
- Cable modem: Zenith
- Services: trial Internet access for 800 participants in Quebec

NEEDHAM, NEWTON, AND WELLESLEY, MASSACHUSETTS

- Sponsor: Continental Cablevision. Launched commercial "Highway 1" service to 100,000 homes passed by two-way HFC plant in suburban Boston.
- Cable modems: 50,000 LANcity
- Services: modem rental and unlimited Internet access priced at $49.95 per month for cable subs.

NEWMARKET, ONTARIO, CANADA

- Sponsor: Rogers Cablesystems
- Cable modem: Zenith
- Services: commercially deployed WAVE service offers unlimited Internet access for $39.95 per month, plus AOL and CompuServe by request. There are 650 paying subscribers.

OCEAN COUNTY, NEW JERSEY

- Sponsor: Adelphia. Adelphia plans to launch offerings to Grand Island, New York; Mt. Lebanon, Pennsylvania; Lansdale, Pennsylvania; Plymouth, Massachusetts; and Adams/North Adams, Massachusetts.
- Cable modem: LANcity
- Services: Power Link high-speed Internet service commercially available to 10,000 homes on a two-way-active portion of the cable system.

The price is $39.95 per month for unlimited access and cable modem rental.

OMAHA, NEBRASKA

- Sponsor: U.S. West TeleChoice
- Cable modem: LANcity
- Services: 13,000 homes passed by two-way HFC plant. Priced at $49.95 per month for basic cable subs.

PALO ALTO, CALIFORNIA

- Sponsor: PaloAlto Cable Co-op. Technical trial offering Internet access to schools and telecommuting for cable employees.
- Cable modem: Com21
- Services: total of 20 participating sites

PHOENIX, ARIZONA

- Sponsor: Cox Communications with plans to launch commercial service to 50,000 homes passed by two-way HFC
- Cable modem: LANcity
- Services: 150 participants in trial of business connectivity, telecommuting, and Internet access

PLYMOUTH, CANTON, AND NORTHVILLE, MICHIGAN

- Sponsor: Continental Cablevision. Plans to launch "Highway 1" service commercially to 50,000 homes.
- Cable modem: LANcity
- Services: technical trial of high-speed Internet access with six users in suburban Detroit

SAN DIEGO, CALIFORNIA

- Sponsor: Cox Communications
- Cable modem: Zenith
- Services: trial of high-speed Internet access with 150 participants

SUNNYVALE, CALIFORNIA

- Sponsor: TCI/@Home
- Cable modem: LANcity
- Services: commercial deployment of @Home service priced at $34.95 per month for unlimited Internet access, and cable modem rental.

14.4 List of MSOs in North America

MSO	Territory/Region
Adelphia Communication	America's 7th largest MSO
AGI Cablevision	Canadian MSO with systems in south-western Ontario
Belleville Cable	Middleton, WI
Bresnan Communications	Mankato, MN
CableAmerica	MSO with systems in: Alabama Arizona California Michigan Missouri New Mexico
Cable Alabama	Huntsville, AL
Cable Atlantic	Newfoundland, Canada
Cable Online	Hertfordshire South Wales Glasgow Northern Ireland and West Yorkshire
Cable Regina	Dallas
Cable TV of the Kennebunks	Kennebunk, ME
Cablevision	Sunflower Cablevision Tennessee Cablevision Midwest Cablevision
Cambridge Cable Group	UK MSO with systems in: Cambridge area Anglia area The east coast

Chamber Cable	MSO with systems in: Edmonds, WA Eugene, OR Sunriver, OR Payette, ID Chico, CA Novata, CA
Classic Cable	MSO with systems in: Colorado New Mexico Nebraska Kansas Missouri Texas Oklahoma Arkansas
CMI Cable	Irish MSO with systems in: Swords/Malahide Maynooth Mullingar Navan and Sligo
Cogeco Cable	Canadian MSO
Comcast	Comcast of the Carolinas Tallahassee System Sarasota System
Continental Cablevision	
Cox Communications	Phoenix System
Foxtel Cable	Australian MSO with systems over the Telstra Cable Network on the east coast and in Adelaide
Fundy Cable	Canadian MSO with systems in: St. John, Moncton Fredericton Edmundston Bathurst
Garden State Cable	New Jersey
Harron Communications	MSO with systems in: Michigan

	Texas
	Delaware Valley
	New York and throughout New England
	Utica System
Inter Mountain Cable	Harold, KY
Jones Intercable	MSO with 41 systems in 20 states
Kyosai CATV	Japanese system serving Kushigata and Wakakusa in Yamanashi
Le Groupe Videotron Ltee	MSO serving Canada and England
Monterey Peninsula TV Cable	Serving:
	Monterey
	Pacific Grove
	Pebble Beach
	Carmel
	Valley
	Seaside
	Marina
	Spreckels
	Salinas, CA
	(all on TCI Cablevision)
MultiVision Cable TV	Anaheim, CA
Nynex Cable Communications	British MSO
Oceanic Cable	Honolulu, HI
OTV Cablevision	Canadian system serving Oliver, Osoyoos, and Keremeos in British Columbia
Paragon Cable	Portland, OR
Post Newsweek Cable	
Rogers Cablesystems	Canadian MSO
Service Electric	MSO with systems in Pennsylvania and New Jersey
Shaw Cable	Canadian MSO
State Cable	MSO with systems in Maine and New Hampshire
Suburban Cable	MSO with systems in Pennsylvania, New Jersey, and Delaware

TCI	Largest MSO Reno System TCI Media Services
TeleScripps Cable	Serving Greater Bluefield, WV and Bluefield, VA
TKR Cable	MSO with systems in New York, New Jersey, and Kentucky
Time Warner	Milwaukee system
Tri County Communications	Serving: Linden New Richmond Romney Wingate Colfax Other rural areas in Virginia
Viacom	
Co-Axial Ltd.	Canadian MSO serving: Hamilton Flambourogh Dundas Ancaster Grimsby Beamsville Jordan Stoney Creek Halton

14.5 List of Cable Modem Vendors

ADC TELECOMMUNICATIONS.   ADC is delivering customer-driven communications systems, including the *homeworx* cable modem product. ADC Telecommunications was founded in 1935 in Minneapolis, Minnesota. The company currently has over 3000 employees worldwide. The cable modem is part of an integrated system solution in fiber-optics, broadband, video, and wireless technologies.

ADC Telecommunications
4900 West 78th Street
Minneapolis, MN 55435
Telephone: 1-612-938-8080

ALCATEL TELECOM. Alcatel is the worldwide leader in the telecom and power cable market. Alcatel Cable offers leading-edge technology in the fields of high-frequency products, electronics, classic copper, and fiber-optic cables.

Alcatel 1570 BB is an optical transmission system for distribution and cable phone that brings the fiber very close to the subscriber. It is a digital video broadcast system compatible to the most relevant standards of DVB and MPEG.

Alcatel Telecom
33 Rue Emeriau, 75015
Paris, France
Telephone: +1 40 585858

COCOM—ENABLING BROADBAND TECHNOLOGY. COCOM is supplying high-performance CATV equipment to end users to enable high-performance behavior on existing communication networks. COCOM is also involved in the development of WWW caching techniques, CATV, CATV modem for end users, cable controller for CATV operators, COCOM cache for end users, and COCOM cache feeder for Internet providers.

COCOM
Lundtoftevej 80
DK-2800 Denmark
Telephone: +45 45 96 95 94
Fax: +45 45 96 95 95

COM21. Com21 began development of the technology and products needed to accommodate the transmission requirements of the emerging broadband communications to the home industry.

Com21 is involved in several technology trials to test bidirectional CATV network and technology feasibility. In Palo Alto, California, several schools are utilizing Com21 cable modems for high-speed Internet and Email access.

Com21
Tasman Drive
Milpitas, CA 95035
Phone: 1-408-953-9100
Fax: 1-408-953-9299

DIGITAL EQUIPMENT CORP. (DEC).   DEC is the leading worldwide supplier of networked computer systems, software, and services. Digital is also supplying the industry in interactive, distributed, and multivendor computing. Digital does more than half its business outside the United States, developing and manufacturing products and providing customer services in the Americas, Europe, Asia, and the Pacific Rim.

Contact General Inquiries and Literature, 1-800-344-4825 (voice) or 1-800-676-7517 (fax).

FIRST PACIFIC NETWORKS, INC. (FPN).   FPN, based in San Jose, California, supplies equipment for two-way communications over hybrid fiber/coaxial cable networks. FPN was founded in 1988.

The FPN3000 cable modem is under development and supplies telephone-over-cable products as well as data and video transmitted over a single wiring system to large numbers of customers.

FPN has licensed its technology, manufacturing, and products in certain market places as part of this strategy. The focus markets include cable telephony applications for worldwide cable television operators and telephone service companies and domestic (United States) electric power utilities.

First Pacific Networks
871 Fox Lane
San Jose, CA 95131
Telephone: 1-408-943-7600
Fax: 1-408-943-7666

GENERAL INSTRUMENT CORPORATION (GI).   GI is a leader in developing technology and systems for the interactive delivery of video, voice, and data. GI is actively developing technology through intensive, high-quality, low-cost manufacturing. The GI product portfolio includes the family of SURFboard products (one-way cable modems). This company is also pioneering research in satellite television encryption and broadband digital compression technologies. Other areas of expertise are radio frequency and fiber-optic distribution electronics.

Fore Systems Inc. and General Instrument Corp. have embarked upon a joint venture, developing an ATM-based cable modem that could give cable TV operators control of service, billing, and security. Fore is contributing its 2.5–10-Gbit switch at the headend, a PC adapter card, and network management equipment.

> General Instrument Corporation
> 8770 W. Bryn Mawr Avenue
> Suite 1300
> Chicago, IL 60631 773.695.1000
> Telephone: 1-516-847-3164

HEWLETT-PACKARD COMPANY (HP). HP designs, manufactures, and services electronic products and systems for measurement, computing, and communication used by people in business, engineering, science, medicine, and education.

HP is one of the largest industrial companies in the United States and one of the world's largest computer companies. More than 55 percent of its business is generated outside the United States; two-thirds of that is in Europe. In telecommunications, HP is actively developing the QuickBurst cable modem and other related products.

> HP Corporate Office
> 3000 Hanover Street
> Mailstop 20BR
> Palo Alto, CA 94304-1185

HYBRID NETWORKS, INC. Hybrid Networks was founded in June 1990 to develop and market high-speed network connectivity products and services. The company's Hybrid Access System (HAS) was developed for a high-speed digital signal to be received by every home, school, or office using existing cable TV and telephone transmission facilities. Hybrid has been testing its high-speed network on TCI, Viacom, Comcast, and Bell-South cable systems. Contact Hybrid Networks, Inc. at 1-408-725-3250.

IBM. IBM is developing a cable online access system that consists of cable modems working together with an access hub. This system enables cable TV operators to deliver high-speed data services such as Internet access, local community information, and telecommuting for work at home applications—all over hybrid fiber coax networks. The cable online access system is capable of providing different grades of service. Cost-sensitive subscribers, for example, can subscribe to best-effort service for cybersurfing, whereas business subscribers can subscribe to premium, guaranteed bit-rate grades of service. The cable online access system uses ATM as the underlying multiplexing and transport technology as a powerful multiplexing technology. The cable online modem implements dynamic IP address assignment to minimize the need for configuration, thus minimizing operator and subscriber expense. Contact IBM at 1-800-IBM-3333.

INTEL.   Intel is the world leader in developing and supplying microprocessor technologies. This firm is actively developing an NB and Cable-Port modem product.

> Intel Corporation, Chandler
> 5000 W. Chandler Blvd
> Chandler, AZ 85226-3699
> Telephone: 1-602-554-8080
> Fax: 1-916-956-5427

LANcity.   LANcity is active in the development of modem technologies. More than 31,000 cable modems have been installed in more than 271 different sites worldwide. In 1995, LANcity unveiled the personal product line for cable TV modem. Currently, LANcity is developing the first industry-standard cable modem, incorporating IEEE 802.14 standards, MCN initiatives, IETF, DAVIC, and installed-base customer demands. The new product, named LCX, is anticipated to be released in 1997.

> LANcity
> Bay Networks, Inc.
> Internet Telecom Business Units
> LANcity Cable Modem Division
> Andover Business Park
> 200 Bulfinch Drive
> Andover, MA 01810-1140
> Telephone: 1-800-LANcity or 508-682-1600 x201.
> Fax: 1-508-682-3200

LUCENT.   Lucent is a world leader in telecommunications networking. For the central offices of the service provider, there's the Exchange MAX® distribution system. An end-to-end wiring network for the home is available under the trade name "HomeStar® Wiring System." Contact 1-317-322-6848.

MOTOROLA.   Motorola Multimedia Group is actively working on CableComm technology. The CableComm family of products is targeted for market to the MSOs, telcos, and end users.

The Motorola Multimedia Group designed Motorola's CyberSurfr cable modems released in 1995. Motorola has delivered CyberSurfr cable data modems, CableComm units for voice over cable, and infrastructure products to Time Warner, Tele-Communications Inc. (TCI), Comcast Corporation, Optus Vision, Shaw Communications, Generale des Eaux, Videotron, and two RBOCs.

Motorola
Corporate Offices
1303 E. Algonquin Road
Schaumburg, IL 60196

NEC. The NEC strategy is to work in stages. The first stage is the development of multimedia equipment; the second stage sees the growth of networks, in which individual multimedia equipment is linked through multimedia networks; and the third stage will see the formation of a new multimedia culture of multimedia societies through the proliferation of multimedia networks. NEC announced a cable modem to address the above plan.

Multimedia Group
Telephone: 1-800-2-WAY-HFC

NEC Corporation
7-1, Shiba 5-chome, Minato-ku,
Tokyo, Japan

PHASECOM INC. Phasecom is developing solutions that enable high-speed data communications over broadband network infrastructures for residential and commercial applications. Phasecom, founded in 1981, is a Cupertino, California—based company developing products for hybrid fiber coax (HFC) services. These products are targeted to interactive multimedia applications over hybrid fiber coax networks—primarily used today for cable TV applications.

Phasecom, Inc.
20700 Valley Green Drive
Cupertino, CA 95014
Telephone: 1-408-777-7784
Fax: 1-408-777-7787

SCIENTIFIC-ATLANTA (SA). SA is at the forefront of today's telecommunications revolution, because cable operators, broadcasters, telephone and utility companies, governments, and corporations worldwide rely on SA's advanced broadband and satellite networks to deliver communications services. Scientific-Atlanta is a global company that conducts business from Argentina to Zaire through 16 worldwide offices and representatives in more than 70 countries.

Scientific-Atlanta is actively involved in tests and deployment of interactive cable networks with Time Warner, U.S. WEST, Ameritech, BellSouth, SNET, and the 1996 Olympic Games.

> Scientific-Atlanta
> Cablemart
> 4311 Communications Drive
> Norcross, GA 30093
> Telephone: 1-800-241-5787 (in Georgia, 1-770-903-5800)

14.6 List of Cable Modems on the Market

Vendor	Cable modem name	Data rate, down/ upstream	Frequency agility	Carrier	Modu- lation	Release date
ADC/ NetComm	Home- works	8.2 Mbps/ 8.2 Mbps	N/A	6 MHz/ 6 MHz	32 QAM/ OFDM	
Alexon	Cable Master	2.016 Mbps/ 64 Kbps	72–800 MHz/ 10–57 MHz	435 kHz/ 108.57 kHz	FSK/FSK	1997
COCOM		30 Mbps/ 1 Mbps	47–862 MHz/ 5–42 MHz	8 MHz/ 500 kHz	64 QAM/ QPSK	
Com21/ 3Com	Com- Port	30 Mbps/ 1.9 Mbps	88–800 MHz 5–40 MHz	upstream = 2 MHz	64 QAM/ QPSK	1996
Digital	LCB	10Mbps/ 10 Mbps			QPSK/ QPSK	1996
General Instrument	SURF- Board	27 Mbps/ 1.7 Mbps	50–860 MHz 5–20 MHz	upstream = 1 MHz	64 QAM/ QPSK	1997
Fore Systems	ATM	1.7 Mbps	5–20 MHz	upstream = 1 MHz		
Hewlett- Packard	Quick- Burst	30 Mbps/ 3 Mbps	88–806 MHz/ 10–32 MHz	upstream = 2 MHz	64 QAM/ QPSK	1996

Vendor	Cable modem name	Data rate, down/ upstream	Frequency agility	Carrier	Modu- lation	Release date
Hughes Network					64 QAM/ QPSK	
Hybrid Networks	Remote Link	30 Mbps/ 512 Kbps	50–750 MHz/ 5–40 MHz	upstream = 300 kHz	64 QAM/ 2,4.8. VSB	
IBM		30 Mbps/ 2 Mbps	54–750 MHz/ 5–40 MHz	upstream = 1.8 MHz	64 QAM/ QPSK	
LANcity	LCP	10 Mbps/ 10 Mbps	54–550 MHz/ 5–40 MHz	6 MHz/ 6 MHz	QPSK/ QPSK	1996
Motorola	Cyber- Surfr	10 Mbps/ 768 Kbps	50–750 MHz/ 5–42 MHz	upstream = 600 kHz	64 QAM/ 4-DQPSK	1996
NetComm Boca	Net- Rocket	30 Mbps/ 64–512 Kbps				1997
NetGame	NeMo	6 Mbps/ 1.5 Mbps	50–750 MHz/ 5–42 MHz	upstream = 1.5 MHz	MSK/MSK	1996
Nortel/Arris Interactive	Corner- stone DataPort	27 Mbps/ 2 Mbps				
Panasonic	TM-1996	30 Mbps			64 QAM/ QPSK	1997
Phasecom	Speed Demon	30 Mbps/ 3 Mbps		6 MHz/ 2 MHz	64 QAM/ QPSK	
Pioneer	Speed Station	28.5 Mbps/ 2.5 Mbps			64 QAM/ 4-DQPAK	1997
Scientific Atlanta	The Godfather	6 Mbps/ 1.5 Mbps		upstream = 1 MHz	QPR/ QPSK	1997
Terayon	TeraPro	10 Mbps/ 10 Mbps	54–750 MHz/ 5–42 MHz	6 MHz/ 6 MHz		1996
Toshiba	PCX	8.192 Mbps/ 2.048 Mbps		upstream = 1.5 MHz	QPSK/ QPSK	1997
Tytec	LinCable				BPSK	1996
West End Systems	Westbond 9604 Universal	256 Kbps/ 256 Kbps			OFDM/ OFDM	
Zenith	Home- works	4 Mbps/ 4 Mbps	5–750 MHz/ 12–108 MHz	6 MHz/ 6 MHz	BPSK/ BPSK	1996

GLOSSARY

access channels Channels set aside by the cable operator for use by the public, educational institutions, municipal government, or for lease on a nondiscriminatory basis.

aerial plant Cable that is suspended in the air on telephone or electric utility poles.

alternative access provider A telecommunications provider other than the local telephone company that provides a connection between a customer's premises (usually a large business customer) to the point of presence of the long distance carrier.

amplifier A device that boosts the strength of an electronic signal. In a cable system, amplifiers are spaced at regular intervals throughout the system to keep signals picture-perfect no matter where you live.

asynchronous transfer mode (ATM) The transfer mode in which the information is organized into cells. It is asynchronous in the sense that the recurrence of cells containing information from an individual user is not necessarily periodic.

ATM cell A digital information block of fixed length (53 octets) identified by a label at the ATM layer.

available bit rate (ABR) An ATM layer service where the limiting ATM-layer transfer characteristics provided by the network may change subsequent to connection establishment.

bandwidth Frequency spectrum used to transmit pictures, sounds, or both. The average television station uses a bandwidth of six million cycles per second (6 megahertz). Also, a measurable characteristic defining the available resources of a device in a specific time period (typically in one second).

broadband A service or system requiring transmission channels capable of supporting rates greater than 1.5 Mb/s.

broadband communications system Frequently used as a synonym for cable television. It can describe any system capable of delivering wideband channels and services.

broadcast addresses A predefined destination address that denotes the set of all service access points.

burst error second Any errored second containing at least 100 errors.

cable-ready television set A television set with an improved tuner that is more resistant to cable service without a franchise.

cable television Communications system that distributes broadcast and nonbroadcast signals as well as a multiplicity of satellite signals, original programming, and other signals by means of a coaxial cable and/or optical fiber.

call An association between two or more users or between a user and a network entity that is established by the user of the network capabilities. This association may have zero or more connections.

CATV (community antenna television or cable television) A communication system that simultaneously distributes several different channels of broadcast programs and other information to customers via a coaxial cable.

cell ATM layer protocol data unit.

central office (CO) The central location in a traditional public network telecommunication environment wherein access is available to signals traveling in both the forward and reverse directions.

channel A communication path. Multiple channels can be multiplexed over a single cable in the cable television environment.

channel capacity Maximum number of channels that a cable system can carry simultaneously.

coaxial cable Actual line of transmission for carrying television signals. Its principal conductor is either a pure copper or copper-coated wire surrounded by insulation and then encased in aluminum.

communications common carrier General name for any medium that carries messages prepared by others for a fee and is required by law to offer its services on a nondiscriminatory basis. Common carriers are regulated by federal and state agencies, and they exercise no control over the message content carried.

competitive access provider A telecommunications entity engaged in providing competitive access service.

connection An association established by a layer between two or more users of the layer service for the transfer of information.

connectionless service A service that allows the transfer of information among service subscribers without the need for end-to-end call establishment procedures.

constant bit rate (CBR) A service class intended for real-time applications, or those requiring tightly constrained delay and delay variation, as would be appropriate for voice and video applications. The consistent availability of a fixed quantity of bandwidth is considered appropriate for CBR service.

converter A device that is attached between the television set and the cable system that can increase the number of channels available on the TV set, enabling it to accommodate the multiplicity of channels offered by cable TV.

data bandwidth A measurable characteristic defining the bandwidth of a device (measured in bits per second).

data link layer In open system interconnection (OSI) architecture, the layer that provides services to transfer data over the transmission link between open systems.

DBS (direct broadcasting satellite) The system in which signals are transmitted directly from a satellite to a home rooftop receiving dish.

delay The elapsed time between the instant when user information is submitted to the network and when it is received by the user at the other end.

demographics Breakdown of television viewers by such factors as age, sex, income levels, education, and race. These figures are used in selling advertising time.

descrambler An electronic circuit that restores a scrambled video signal to its standard form.

digital compression An engineering technique for converting a cable television signal into digital format (in which form the signal can easily be stored).

distant signals A television channel from another market imported and carried locally by a cable television system.

distribution system Part of a cable system consisting of trunk and feeder cables used to carry signals from headend to customer terminals.

downstream Flow of signals from the cable system headend through the distribution network to the customer.

drop cable Coaxial cable that connects a residence or service location from a tap on the nearest feeder cable.

dual cable Two independent distribution systems operating side by side, providing double the channel capacity of a single cable.

earth station Structure referred to as a *dish* used for receiving.

end user A human being, organization, or telecommunications system that accesses the network in order to communicate via the services provided by the network.

errored second   Any one-second interval containing at least one bit error.

extended subsplit   A frequency division scheme that allows bidirectional traffic on a single cable. Reverse path signals come to the headend from 5 to 45 MHz. Forward-path signals go from the headend from 54 to the upper frequency limit.

feeder cable   Cables that run down streets in the served area and distribute the signal to the individual taps.

feeder line   Cable distribution lines that connect the main trunk line or cable to the smaller drop cable.

fiber optics   Very thin and pliable tubes of glass or plastic used to carry wide bands of frequencies.

FM cable system   FM radio signals offered by a cable system (the cable must be connected to the customer's FM stereo receiver).

forward channel   The direction of RF signal flow away from the headend toward the end user.

frequency bandwidth   A measurable characteristic defining the number of cycles that can be conveyed or transported by a device in one second, measured in Hertz (cycles per second).

guardband   Provides for slot timing uncertainty due to inaccuracy of the ranging.

hardware   Equipment involved in production, storage, distribution, or reception of electronic signals, such as the headend and the coaxial cable.

HDTV   A television signal with greater detail and fidelity than the one in current use.

headend   The central location in an MSO environment that has access to signals traveling in both the forward and reverse directions.

header   Protocol control information located at the beginning of a protocol data unit.

HFC system   Hybrid fiber coax cable television system. Fiber is used for the trunks and coax is used to access the end nodes.

high split   A frequency division scheme that allows bidirectional traffic on a single cable. Reverse-path signals propagate to the headend from 5 to 174 MHz. Forward-path signals go from the headend from 234 MHz to the upper frequency limit. A guardband is located from 174 to 234 MHz.

homes passed   Total number of homes that have the potential for being hooked up to the cable system.

HUBS   Local distribution centers where signals are taken from a master feed and transmitted over cable to customers.

independent   Individually owned and operated cable television system not affiliated with an MSO.

interconnect   Connection of two or more cable systems by microwave, fiber, coaxial cable, or satellite, so that programming or advertising may be exchanged, shared, or simultaneously viewed.

interdiction   A method of receiving TV signals by jamming unauthorized signals but having all other signals received in the clear. Because the jamming is accomplished outside the home and does not require a set-top terminal in the home, interdiction is receiving more operator interest, especially in light of recent FCC actions encouraging more consumer-friendly approaches.

interexchange carriers   A long-distance carrier between serving areas of LATAs.

INTER-LATA   The provision of telecommunications services between LATAs. Pursuant to the AT&T Consent Decree, the RBOCs prior to the 1996 Reform Act were prohibited from providing telecommunications services between LATAs.

intra-LATA   The area within a LATA in which, pursuant to the AT&T Consent Decree, the RBOCs are permitted to offer local telephone service.

isochronous   The time characteristics of an event or signal recurring at known, periodic time intervals.

layer   A subdivision of the open system interconnection (OSI) architecture, constituted by subsystems of the same rank.

leased channels   Any channels made available by the operator to potential programmers for a fee.

local area network (LAN)   A nonpublic data network in which serial transmission is used without store and forward techniques for direct data communication among data stations located on the user's premises.

local exchange carrier (LEC)   A local telephone company within a serving area or LATA.

local loop   The set of facilities used by a telephone company to transport signals between a central office, roughly similar to a cable TV headend, and a customer location. The LOCAL LOOP using twisted-pair copper wire typically stretches a maximum of 18,000 feet between the central office and customer premises.

local originating programming   Programming developed by an individual cable television system specifically for the community it serves.

logical link control (LLC) procedure   In a local area network (LAN) or a metropolitan area network (MAN), that part of the protocol that governs the assembling of data link layer frames and their exchange between data stations, independent of how the transmission medium is shared.

MAC address   An address that identifies a particular medium access control (MAC) sublayer service access point.

medium-access control (MAC) procedure   In a subnetwork, that part of the protocol that governs access to the transmission medium, independent of the physical characteristics of the medium, but taking into account the topological aspects of the subnetworks in order to enable the exchange of data between nodes. MAC procedures include framing, error protection, and so on.

medium-access control (MAC) sublayer   The part of the data link layer that supports topology-dependent functions and uses the services of the physical layer to provide services to the logical link control (LLC) sublayer.

metropolitan area network   A network that spans a metropolitan area. Generally, a network spans a larger geographic area than a LAN but a smaller geographic area than a WAN.

microwave   One method of interconnecting a cable system with a series of high-frequency receive-and-transmit antennas mounted on towers spaced up to 50 miles apart.

midsplit   A frequency-division scheme that allows bidirectional traffic on a single cable. Reverse-channel signals propagate to the headend from 5 to 108 MHz. Forward-path signals go from the headend from 162 MHz to the upper frequency limit. The guardband is located from 108 to 162 MHz.

MSO (multiple system operators)   A company that owns and operates more than one cable television system.

multicast addresses   A predefined destination address that denotes a subset of all service access points.

multiplexing   The potential transmission of several feeds of the same cable network with the same programming available at different times of the day. This is seen as one possible use of the additional channel capacity that may be made available by digital compression. Multiplexing is also used by some cable networks to mean transmitting several slightly different versions of the network, like several MTV channels carrying different genres of music.

multipoint access   User access in which more than one terminal equipment is supported by a single network termination.

multipoint connection   A connection with more than two end points.

NCTA (National Cable Television Association)   The major trade association for the cable television industry.

near video on demand (NVOD)   An entertainment and information service that broadcasts a common set of programs to customers on a scheduled basis. At least initially, NVOD services are expected to focus on delivery of movies and other video entertainment. NVOD typically features a schedule of popular movies and events offered on a staggered-start basis (every 15 to 30 minutes, for example).

network layer   In open system interconnection (OSI) architecture, the layer that provides services to establish a path between open systems with a predictable quality of service.

network management   Within IEEE 802, the functions related to the management of the data link layer and the physical layer resources and their stations across the IEEE 802 local area network (LAN) or metropolitan area network (MAN).

node   A device that consists of an access unit and a single point of attachment of the access unit for the purpose of transmitting and receiving data.

number portability   A capability that permits telecommunications users to maintain the same telephone access number as they change telecommunications suppliers.

on-demand service   A type of telecommunication service in which the communication path is established almost immediately in response to a user request brought about by means of user-network signaling.

pay-per-view   Cable programming for which customers pay on a one-time basis (e.g., for prize fights, Broadway shows, and movie premieres).

pay programming   Movies, sports, and made-for-cable specials that are available to the cable customer for a charge in addition to the basic fee.

peer entities   Entities within the same layer.

penetration   Ratio of the number of cable customers (or pay-TV customers) to the total number of households passed by the system.

personal communications services (PCSs) A new wireless communications service that allows users to communicate through the use of miniature handheld devices transmitted over radio waves. The technology uses a network of transmission towers or minicells to relay the signal from one point to another.

physical layer In open system interconnections (OSI) architecture, the layer that provides services to transmit bits or groups of bits over a transmission link between open systems.

point of presence (POP) The place at which, pursuant to the AT&T consent decree, a long-distance carrier interconnects with a local telephone company.

pole attachment Cable television hookups to telephone or electric utility poles.

protocol A set of rules and formats that determines the communication behavior of layer entities in the performance of the layer functions.

quality of service (QOS) The accumulation of the cell loss, delay, and delay variation incurred by the cells belonging to a particular ATM connection.

reverse channel The direction of signal flow towards the headend, away from the subscriber.

satellite (domestic communications) Device located in geostationary orbit above the earth that receives transmissions from separate points and retransmits them to cable systems, DBSs, and others over a wide area.

satellite master antenna television system (SMATV) Systems that serve a concentration of TV sets such as an apartment building, hotel, and so on, utilizing one central antenna to pick up broadcast and/or satellite signals.

scrambling A signal security technique for rendering a TV picture unviewable while permitting full restoration with a properly authorized decoder or descrambler.

service data unit (SDU) Information that is delivered as a unit between peer service access points (SAPs).

set-top box Any of several different electronic devices that may be used in a customer's home to enable services to be on that customer's television set. If the set-top device is for extended tuning of channels only, it is called a *converter*. If it restores scrambled or otherwise protected signals, it is a *descrambler*.

shop at home Programs allowing customers to view products and/or order them by cable television, complete with catalogs, shopping shows, and so on.

sublayer   A subdivision of a layer in the open system interconnection (OSI) reference model.

subnetwork   Subnetworks are physically formed by connecting adjacent nodes with transmission links.

subscriber   A customer who pays a monthly fee to cable system operators for the capability of receiving a diversity of programs and services.

subsplit   A frequency division scheme that allows bidirectional traffic on a single cable. Reverse-path signals come to the headend from 5 to 30 (up to 42 on newer systems) MHz. Forward path signals go from the headend from 54 to the upper frequency limit.

subsystem   An element in a hierarchical division of an open system that interacts directly with elements in the next higher division or the next lower division of that open system.

switched connection   A connection established via signaling.

systems management   Functions in the application layer related to the management of various open systems interconnection (OSI) resources and their status across all layers of the OSI architecture.

traffic parameter   A parameter for specifying a particular traffic aspect of a connection.

transit delay   The time difference between the instant at which the first bit of a PDU crosses one designated boundary and the instant at which the last bit of the same PDU crosses a second designated boundary.

translator   Relay system that picks up distant television signals, converts the signals to another channel to avoid interference, and retransmits them into areas the original television station could not reach.

transmission link   The physical unit of a subnetwork that provides the transmission connection between adjacent nodes.

transmission medium   The material on which information signals may be carried, such as optical fiber, coaxial cable, and twisted-wire pairs.

transmission system   The interface and transmission medium through which peer physical-layer entities transfer bits.

transponder   The part of a satellite that receives and transmits a signal.

trunk cable   Cables that carry the signal from the headend to groups of subscribers.

trunking   Transporting signals from one point (an antenna site for TV systems) to another.

unbundling   The separation and discrete offering of the components of the local telephone service. Unbundling of network components facilitates the provision of pieces of the local network, such as local switching and transport, by telephone company competitors.

unicast   An address specifying a single network device.

unspecified bit rate (UBR)   The UBR service class is intended for delay-tolerant or non-real-time applications, or those which do not require tightly constrained delay and delay variation, such as traditional computer communications applications. Sources are expected to transmit noncontinuous bursts of cells.

upstream   The flow of any information from the customer, through the cable system, to the headend.

variable bit rate (VBR)   A type of telecommunication service characterized by a service bit rate specified by statistically expressed parameters that allow the bit rate to vary within defined limits.

variable bit rate (VBR), non-real-time users   The non-real-time VBR service class is intended for non-real-time applications that have bursty traffic characteristics and can be characterized in terms of a generic cell-rate algorithm (GCRA).

variable bit rate (VBR), real-time   The real-time VBR service class is intended for real-time applications, or those requiring tightly constrained delay and delay variation, as would be appropriate for voice and video applications. Sources are expected to transmit at a rate that varies in time; the source can be described as bursty. VBR service may support statistical multiplexing of real-time sources, or may provide a consistently guaranteed QOS.

video dialtone   The means by which telephone companies may provide transmission facilities and for on-telco video programming as well as certain enhanced services to third-party programmers.

video on demand (VOD)   An entertainment and information service that allows customers to order programs from a library of material at any time.

virtual channel (VC)   The communication channel that provides for the sequential unidirectional transport of ATM cells.

ACRONYMS

AAL	ATM adaptation layer
ABR	Available bit rate
ADSL	Asymmetric digital subscriber line
ANM	Answer message
ANSI	American National Standards Institute
API	Application program interface
ATM	Asynchronous transfer mode
BECN	Backward explicit congestion notification
BER	Bit error rate
B-ICI	BISDN intercarrier interface
BISDN	Broadband integrated services digital network
BSS	Broadband switching system
BW	Bandwidth
CAC	Connection admission control
CATV	Community antenna television
CBR	Constant bit rate
CCITT	International Telephone and Telegraph Consultative Committee
CdPN	Called party number
CDV	Cell delay variation
CES	Circuit emulation service
CgPN	Calling party number
CLP	Cell loss priority
CDMA	Code division multiple access
CPCS	Common part convergence sublayer
CPE	Customer premises equipment
CRC	Cyclic redundancy check
CSI	Carrier selection information
CSMA/CD	Carrier sense multiple access with collision detection
DS-0	Digital signal level 0 (DS0 = 64 Kb/s)
DTE	Data terminal equipment

EFCI	Explicit forward congestion indication
ETSI	European telecommunications standards institute
FEBE	Far-end block error
FECN	Forward explicit congestion notification
FERF	Far-end receive failure
FRS	Frame relay service
E1	ETSI digital signal level 1 (E1 = 2.048 Mb/s)
HE	Headend
HEC	Header error control
HFC	Hybrid fiber coax
HDTV	High-definition TV
ICIP	Intercarrier interface protocol
IEEE	Institute of Electrical and Electronics Engineers
IETF	Internet engineering task force
ILEC	Independent local exchange carrier
IN	Intelligent network
ITU	International Telecommunications Union
IP	Internet protocol
ISDN	Integrated services digital network
IXC	Interexchange carrier
LAN/WAN	Local/wide-area network
LATA	Local access and transport area
LEC	Local exchange carrier
LLC	Logical link control
MAC	Medium-access control
MIB	Management information base
MMDS	Multichannel/multipoint distribution system
MPEGx	Motion picture editors' group compression algorithm x
NNI	Network node interface
NPC	Network parameter control
NT	Network terminal
OSI	Open system inerconnect
PV	Personal computer

PCR	Peak cell rate
PDH	Pleisiochronous digital hierarchy
PLCP	Physical layer convergence procedure
POTS	Plain old telephone service
PMD	Physical-medium-dependent
PVC	Permanent virtual connection
QAM	Quadrature amplitude modulation
QPSK	Quaternary phase shift keying
QOS	Quality of service
RF	Radio frequency
SS7	Signaling system number 7
SAAL	Signaling ATM adaptation layer
SCR	Sustained cell rate
SDH	Synchronous digital hierarchy
SMDS	Switched multimegabit data service
SNMP	Simple network management protocol
SONET	Synchronous optical network
SPE	Synchronous payload envelope
SRTS	Synchronous residual time stamp
SSCOP	Service-specific connection-oriented protocol
SSCS	Service-specific convergence sublayer
STB	Set-top box
STM	Synchronous transfer mode
SVC	Switched virtual connection
TDMA	Time division multiple access
UNI	User network interface
UPC	Usage parameter control
VBR	Variable bit rate
VCC	Virtual channel connection
VCI	Virtual channel identifier
VPC	Virtual path connection
VPCI	Virtual path connection identifier

BIBLIOGRAPHY

Abbott, C. "Proposal for reservation-based MAC protocol." Contribution number IEEE 802.14-95/140, November 1995.

Abbott, C., S. Behtash, and L. Yamano. "Proposal for multirate QPSK/QAM PHY protocol for 802.14." Contribution number IEEE 802.14-95/141, November 1995.

Agarwal, R. "A simple and efficient multiple-access protocol." Contribution number IEEE 802.14-95/136, November 1995.

"Algorithm for a Random Access Broadcast Channel." *IEEE Trans. on Inform. Theory,* vol. 34.

ALOHA. "Packet System with and without Slots and Capture." *Computer Communications Review,* April 1975.

Angelopoulos, J. and G. Stassinopoulo. "A proposal for the support of ABR services in HFC." National Tech. U. of Athens, Contribution Number IEEE 802.14-96/166, July 1996.

ANSI/IEEE. "Carrier sense multiple access with collision detection (CSMA/CD) access method and physical layer specifications." ANSI/IEEE Std. 802.3-1985, 1985.

ATM Forum. "BISDN intercarrier interface (B-ICI) specification version 2.0."

ATM Forum. "Traffic management specification, draft version 4.0."

ATM Forum. "Traffic control and congestion control." UNI specification version 3.0, section 3.6, September 10, 1993.

ATM Forum. "User-network interface specification version 3.1," September 1994.

ATM Forum. "User-network interface (UNI) signaling specification version 4.0," July 1996.

Azzam, A. et al. "ATM over ADSL," *ATM Forum Magazine,* 1997.

Azzam, A. et al. "IEEE P 802.14 Cable TV functional requirements and evaluation criteria." Contribution number IEEE 802.14-94/002R3, November 1993.

Azzam, A. et al. "Prospectus on voice/POTS over ATM." Internal Alcatel white paper, August 1996.

Azzam, A. et al. "Video on demand architecture," July 1995.

Bertsekas, D. and R. Gallager. *Data Networks,* 2d ed. Prentice Hall, 1992.

Bingham, J. and K. Jacobsen. "A proposal for an SDMTPHY layer." Contribution number IEEE 802.14-95/138, November 1995.

Bingham, J. and K. Jacobsen. "A proposal for MAC protocol to support both QPSK and SDMT." Contribution number IEEE 802.14-95/137, November 1995.

Bingham, J. "CATV rev. channel using synchronized DMT (part 1)." Contribution number IEEE 802.14-95/001, January 1995.

Bingham, J. "CATV rev. channel using synchronized DMT (part 3)." Contribution number IEEE 802.14-95/003, January 1995.

Bingham, J. "Multicarrier Modulation for Data Transmission: An Idea Whose Time Has Come." *IEEE Communications,* May 1990.

Bisdikian, C. "A review of random access algorithms." Contribution number IEEE 802.14-96-019, January 1996.

Bisdikian, C. "Performance analysis of the multislot *n*-ary stack random access algorithm (msSTART)." Contribution number IEEE 802.14-96-117, IEEE 802.14 working group (WG) meeting, May 1996.

CableLabs. "A Successful Return Path on Cable Plant is Crucial for the Implementation of New Services." *SPECS Technology Newsletter* **8**(2), March/April 1996.

CableLabs. "Cable Television's Entree into Personal Communications Services." *SPECS Technology Newsletter* **7**(2), March 1995.

"CableLabs." *SPECS Technology Newsletter* 3(7), May 1996.

CableLabs. "Two-way cable television system characterization: final report," April 12, 1995.

Cable Television Laboratory, Inc. *Digital Transmission Characterization,* November 1994.

Capetanakis, T. "Tree Algorithm for a Packet Broadcasting Channel." *IEEE Transaction Theory,* vol. IT-25.

"Channelized solutions." FPN internal memorandum.

Chaudhry, T. "IEEE 802.14 MAC layer." Contribution number IEEE 802.14-95/164, November 1995.

Chen, K. C. "Implementation of an Experimental Wireless LAN Based on Randomly Addressed Polling." *Proc. 1995 Communication Networks Workshop,* 1995.

Chen, K. C. and D. C. Twu. "General multilayer collision resolution with reservation: A MAC protocol for broadband communication network." Contribution number IEEE 802.14-95/167, November 1995.

Chen, W. "Upstream ingress noise model." Contribution number IEEE 802.14-95/110, September 1995.

Chiu, Ran-Fun and Robert C. Hutchinson. "Multimedia transmission link protocol: A proposal for the 802.14 MAC layer." Contribution number IEEE 802.14/95-160, November 6, 1995.

Cioffi, J. and K. Jacobsen. "15 Mbps on the mid-CSA." T1E1.4 contribution number 94-088, Palo Alto, April 1994.

Citta, R. "Phase-2 simulation results for adaptive random access protocol." Contribution number IEEE 802.14-96/114, IEEE 802.14 working group (WG) meeting, May 1996.

Citta, R., G. Sgrignoli, and D. Mutzabaugh. "Performance history in two-way cable parts unitizing a PSK communication system." Contribution number IEEE 802.catv-94/009, 1994.

Citta, R. and D. Willming. "Proposal for a spectrum upstream modem for CATV networks." Contribution number IEEE 802.14-95/143, October 1995.

Clanton, C. "Clarification of security requirements for shared media access." Contribution to ATM_Forum/96-1256, October 1996.

Cruickshank, R. "Request for performance measurements of high-speed cable data services." Contribution number IEEE 802.14-95/150, October 30, 1995.

Currivan, B. and W. Chen. "Cable channel model parameters." Contribution number IEEE 802.14-95/42, May 1995.

Currivan, B. "CATV upstream channel model, rev. 1.0." Contribution number IEEE 802.14-95/133, March 1997.

Currivan, B. "Proposal for QPSK upstream modulation proposal update #2." Contribution number IEEE 802.14-95/132, November 1995.

David Sarnoff Research Center. "Resolving noise problems in the demodulation of downstream (64/256 QAM) and signal ingress in the return channel for 16 QAM signals." Contribution number IEEE 802.14 96/124, May 1, 1996.

Dawson, F. "Cable Consensus on Modems Begins to Crack." *Inter@ctive Week,* October 1996.

Dawson, F. "Investigating before Investing: Filling the Potholes on the Road to Digital." *Communications Engineering & Design,* February 1996.

"Delivery system architecture and interfaces." DAVIC 1.0 specification part 4, technical report, revision 5.0.

Deloddere, D. et al. "Interactive Video On Demand." *IEEE Communications Magazine*, May 1994.

Deloitte and Touche. "Independent survey of corporate telecommunications managers," 1996.

Denny, L. "Why 802.14 should support telephony." Contribution number IEEE 802.14-96/199, July 1996.

De Prycker, Martin. "Asynchronous transfer mode: solution for B-ISDN," 1993.

Desmet, E. "The ATM layer available bit rate service category." Alcatel internal position paper, 1996.

Digital Video Broadcasting (DVB) Project. "Background documents on digital video broadcasting," April 1994.

Eldering, C., M. Kauffman, D. Grubb, S. Warwick, and G. Hawley. "Engineering requirements for hybrid fiber-coax telephony systems." Contribution number IEEE 802.catv/021, 1994.

Eng, J. "Standards for HFC-Based Residential Broadband: IEEE Project 802.14, Its Mission, Charter, and Status." *SPIE Proceedings*, 1995.

Enns, R. "Multimedia MAC (M3) proposal." Contribution number IEEE 802.14-95/162, November 1995.

European Telecommunication Standard. Draft proposal ETS 300 429, August 1994.

Fernandez, J. M. "Downstream physical layer recommendations." Contribution IEEE 802.14-96/179, July 1996.

Fernandez, J. and L. Montreuil. "A proposal to use QAM for the hybrid fiber coax downstream." Contribution number IEEE 802.14-95/069, July 1995.

Gatherer, Alan. "CATV downstream channel model, rev. 1.2." Contribution number IEEE 802.14-96/175, March 1997.

Gatherer, A. and T. Wright, "Spectral inversion in the downstream." Contribution number IEEE 802.14-97/036, March 1997.

Gingold, D. "Integrated digital services for cable: economics, architecture, and the role of standards—MIT research program on communications policy." Contribution number IEEE 802.14-96/230, September 1996.

Golmie N. et al. "Additional test scenarios for the evaluation of the MAC protocol proposals." Contribution number IEEE 802.14/96-099, IEEE 802.14 working group (WG) meeting, March 1996.

Golmie N. et al. "On the issue of frame format (size and contention/data slots ratio)." Contribution number IEEE 802.14/96-159, IEEE 802.14 working group (WG) meeting, May 1996.

Golmie N. et al. "Performance evaluation of contention resolution algorithms: Ternary-tree vs. p-persistence." Contribution number IEEE 802.14/96-241, IEEE 802.14 working group (WG) meeting, September 1996.

Golmie N. et al. "Performance evaluation of MAC protocol components for HFC networks." *Broadband Access Systems Proc. SPIE 2917,* Boston, Massachusetts, 1996, pp. 120–130.

Gosh, M. and S. Hulgelker. "Method for combating ingress noise for multipath in a CATV channel." Contribution number IEEE 802.14-96/111, May 1996.

Graham, C. and Department of Computer Sciences, Illinois Institute of Technology. "Criteria for a MAC protocol standard for the cable television medium." Contribution number IEEE 802.catv-94/03, 1994.

Graham, C. and Department of Computer Sciences, Illinois Institute of Technology. "DQRAP: A proposed standard for the 'last mile' of an IEEE 802.6 MAN." Contribution number IEEE 802.6-93/13, 1993.

Grossman, D. "Security overview." Contribution number IEEE 802.14-97/030, March 1997.

Grossman, D. "Signaling requirements for ATM over HFC." Contribution number IEEE 802.14-97/008, January 1997.

Haruyama, H. "PHY/MAC proposal adopting accurate delay/level control." Contribution number IEEE 802.14-95/170, November 1995.

Heegard, C. "Comparing J.83 annex A and B and flexible interleaving for J.83B." Contribution number IEEE 802.14-96/190, July 1996.

Heidemann, R. "The IVOD Berlin Project: Access Technology for Service Provisioning." *Alcatel Telecommunications Review,* 3rd quarter, 1996.

Hilton, Rhonda. "Cable industry requests ATM compatibility." Contribution number IEEE 802.14-96/228, September 1996.

Hilton, Rhonda for CableLabs. "ATM selected topics and specs technology," January–February 1995 and September–October 1994.

Hilton, Rhonda and R. Prodan. "Evolving cable network architecture cable televisions laboratory." Contribution number ATM Forum/95-0075, February 1995.

Hogue, M. "Inside residential ingress model." Contribution number IEEE 802.14-95/40, May 1995.

Horwood, Ellis, D. Deloddere, W. Verbiest, and H. Verhille. "Interactive Video on Demand," *IEEE Communications,* May 1994.

Hou, V. "802.14 MAC layer proposal." Contribution number IEEE 802.14-95/129, November 1995.

Hundt, Reed E. "Federal Communications Commission on Implementation of the Telecommunications Act of 1996 the Subcommittee on Telecommunications and Finance Committee on Commerce, U.S. House of Representatives." July 18, 1996.

Hutchinson, B. "Multimedia transmission link protocol MXL: A proposal for 802.14 MAC layer." Contribution number IEEE 802.14-95/160, November 1995.

Hutchinson, B. "Multimedia transmission link protocol MXL: A proposal for 802.14 MAC layer." Contribution number IEEE 802.14-95/160, November 1995.

ISO/IEC 13818-1. "Generic coding of moving pictures and associated audio," June 1994.

ISO-IEC/JTCI/CD standard organization, Std 802.3-1985. ISO Draft International Standard 8802/3.

ITU-T Recommendation I.113. "Vocabulary of Terms for Broadband Aspects of ISDN," 1992.

ITU-T Recommendation I.150. "BISDN ATM Functional Characteristics," 1992.

ITU-T Recommendation I.211. "BISDN Services Aspects," 1992.

ITU-T Recommendation I.311. "BISDN General Network Aspects," 1992.

ITU-T Recommendation I.321. "BISDN Protocol Reference Model," 1990.

ITU-T Recommendation Q2761. "Functional Description of the BISDN User Part of SS#7," Feb., 1995.

ITU-T Recommendation Q2762. "General Functions of Messages and Signals of the BISDN User Part," Feb. 1995.

ITU-T Recommendation Q2931. "BISDN User-Network Interface Layer 3 Protocol," Oct. 1995.

Jacobsen, K. "CATV rev. channel using synchronized DMT (part 2)." Contribution number IEEE 802.14-95/002, January 1995.

Jacquet, P., P. Muhlelhaler, and P. Robert. "CATV slotted multiple-access MAC." Contribution number IEEE 802.14-95/166, November 1995.

Jones, L. "T1S1.5 traffic management/congestion control baseline document." Contribution number T1S1.5/95-006R1, 1995.

Kalet, Irving. "The Multitone Channel," *IEEE Transactions on Communications* **37**(2), February 1989.

Karaoguz, J. and Gottfried Ungerboeck. "Formal proposal: Frequency-agile multimode (FAMM) single-carrier modems for upstream transmission in HFC systems." Contribution number IEEE 802.14-95/131, November 7, 1995.

Kidambi, S. and R. Gross. "Formal proposal for a DWMTPHY layer." Contribution number IEEE 802.14-95/096, September 1995.

Klasen, Wolfgang, Michael Munzert, and Bernhard Nauer. "Security framework for ATM networks." Contribution number ATM Forum 97-0068, June 1996.

Kolze, T. "802.14 PHY layer proposal." Contribution number IEEE 802.14-95/130R1, November 1995.

Koperda, K. "Providing CBR service over XDQRAP." Contribution number IEEE 802.14-95/147, November 1995.

Koperda, F. "IEEE 802.14 reference model." Contribution number IEEE 802.14-95/012, January 1995.

Kumar, S. et al. "Service requirements and network model for work-at-home/telecommuting, file transfer, infotainment (information-on-demand), and distance learning." Contribution number ATM Forum/94M-011, 1994.

Kuska, M. "MCNS—radio frequency interface physical layer specification." Contribution number IEEE 802.14-97/017, January 1997.

Kwok, T. "Communications Requirements of Multimedia Applications: A Preliminary Study." *Proc. International Conference on Selected Topics in Wireless Communications*, Vancouver, Canada, 1992.

Lam, S. et al. "A MAC layer proposal for IEEE 802.14 WG." Contribution number IEEE 802.14-95/164, October 1995.

Lam, S. "Packet Broadcast Networks: A Performance Analysis of the R-ALOHA Protocol." *IEEE Transactions on Computers* **C-29**(7), July 1980.

Laubach, M. "ATM HFC overview." Contribution number IEEE 802.14-95/022, March 1995.

Laubach, M. "MAC <> PHY-TC sublayer interface." Contribution number IEEE 802.14-97/025, January 1997.

Laubach, M. "The UPSTREAM protocol for HFC networks." Contribution number IEEE 802.14-95/152, November 1995.

Lee, Kyoo. "Physical layer protocol for HFC networks in support of ATM and STM integration." Contribution number IEEE 802.14-95/145, November 1995.

Lee, R. "16-VSB cable modem." Contribution number IEEE 802.catv 94/011, 1994.

Limb, J. "Performance evaluation process for MAC protocols." Contribution number IEEE 802.14-96/083R2, May 1996.

Limb, J. "Simulation of the performance of XDQRAP under a range of conditions." Contribution number IEEE 802.14/95-049, 1995.

Lin, D. "MAC proposal for upstream and downstream." Contribution number IEEE 802.14-95/144, November 1995.

Lin, D. "Toward a framework for high-speed communications in cable TV networks." Contribution number IEEE 802.catv-94/03, January 1994.

Lindberg, Bertil C. *Digital Broadband Networks and Services,* 1995.

"Lower-layer protocols and physical interfaces." DAVIC 1.0 specification part 8, technical report, revision 5.0.

Martin, James. *Telecommunications and the Computer,* 1990.

Massey, J. L. *Collision-Resolution Algorithms and Random-Access Communications: Multiuser Communications Systems,* 1981.

Mathys, P. and P. Flajolet. "Q-ary Collision Resolution Algorithms in Random-Access Systems with Free or Blocked Channel Access." *IEEE Trans. on Inform. Theory* 3(2), March 1995.

McDysan, D. E. and D. L. Spohn. *ATM.* McGraw-Hill, 1994.

McNamara, R. P. "Ingress Noise." *TDMA Transport,* October 2, 1995.

McNeil, B., R. Norman, and C. Bisdikian. "Formal proposal for 802.14 MC protocol (part 2 of 2) MLAP (MAC layer access protocol)." Contribution number IEEE 802.14-95/158, November 1995.

Miller, B. "INTRA modem PHY proposal." Contribution number IEEE 802.14-95/128, November 1995.

Mollenauer. J. "Dynamic transfer mode." Contribution number IEEE 802.14-96/028, January 1996.

Momona, Morihisa and Shuntaro Yamazaki. "Framed pipeline polling for cable TV networks (rev. 2.0)." Contribution number IEEE 802.14-95/100, September 1995.

Nielsen, Per Moller. "BATMAN—Danish ATM Cooperation" *Teleteknik* (English ed.) 38(1), 1994.

Norman, R. "Proposed relationship of 802.2 LLC to 802.14 MAC data units." Contribution number IEEE 802.14-95/105, September 1995.

Norman, R. "Protocol stack and topology assumptions for MLMP and MLAP proposals." Contribution number IEEE 802.14-95/156, November 1995.

Norman, R. "Timers synch and registration message flow." Contribution number IEEE 802.14-97/024, January 1997.

Prodan, R. "Additional physical layer considerations." Contribution number IEEE 802.14-96/227, September 1996.

Prodan, R. "Letter from CableLabs regarding annex A/B." Contribution number IEEE 802.14-96/205, July 1996.

Prodan, R. "Physical layer considerations." Contribution number IEEE 802.14-96/163, July 1996.

Quilici, J. "Management of noise in upstream channels." Contribution number IEEE 802.14-95/38, May 1995.

Quinn, S. "Requirements on HFC access networks: A public network operator's perspective." Contribution number IEEE 802.14-95/099, September 1995.

Ratta, G. "Report of July 26, 1996 T1S1.5 conference call." Contribution number T1S1.5/96-122R1, September 1996.

Reed, A. *Broadcasting and Cable Yearbook 1996*. Reference Publishing Company, 1996.

Rogers Cable Systems. "Appendix F: Two-Way Transmission Characteristics of Cable Systems" in *Request for Proposals for a Telecommunications System*, 1994.

Samueli, H., C. Reames, W. Wall, and I.E.O. Montreuil. "Performance results of a 64/256 QAM chip set." Contribution number IEEE 802.6-94/016, May 1994.

Samueli, H., C. Reames, L. Montreuil, and W. Wall. "Performance results of a 64/256-QAM CATV receiver chip set." Contribution number IEEE 802.catv-94/106, 1994.

Schneier, Bruce. *Applied Cryptography, Protocols, Algorithms, and Source Code.* John Wiley & Sons, 1994.

Scientific-Atlanta, Inc. "Broadband transport for interactive video service." Publication number 69L007A, January 1994.

Sierens, C. et al. "ATM Access Trial on CATV Networks in the Framework of RACE BISIA," *ATNAC'95 (Australian Telecommunication Networks & Applications Conference)*, Sydney, December 1995.

Sierens, C. et al. "ATM-based PHY for both in-band and out-of-band systems." Contribution IEEE 802.14-95/165, November 1995.

Sierens, C. et al. "Method for minislot alignment and granting in DVB-based frames." Contribution IEEE 802.14-96/185, July 1996.

Sierens, C. et al. "Working document on the comparison of burst QPSK specifications." Contribution IEEE 802.14-96/129, May 1996.

Sriram, K. et al. "An Adaptive MAC-Layer Protocol for Multiservice Digital Access via Tree-and-Branch Communication Networks." *SPIE Proceedings 2609,* 1995.

Sriram, P. "Adaptive digital access protocol (ADAPt): A MAC protocol for multiservices broadband access networks." Contribution number IEEE 802.14-95/142, November 1995.

Sriram, P. "ADAPt MAC PDU and MAC-PHY services in support of ATM." Contribution number IEEE 802.14-95/168, November 1995.

"STAck Algorithm for Random Multiple-Access Communication." *IEEE Trans. on Inform Theory* **31**(2), 1993.

Stanwyck, D. "A reference architecture for enhanced CATV networks." Contribution number IEEE 802.catv/12, 1994.

SWP on Q5/11. "Baseline document for LTA (version 5.1)." ITU Telecommunication Standardization Sector Study Group 11, Working Party 4/11, November 1996.

Teboul, Guillene. "Flexible MAC algorithm." Contribution number IEEE 802.14-95/155R1, November 1995.

Teboul, Guillene. "Physical layer for multiservice HFC network." Contribution number IEEE 802.14-95/154, November 1995.

Time Warner Excalibur Group. *Broadband Commerce and Technology Newsletter,* July 1996.

Tracey, D. "Support of communicative services for IEEE 802.14 HFC network." Contribution number IEEE 802.14-97/014, January 1997.

"Tracking the Development of High-Speed Cable Modem." *Cable Datacom News,* 1996.

Ulm, J. "A MAC protocol for 802.14." Contribution number IEEE 802.14-95/134, November 1995.

Ulm, J. "MCNS MAC downstream framing." Contribution number IEEE 802.14-97/023, January 1997.

Van der Plas, G. "Access Network Alternatives and Associated Trial Experiences." *ISSLS'96,* February 1996.

Van der Plas, G. "APON: An ATM-Based FITL System." *EFOC&N'93*, 1993.

Van Driel, C. J. L. "An Access Platform for CATV Networks to Support Multimedia Services." *Proc. 10e Kabel Congres*, Den Haag, November 1994.

Van Driel, C. J. L. and W. A. M. Snijders. "Network Evolution for End-User Access to Interactive Digital Services: The Last Mile of the Information Superhighway." *Proc. IBC Conference*, Sydney, August 1994.

Vandenameele, J. "How to Upgrade CATV Networks to Provide Interactive ATM-Based Services." *Proc. GLOBECOM 95*, Singapore, November 1995.

Van Grinsven, P. et al. "Accurate ranging for synchronized burst QPSK." Contribution number IEEE 802.14-96/183, July 1996.

Van Hauwermeiren, L. et al. "Offering Video Services over Twisted-Pair Cables to the Residential Subscriber by Means of an ATM-based ADSL Transmission System." *Proc. International Switching Symposium—ISS'95*, Berlin, 1995.

Verbiest, W. "Integrated Broadband Access," *Proc. Fourth IEE Conference on Telecommunications*, Manchester, 1993.

Verbiest, W. et al. "FITL and B-ISDN: A Marriage with a Future," *IEEE Communications Magazine*, June 1993.

Wall, B. and J. M. Fernandez. "Preliminary report on the effect of interleave depth and frequency offset on the effect of QAM transmission over cable systems." Contribution number IEEE 802.14/95-071, 1995.

Wardani, L. "Discussion of trellis coding." Contribution number IEEE 802.14-96/210, July 1996.

Wesley, P. and Weldon E. J. *Error Correcting Codes*, 2d ed. The MIT Press, 1972.

Wojnaroski, L. (ed.). "ATM Forum/RBB living list." Document ATM_FORUM/96-0006.

Wojnaroski, L. (ed.). "ATM Forum/Baseline text for the residential broadband working group." ATM Forum 95-1416R3.

Wu, Chien-Ting and Graham Campbell. *Extended DQRAP (XDQRAP)*. Georgia Tech, in press.

Wu, Chien-Ting and Graham Campbell. "Extended DQRAP (XDQRAP): A Cable TV Protocol Functioning as a Distributed Switch." *Proc. 1st International Workshop on Community Networking*, San Francisco, July 1994.

X, Yang Lee. "IEEE 802.14 physical layer." Contribution number IEEE 802.14-95/163, November 1995.

Xu, Wenxin and Graham Campbell. "A Distributed Queuing Random Access Protocol for a Broadcast Channel." *Proc. ACM SIGCOMM,* 1993.

Zeisz, R. "Formal proposal for 802.14 MC protocol (part 1 of 2) MLMP (MAC layer management protocol)." Contribution number IEEE 802.14-95/157, November 1995.

INDEX

ABOUT THE AUTHOR

Albert Azzam (Raleigh, NC) is a member of the IEEE 802.14 working group that is finalizing the High-Speed Cable Modem Standard. A 26-year telecommunications veteran (BSEE, Syracuse) and holder of four patents, presently with Alcatel, he has served as secretary of the IEEE 802.14 group since its inception. He also served as vice-chair in the signaling working group of T1S1 and has been active in the ATM standardization process for the last 9 years.

Learning Resources
Centre